The Building Blocks

	Visual Representation	Textual Form	Definition	Description
Entities	Object	Nouns; capitalized first letter in every word; if ending with "ing", "Object" is placed as a suffix	*An **object** is a thing that has the potential of stable, unconditional physical or mental existence.*	Static things. Can be changed only by processes.
	Process(ing)	Nouns in gerund form; capitalized first letter in every word; if not ending with "ing", "Process" is placed as a suffix	*A **process** is a pattern of transformation that an object undergoes.*	Dynamic things. Are recognizable by the changes they cause to objects.
	Object state	Nouns, adjectives or adverbs; non-capitalized	*A **state** is a situation an object can be at.*	States describe objects. They are attributes of objects. Processes can change an object's state.
OPL Conventions	**Non-Reserved Words**	Arial bold by default	*Names of entities.*	Words used by the system architect, unique to the system.
	reserved words	Arial by default; non-capitalized	*Object-Process Language (OPL) words.*	Words or phrases used by OPL, the same in every sentence of a certain type.

Links: The Mortar

Tagged Structural Links

Generally used between objects, but may also be used between processes.
Cannot be used to link an object to a process.

Link Name	Object Process Diagram (OPD) Symbol	OPL Sentence	Description
Tagged	R Object — refers to → S Object	**R Object refers to S Object.**	Relation from source object to destination object; relation name is entered by architect, and is recorded along link.
(Null)	R Object — → S Object	**R Object relates to S Object.**	Relation from source object to destination object with no tag.
Bi-directional Tagged	R Object — precedes / follows — S Object	**R Object precedes S Object. S Object follows R Object.**	Relation between two objects; relation names are entered by architect, and are recorded along link.
(Null) Bi-directional	R Object ← → S Object	**R Object and S Object are related.**	Relation between two objects with no tag.

The Four Fundamental Structural Relations

Shorthand Name	Aggregation	Exhibition	Generalization	Instantiation
Symbol	▲	◮	△	◬
Meaning	Relates a whole to its parts	Relates an exhibitor to its attributes	Relates a general thing to its specializations	Relates a class of things to its instances

A fundamental structural relation can have many descendants.
The different OPL sentences and OPD pictures are listed below.

Structural Relation Name and **Shorthand Name**	Number of Descendants						Description
	One		Two		Three or more		
	OPD	OPL	OPD	OPL	OPD	OPL	
Aggregation- Participation	A▲B	**A** consists of **B**.	A▲B C	A consists of **B** and **C**.	A▲B C D	**A** consists of **B, C,** and **D**.	B, C and D are parts of the whole A.
Exhibition- Characterization	A◮B	**A** exhibits **B**.	A◮B C	A exhibits **B** and **C**.	A◮B C D	A exhibits **B, C,** and **D**.	B, C and D are attributes of A. If B is a process, it is an operation of A.
Generalization- Specialization	A△B	**B** is an **A**.	A△B C	**B** and **C** are As.	A△B C D	**B, C,** and **D** are As.	B, C and D are types of A.
Classification-**Instantiation**	A◬B	**B** is an instance of **A**.	A◬B C	**B** and **C** are instances of **A**.	A◬B C D	**B, C,** and **D** are instances of **A**.	B, C and D are unique objects of the class A.

The four fundamental relations are also applicable to processes. Only exhibition can link objects with processes. Instantiation cannot generate a hierarchy while the other three can. Any number of things can be linked to the root.

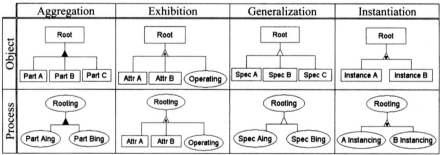

Object-Process Methodology

Springer
Berlin
Heidelberg
New York
Barcelona
Hong Kong
London
Milan
Paris
Tokyo

Dov Dori

Object-Process Methodology

A Holistic Systems Paradigm

Foreword by Edward F. Crawley
With 246 Figures and CD-ROM

 Springer

Professor Dov Dori

Technion, Israel Institute of Technology
Faculty of Industrial Engineering and Management
Haifa 32000, Israel

dori@ie.technion.ac.il

and

Massachusetts Institute of Technology
Engineering Systems Division, Building E40-347
77 Massachusetts Avenue
Cambridge, MA 02139-4307, USA

dori@mit.edu

Library of Congress Cataloging-in-Publication Data applied for
Die Deutsche Bibliothek – CIP-Einheitsaufnahme
Dori, Dov: Object process methodology: a holistic systems paradigm/
Dov Dori. – Berlin; Heidelberg; New York; Barcelona; Hong Kong;
London; Milan, Paris; Tokyo: Springer, 2002

ADDITIONAL MATERIAL TO THIS BOOK CAN BE DOWNLOADED FROM HTTP://EXTRA.SPRINGER.COM

ISBN 3-540-65471-2

ISBN 3-540-65471-2 Springer-Verlag Berlin Heidelberg New York

ACM Computing Classification (1998):
D.2.2, D.2.9–10, D.3.2–4, H.1.1, I.2.4

Springer-Verlag Berlin Heidelberg New York,
a member of BertelsmannSpringer Science+Business Media GmbH
http://www.springer.de

© Springer-Verlag Berlin Heidelberg 2002
Printed in Germany

Material in this book is patent pending.
A patent titled "Modeling System" was filed with the US Patent & Trademark Office,
Document Number 20020038206.

Cover design: KünkelLopka, Heidelberg
Data conversion: G&U e.Publishing Services, Flensburg
Printed on acid-free paper   SPIN: 10560832   45/3142 GF– 5 4 3 2 1 0

To my family

Foreword

The contemporary technological world is made up of complex electro-mechanical-information products that intimately involve human operators. Such a description includes products that range from a simple kitchen appliance to an air-traffic control system. Increasingly, the value these products deliver is based on their system nature; they have many interacting components that combine to produce emergent features or services that are desirable to the customer.

The development of more rigorous approaches to the engineering of these systems is seen by industry as a key area for process improvement, and should be recognized by academia as an important area of scholarly development. Since the engineering process dominates development time and cost, and the system engineering process defines value delivered by a product, improvements in system engineering and product modeling will have direct influence on product utility and enterprise competitiveness. The task of engineering a new system has become more complicated by the vastly increasing number of components involved, the number of disparate disciplines needed to undertake the task, and the growing size of the organizations involved. Despite the common experience that members of many organizations share, they often lack a common product development vocabulary or modeling framework. Such a framework should be rigorously based on system science; be able to represent all the important interactions in a system; and be broadly applicable to electrical, informational, mechanical, optical, thermal, and human components.

Object Process Methodology provides such a framework. OPM includes a clear and concise set of symbols that form a language enabling the expression of the system's building blocks and how they relate to each other. It is a symbolic representation of the objects in a system and the processes they enable. Considering the historical development of engineering disciplines, it is an appropriate time for such a rigorous framework to emerge. Disciplines often move through a progressive maturation. Early in the history of an intellectual discipline, we find observation of nature or practice, which quickly evolves through a period in which things are classified. A breakthrough often occurs when classified observations are abstracted and quantified. These phases characterize much of the work done to date in system engineering and product/system development.

We are now entering a phase, in which symbolic representation and manipulation can be developed. Such a symbolic system can lay the foundation for the ultimate step in the evolution of a discipline, the ability to predict the outcome beforehand. Mature disciplines, such as mechanics, are well into the era of symbolic manipulation and prediction. Maturing disciplines, such as human genomics, are in the phase of symbolic representation. OPM is a parallel development in symbolic representation of systems.

OPM represents the two things that are inherent in a system: its objects and its processes. This duality is recognized throughout the community that studies systems, and sometimes goes by labels such as form/function, structure/function, and functional requirements/design parameters. Objects are what a system or product is. Processes are what a system does. Yet, it is remarkable that so few modeling frameworks explicitly recognize this duality. As a result, designers and engineers try to jump from the goals of a system (the requirements or the "program") immediately to the objects. Serious theory in such disparate disciplines as software design, mechanical design and civil architectural design recognizes the value of thinking about processes in parallel with objects. Not only does OPM represent both objects and processes, but it also explicitly shows the connections between them.

Object Process Methodology has another fundamental advantage – it represents the system simultaneously in a graphic representation and in a natural language. The two are completely interchangeable, and represent the same information. The advantage in this approach lies in appreciating the human limitation to the understanding of complexity. As systems become more complex, the primary barrier to success is the ability of the human designers and analysts to understand the complexity of the interrelationships. By representing the system in both textual and graphical form, the power of "both sides of the brain" – the visual interpreter and the language interpreter – is engaged. These are two of the strongest processing capabilities hard-wired into the human brain.

OPM allows a clear representation of the many important features of a system: its topological connections, its decomposition into elements and subelements, the interfaces among elements, and the emergence of function from elements. The builder or viewer of the model can view abstractions or zoom into some detail. One can see how specification migrates to implementation. These various views are invaluable when pondering the complexity of a real modern product system.

I have used OPM in my System Architecture course at MIT. It has proved an invaluable tool to professional learners in developing models of complex technical systems, such as automobiles, spacecraft and software systems. It allows an explicit representation of the form/function duality, and provides an environment in which various architectural options can be examined. Incorporating OPM into my subject has added the degree of rigor of analysis necessary to move the study of technical system architecture towards that of an engineering discipline.

One can anticipate that there will be many academic applications of OPM. I would consider using it in intermediate or advanced subjects in system engineering, product development, engineering design and software engineering. It is ideal for courses that demonstrate how various disciplines come together to form a multi-disciplinary product.

Likewise, OPM can form the backbone of a corporate or enterprise modeling system for technical products. Such a representation would be especially valuable in conceptual and preliminary design, when much of the value, cost and risk of a product are established and only a few other modeling frameworks are available for decision support.

Massachusetts Institute of Technology   Edward F. Crawley
Cambridge, Massachusetts
May 2001

Preface

Before Fortran, putting a human problem ... on a computer was an arduous and arcane task. ... Like high priests in a primitive society, only a small group of people knew how to speak to the machine. Yet there were some heretics in the priesthood, and Mr. Backus was one of them. "I figured there had to be a better way," he recalled nearly five decades later, "you simply had to make it easier for people to program."

S. Lohr (2001)

This book is about how to make it easier for people to understand and develop systems. Systems are all around us. Understanding and developing complex systems involve a host of disciplines that need to be brought together in a unifying framework. Systems evolve over time, and they keep growing and getting more complex. Hundreds, perhaps thousands, of individuals are required to maintain and operate systems such as electrical power grids or globally interconnected telecommunication networks. Nature's global climate, the solar system and biological systems are at least as intricate. Systems science and engineering are emerging as new interdisciplinary fields of study that identify and utilize important commonalties among systems of all types.

A fundamental requirement of any science and engineering domain is that its intellectual underpinnings be formulated and well grounded. Such formulation, in turn, mandates that a set of elementary building blocks is agreed upon and relations among these building blocks are studied and understood. A unifying approach is indispensable for developing, communicating, supporting, and evolving systems of various domains, types, magnitudes and complexities. To this end, a clear and concise language must be developed. Pioneers in established science and engineering disciplines, like physics, chemistry, mechanical engineering, and electrical engineering, have long agreed on such languages. More recently, biology has joined this list and is now establishing the molecular foundations of life. Being a young filed, systems science and engineering is just beginning to contemplate this problem and sort it out. This book will hopefully advance the state of affairs regarding our ability to model and understand systems.

The book is intended for people interested in modeling systems, and reading it does not require special background in either mathematics or programming. Specifically, system integrators, analysts, designers, modelers, executives, and project leaders in a variety of industry and service domains, private as well as public, will benefit from reading the book and applying Object-Process Methodology (OPM) for the purpose of developing better systems faster and more reliably. An equally important point is that developing systems with OPM is fun. The combination of graphics and natural language that automatically complement each other when an automated tool, such as Systemantica® (Sight Code, 2001) is employed, is an enjoyable way of analysis and design that increases one's confidence in the quality of the resulting system specification. Information technology professionals, including computer scientists, software engineers, information systemmanagers, and database administrators will gain insight into possible extensions of Object technology. Scientists and engineers, along with science and engineering educators and students of all ages will hopefully find OPM useful to model, communicate, and explain systems they research and design.

The book, which consists of 15 chapters arranged in three parts, is designed as a textbook for graduate or advanced undergraduate courses. Typical course titles include engineering systems, specification and analysis of information systems, systems architecture, information systems engineering, systems design and management, or Object-Process Methodology. Indeed, various versions of this book have been used and tested in graduate and undergraduate courses at the Technion and MIT during the past five years. Each chapter concludes with a summary of the chapter's highlights and problems that enable hands-on experience, elaborate on concepts discussed in the chapter and provide food for further thought.

Part I, *Foundations of Object-Process Methodology*, is a gradual exposure to OPM. Chapter 1 is a gentle introduction to basic principles of OPM through a simple example of a wedding. Walking through a comprehensive case study of an automated teller machine (ATM), Chapter 2 exposes the reader to Object-Process Diagrams as the graphic representation of the single object-process model. Continuing with the ATM case study, Chapter 3 introduces Object-Process Language as the complementary modality to the graphic one and shows how the two synergistically reinforce each other. Chapter 4 discusses in depth the two basic building blocks of OPM, objects and processes. In Part II, *Concepts of OPM Systems Modeling*, the dynamic and static system aspects are the focus of Chapters 5 and 6, respectively. The next two chapters elaborate on the four OPM fundamental structural relations. Aggregation and Characterization are discussed in Chapter 7; Generalization and Classification in Chapter 8. Chapter 9, which concludes Part II, is devoted to complexity management. Part III, *Building Systems with OPM*, shows how OPM is used to develop systems, products and projects. Systems and modeling is the topic of Chapter 10. Chapter 11 provides a comprehensive OPM model of system evolution and lifecycle, and discusses how this model can be used to develop systems in an OPM-based environment. The next two chapters expand upon issues presented in Parts I and II. Chapter 12 elaborates on states, while Chapter 13 presents advanced OPM concepts. Chapter 14 discusses systems theory and Chapter 15 concludes with a survey of object-oriented (OO) methods and their relation to OPM.

I thank Professor Ed Crawley for his encouragement and support throughout my stay at MIT and, in particular, for his enthusiastic Foreword to this book. OPM has become and integral part of the course "Systems Architecture," which he and Professor Olivier de Weck teach at MIT. Thanks to Robert M. Haralick whose draft, written at my request, was the basis for Section 4.1. Thanks for the professional editing as well as the detailed and helpful comments to Dr. Hans Wössner of Springer-Verlag. Thanks to Idan Ginsburg and William Litant for their meticulous proofreading. Thanks to my wife Judy and my daughters, Limor, Tlalit, Shiri and Inbal, who have been really helpful and considerate. Special thanks to Shiri Dori, whose thorough proofreading and wise comments, at both the editorial and content level, as well as the problems she composed, were extremely helpful.

Haifa, Israel, and Cambridge, Massachusetts   Dov Dori
January 2002

Contents Overview

Table of Contents

Part I

Foundations of Object-Process Methodology

Chapter 1
A Taste of OPM

...We're limited only by our imaginations and what we think we can make computers do. Ben Kovitz (1998)

Imagine that an intelligent, friendly extra-terrestrial creature has just landed on planet Earth, without any prior knowledge of earthly physical and societal systems. Trying to figure out what his senses are telling him and how to construct a model of the surrounding reality, this creature is in a situation not unlike that of a human system developer, who is beginning to evolve a new system in an unfamiliar domain. Lacking sufficient knowledge about the domain that hosts the system and its environment, the analyst must start with an arbitrary collection of facts. The collected observations are not overly refined nor are they extremely abstract. The knowledge and understanding of the system is gradually improved through activities such as observing the current state of affairs and practices, inquiring, interviewing professionals in the field, and reading relevant documents.

Soon enough, details about the system begin to mount. An intuitive yet formal way of documenting this growing amount of collected information must be in place. Subsequent, higher order cognitive activities follow, including understanding, modeling, analyzing, designing, presenting and communicating the analysis findings and design ideas. These mental activities rely heavily on a solid infrastructure for organizing the accumulated knowledge and creative ideas. Long before the system becomes particularly intricate, it is very helpful to display these results in a medium other than the brain.

Systems and products are becoming increasingly complicated. Technology has been so pervasive that even commonly used products feature high computational power, embedded within increasingly miniature, precise and involved hardware. Systems of an infrastructure nature, such as air traffic control, the Internet, and electronic economy, are orders of magnitude more complex than products individuals normally use. Understanding natural, artificial, and social systems requires a well-founded, yet intuitive methodology that is capable of modeling these complexities in a coherent, straightforward manner. The same methodology should be useful for designing new systems and improving existing ones. These systems are intricate enough as it is; there is no need to add confusion by using a method that is itself complex.

Artificial systems require development and support efforts throughout their entire lifecycle. Systematic specification, analysis, design and implementation of new systems and products are becoming ever more challenging and demanding, as contradicting requirements of shorter time-to-market, rising quality, and lower cost, are on the rise. These trends call for a comprehensive methodology, capable of tackling the mounting challenges that the evolution of new systems poses. The development of Object-Process Methodology was motivated by this call.

The idea of developing systems in a unified frame of reference is not new. For example, Bauer and Wössner (1981) noted that systematic development of basic concepts leads to methods that cover the entire system's lifecycle. Object-Process Methodology, or OPM for short, takes a fresh look at modeling complex systems that comprise humans, physical objects and information. OPM is a formal paradigm to systems development, lifecycle support, and evolution. It caters to people's intuition and train of thought. Preliminary ideas of OPM were expressed in (Dori, 1995; 1996) and applied in such diverse areas as computer integrated manufacturing (Dori, 1996A), image understanding (Dori, 1996B), modeling research and development environments (Meyersdorf and Dori, 1997), algorithm specification (Wenyin and Dori, 1999), document analysis and recognition (Dori, 1995A; Wenyin and Dori, 1998), and modeling electronic commerce transactions (Dori, 2001).

Natural and artificial systems alike exhibit three major aspects: function (what these systems do), structure (how they are constructed), and behavior (how they change over time). Since OPM does not make any assumptions regarding the nature of the system in question, it can be applied in any domain of human study or endeavor. In the case of artificial systems, it provides a framework for the entire system's lifecycle, from the early stages of requirement elicitation and analysis, through further development and deployment, all the way to termination and initiation of a new generation.

OPM combines formal yet simple graphics with natural language sentences to express the function, structure, and behavior of systems in an integrated, single model. The two description modes OPM uses are semantically equivalent, yet appeal to two different parts of the brain, the visual and the lingual. OPM is a prime vehicle for carrying out the tasks that are involved in system development. It does so in a straightforward, friendly, unambiguous manner. The design of OPM has not been influenced by what current programming languages can or cannot do, but rather, what makes the most sense. Due to the resulting intuitiveness, OPM is communicable to peers, customers and implementers. At the same time, the formality of OPM makes it amenable to computer manipulation for automatically generating large portions of the conceived system, notably program code and database schema.

Where does the name Object-Process Methodology come from? Objects and processes are the two main building blocks that OPM requires to construct models. A third OPM entity is *state*, which is a situation at which an object can be and therefore a notch below object. Objects, processes and states are the only bricks involved in building systems. The links connecting these three entities act as the mortar that holds them together.

As for methodology, Computer Desktop Encyclopedia (2001) defined it as "the specific way of performing an operation that implies precise deliverables at the end of each stage." OPM specifies a way of understanding and developing systems. Its deliverables include a set of Object-Process Diagrams (OPDs) and a corresponding collection of sentences written in Object-Process Language (OPL). Three more general definitions of the term "methodology" are found in The American Heritage Dictionary (1996):

(a) A body of practices, procedures, and rules used by those who work in a discipline or engage in an inquiry; a set of working methods
(b) The study or theoretical analysis of such working methods
(c) The branch of logic that deals with the general principles of the formation of knowledge.

OPM contains important elements of each of these three definitions.[1] These elements are treated and discussed throughout the book. This chapter introduces OPM's basic principles and features and demonstrates its modeling power using an example of a wedding.

1.1 The Wedding Example: A Sneak Preview of OPM

To gain familiarity with the basic ideas of OPM and its two methods of representation, we will describe a wedding system at increasing levels of complexity. This example introduces some of the most important concepts that OPM is built upon, such as the independence of processes, their effect on objects, and the relations among objects and processes. Step by step, we will create an OPD (Object-Process Diagram) of this system and its corresponding set of OPL (Object-Process Language) sentences. In this chapter, each OPD is accompanied by a legend that introduces OPM's commonly used symbols.

1.2 OPM Building Blocks: Objects, Processes, and States

OPM is built of just three types of entities: objects, processes and states, with objects and processes being higher-level building blocks. Objects exist, and processes transform the objects by generating, consuming, or affecting them. States are used to describe objects, and are not stand-alone things.

The symbols for objects and processes are rectangles and ellipses, respectively. The name of the object or process is recorded inside the corresponding symbol.

---

[1] Definition (a) above overlaps with The American Heritage Dictionary (1996) definition of "method," as a means or manner of procedure, especially a regular and systematic way of accomplishing something; the procedures and techniques characteristic of a particular discipline or field of knowledge. However, since OPM contains the additional elements of theoretical analysis methods and formation of knowledge, "methodology" is more appropriate.

Figure 1.1[2] shows **Person** as an example of an object and **Marrying** as an example of a process. The first letter in object and process names is always capitalized. To differentiate objects from processes, process names end with the suffix **ing**, indicating that they are active, dynamic things.

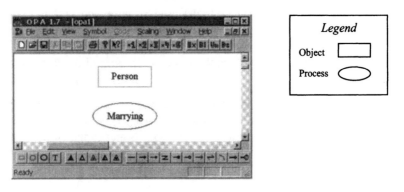

Figure 1.1. The object and process symbols applied to **Person** and **Marrying**, respectively

At any point in time, each object is at some *state*. In our example, we restrict the possible states of **Person** to be **single** or **married**. The OPD in Figure 1.2 shows the symbol of state as a rounded corner rectangle, or "rountangle" for short,[3] enclosing the state name, which starts with a lower-case letter. The state symbol is drawn inside the rectangle symbolizing the object that "owns" the state. The sentence describing Figure 1.2 in Object-Process Language (OPL) lists the possible states of the object:

Person can be single or married.

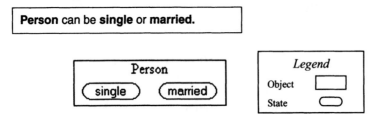

Figure 1.2. The two states of the object **Person**

The **bold** words in the OPL sentence above, such as **Person**, denote *non-reserved words*, whereas the non-bold words, such as can be, are *reserved words*, which are part of the sentence structure. Any OPL sentence consists of non-

---

[2]   The OPDs in Figure 1.1 and in the rest of this book were drawn using OPCAT – Object-Process CASE Tool. This Computer Aided Software Engineering (CASE) tool supports OPM by drawing OPDs and checking the legality of the various links. It is downloadable from http://iew3.technion.ac.il/%7Edori/opcat/index-continue.html. To reduce clutter and save space, subsequent OPDs will be presented without the surrounding OPCAT user interface that appears in Figure 1.1.

[3]   This word is due to D. Harel.

reserved words – domain-specific words, which the system architect uses for the specific system – and reserved words, which link the non-reserved words and provide for creating a natural language sentence. In a way, the reserved words glue the system-specific terms (the non-reserved words) together in a meaningful way. OPL sentences are certainly far more readable than a script of any computer programming language. These sentences may not be the most natural, nor grammatically perfect. However, as we shall see, they are carefully designed to convey a clear and straightforward meaning through well-phrased and humanly understandable constructs.

A process can transform an object in three different ways: by creating it, by destroying it, or by affecting it in some way. When a process affects an object, it changes the state of that object. Links within the OPD express this graphically. Figure 1.3 contains two such links, ⟶▷ , arrows with elongated hollow triangular heads. An *input link* goes from the input state to the affecting process, and an *output link* goes from that process to the output state. In our example, **single** is the input state. It is linked with an input link to the process **Marrying**. From **Marrying**, an output link leads to **married**, the output state. The appropriate OPL sentence follows.

> **Marrying** changes **Person** from **single** to **married**.

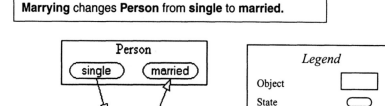

Figure 1.3. The process **Marrying** affects the object **Person** by changing its state from **single** to **married**.

Note that the sentence has the same meaning as the combination of the two links in Figure 1.3, from the **single Person** through **Marrying** to the **married Person**. These links are procedural links. Procedural links connect processes to objects or to states of objects (as is the case here) in a variety of ways.

1.3 Specialization and Inheritance

Specialization is the relation between a general thing and a type of that thing. The process **Marrying** depends upon a **Woman** and a **Man** getting married.[4] **Man** and **Woman** are different types, or specializations, of **Person**. The white equilateral tri-

---

[4]   While the wedding here is traditional in the sense that it is between people of different genders, it does not imply any bias against other engagement types.

angle ($\triangle$) in Figure 1.4 denotes this. This link is a structural link, which is fundamentally different than the procedural links we used in Figure 1.3. Structural links relate objects to other objects, and processes to other processes, but not objects to processes. The same semantics that the new link expresses in Figure 1.4 is expressed in the following OPL sentence:

Man and **Woman** are **Persons.**

This sentence demonstrates the differences between natural, spoken language and OPL. A more natural sentence would be "**Man** and **Woman** are **People.**" But OPL in its basic form does not handle such exceptions. Yet, the sentence is still clear and comprehensible.

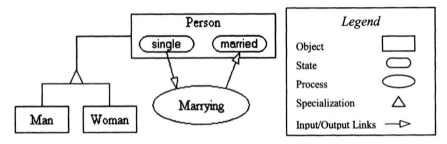

Figure 1.4. Man and **Woman** as specializations of **Person**

Since **Woman** and **Man** are specializations of **Person**, they *inherit* the states of **Person**. We have seen that **Person** can be in one of two states: **single** or **married**. Figure 1.4 expresses that both **Woman** and **Man** are specializations of **Person**. Through inheritance, each can be **single** or **married**. We have seen that **Marrying** affects **Person** by transforming it from **single** to **married**. Since **Woman** and **Man** are specializations of **Person**, they inherit not only the states of **Person**, but also the transformation from **single** to **married** that **Person** undergoes. This effect is inherited to both **Woman** and **Man**: **Marrying** affects each one of them by transforming her or him from **single** to **married**.

1.4 Aggregation and the Result Link

The **Marrying** process yields a new object, **Couple**, which consists of **Woman** and **Man**. The aggregation symbol, shown as a black equilateral triangle (▲) in Figure 1.5, denotes this whole-part relation between the whole, the **Couple**, and its parts, **Woman** and **Man**. Aggregation is another structural link, expressing a different relation than specialization. The arrow from **Marrying** to **Couple** is a *result link*, a procedural link denoting the fact that a new object has been generated as a result of the occurrence of the process. These two links are shown connected to the new object **Couple** in Figure 1.5, and are reflected by these two OPL sentences:

> **Couple** consists of **Man** and **Woman**.
> **Marrying** yields **Couple**.

The reserved words in these two sentences express the relation type. The phrase "consists of" in the first sentence relates the whole, **Couple**, to its parts, **Man** and **Woman**. The same phrase would be used for any other whole-part relationship, for example: **House** consists of **Walls** and **Roof**. In the same manner, the reserved phrase **yields**, which, in this example, is a single word, connects a process and its resulting object, as in **Cutting** yields **Slice**. The link from the process **Marrying** to the object **Couple** denotes that **Couple** is the result of the occurrence of **Marrying**. This link is therefore called the result link. As we progress along the book, we will encounter different types of OPL sentences. We will become familiar with the reserved phrases these sentence types use and with the graphical constructs that are equivalent to these sentences.

The result link, the input link and the output link are similar in their use, but not identical. All three are procedural links, connecting processes to objects or states. The input link connects an object's original state with a process. The output link, in turn, connects that process to a different state of the same object; this state is the output state. The result link denotes the creation of an entirely new object as an outcome of the process. While graphically all three look alike, their distinction is inferable from their context, i.e., the combination of their source and destination. The result link originates from a process and ends at an object, the input link from a state to a process, and the output link from a process to a state. This is an example of the context sensitivity of the graphic symbols in an OPD: depending on its context in the OPD, the same symbol can have more than one meaning. And so, the same symbol may serve a few purposes, albeit related ones. Context sensitivity enables a small set of symbols to express rich semantics.

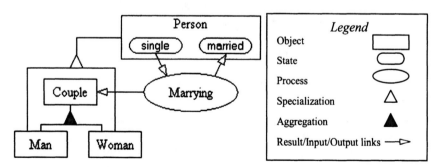

Figure 1.5. Marrying yields **Couple**, which consists of **Woman** and **Man**, each being a specialization of **Person**.

> **Person** can be **single** or **married.**
> **Marrying** changes **Person** from **single** to **married.**
> **Man** and **Woman** are **Persons.**
> **Marrying** yields **Couple.**
> **Couple** consists of **Man** and **Woman.**

Frame 1. The OPL paragraph of the OPD in Figure 1.5

Figure 1.5 is the final OPD of the wedding system. It expresses the same semantics as the OPL paragraph in Frame 1, which is the collection of the OPL sentences in this chapter. Both representations of the wedding system are equally valid, and can be used in conjunction or separately. Moreover, when using the two representations concurrently, one of the modes can compensate for potential misunderstandings in the other. If a reader is not familiar with a particular OPD symbol, for example, the appropriate OPL sentence can explain its meaning.

Summary

Through the wedding example we have been exposed to basic ideas of OPM. The constructed set of Object-Process Language (OPL) sentences and the Object-Process Diagram (OPD) demonstrate that small is beautiful. Using a small set of symbols, we have been able to specify clearly and intuitively, yet formally, exactly what the wedding system is and how it operates. Recognizing processes (**Marrying**, in our example) as independent building blocks alongside objects enables coherent modeling of the system's dynamics. The behavior of the system is integrated with the static object model and complements it.

The two OPM representation modes, the graphic and the textual, provide complementary, crisp representations of the system. The OPL sentences and the OPD complement each other and appeal to the two sides of the brain at once. The single diagram format provides for a concise and economical symbol set with a compact yet semantically rich vocabulary. We saw that three procedural links, the input, output and result links, are expressed by the same symbol.

We have seen the three OPM entities: object, process and state. Objects are things that exist, while processes are things that affect objects. States are situations at which objects can be. Processes transform objects in one of three ways: generating, consuming or affecting them. The effect a process has on an object is manifesting through a change in the object's state. Before the **Marrying** process started, both the **Man** and the **Woman** were **single**, and after it ended, they were **married**. The object **Couple**, which did not exist prior to **Marrying**, was generated as a result of its occurrence. In OPL, the three entity types are also easily distinguishable. Objects and processes are capitalized, states are non-capitalized, and processes end with **ing**.

Two types of links connect entities with each other: *structural links* and *procedural links*. Procedural links, explained in Chapter 5, express the behavior of the system. In the OPD of Figure 1.5, we saw the three types of procedural links discussed above: result link, input link and output link. Structural links, explained in Chapters 6 through 8, express persistent, long-term relations among objects or among processes in the system. Figure 1.5 contains two examples of structural links: aggregation and specialization, which are symbolized by a black triangle and a white triangle, respectively. Since the structural and the procedural links are expressed in the same diagram, they provide a complete picture of the system in a single graphic model, which is complemented by a textual one.

Problems

1. In the wedding example, we defined **Person** as having two states, **single** and **married**. Name two other states that might be applicable for **Person** in this context.
2. Draw an OPD of **Person**, with the states you came up with in problem 1 as well as the ones it had before.
3. Write an OPL sentence expressing the OPD in problem 2.
4. Create a system with **Person** and **Divorcing**. Show how the added process affects the object. You may wish to use your answers to problems 1–3.
5. We may want to hide the states of an object. In Figure 1.6(a) the states of the object **Person** from Figure 1.3 have been suppressed.
   a. Explain the motivation of suppressing the states of an object, as done in Figure 1.6(a) relative to Figure 1.3.
   b. What happened to the edges of the input and output links relative to Figure 1.3?

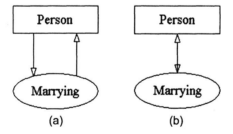

(a)   (b)

Figure 1.6. (a) Figure 1.3 with states hidden. (b) The effect link replaces the two links.

6. In Figure 1.6(b), we have replaced the input and output links by a single, bidirectional effect link. Which is more efficient in your opinion?
7. What is the relationship between the input and output links and the effect link?
8. Compare the semantics of the effect link in Figure 1.6(b) with the links in Figure 1.3. Which conveys more information? Which is more succinct?
9. A person withdraws money from an ATM using his or her cash card.
   a. Identify the objects in this system.
   b. Identify the processes in this system.
   c. Draw an OPD of the system.

Chapter 2

Object-Process Diagrams

*The expert may, in the process of explaining some
idea or description of a behavior, suddenly reach for
pad and draw sketches of what he/she does, and say
"it has to look like this" or "I know just by looking at
the chart if something is wrong."*

M. Firlej and D. Helens (1991)

In Chapter 1, we were exposed to basic OPM principles and features. In this
chapter, OPM's visual capabilities are presented. Diagrams are intuitive and therefore widely used. As Cook (1999) noted, "with diagrams the meaning is obvious,
because once you understand how the basic elements of the diagrams fit together,
the meaning literally stares you in the face." This is one of the premises OPM is
built upon. Diagrams contain symbols for objects, processes and states that are
interconnected with several types of links with precise semantics. These diagrams
are naturally called Object-Process Diagrams (OPDs). Using an ATM system as a
case in point, this chapter presents an overview of OPDs, their properties and the
main symbols they use. An ATM is appropriate as an introductory case study, since
it is a familiar system. Familiarity with the domain and the system being investigated assists the acquaintance with OPM, as the reader can draw on prior knowledge.

2.1 Objects and Aggregation

We first present and consider a small portion of the ATM case study. Gradually,
details of this system are exposed and modeled. Frame 2 contains the part of the
problem statement with which we will be concerned initially. We will follow it
sentence by sentence, show how OPM is employed to model the ATM system, and
illustrate its analysis and design. As we model the system, we encounter characteristics of OPM, which are explained as they are being put to use.

From the beginning of the first sentence in the problem statement in Frame 2,
we extract the first two objects that participate in the ATM system: **Consortium**
and **Bank**. As we have seen in Chapter 1, the symbol of an object is a rectangle that
encloses the name of the object. This object symbol, shown in Figure 2.1, follows
the common object-oriented graphic representation conventions.[1]

---

[1] Adherence to common or standard symbols where possible has been a guideline in devising the OPD symbol vocabulary. For example, the symbols for OPD objects, processes (as
in data flow diagrams, or DFDs) and states (as in Statecharts) follow this principle.

A consortium consisting of five banks operates an ATM system. Each bank holds many accounts, one for each customer. An account is accessible through a cash card. A customer can own one or two cash cards. The main process the ATM system is designed to carry out is the execution of cash dispensing and depositing transactions. Each transaction refers to a specific account. Executing the transaction will result in one of the following two outcomes: it can yield a (successful) transaction or it can issue a denial notice.

Frame 2. The problem statement of the ATM case study

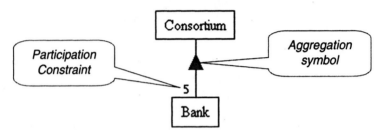

Figure 2.1. An Object-Process Diagram (OPD) showing an Aggregation-Participation (whole-part) relation and participation constraint between **Consortium** (the "whole") and **Bank** (the "part")

The phrase "consisting of" in the above problem statement sentence specifies that a whole-part relation exists between the object **Consortium** and the object **Bank**. The *single* **Bank** is part of the *whole* **Consortium**. This relation is also called the *Aggregation-Participation* relation. We wish to express this whole-part relation graphically in an OPD. The gray balloons on the right hand side of Figure 2.1 are not part of the OPD; they are just used to point at the various components within the OPD. The OPD contains two objects: **Consortium** and **Bank**. These two objects are related through the Aggregation-Participation relation, where **Consortium** is the whole and **Bank** is the part. Following Embley et al. (1992), the Aggregation-Participation relation is denoted by a black triangle, ▲, such that its tip is linked with the aggregate, or whole (**Consortium**, in our example) and its base, to the parts (**Bank**, in our example).

The fact that there are five banks in the **Consortium** is reflected by the *participation constraint*. The label **5**, recorded next to the **Bank** rectangle end of the Aggregation-Participation link, denotes this. No label is attached to the other end of that link, next to **Consortium**, since the participation constraint on this side of the link is one, which is the default value.

2.2 Structural Relations and Structural Links

We continue with modeling our system, and from the problem statement we read:

Each bank holds many accounts, one for each customer.

We model these facts in two stages. First, we model the fact that a bank holds many accounts. This is reflected in the OPD of Figure 2.2, where the object **Account** has been added and associated with the object **Bank** by a *tagged structural link*, labeled **holds**.

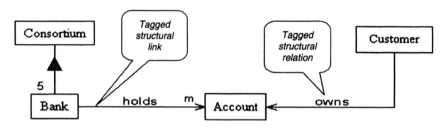

Figure 2.2. A structural link labeled **holds** associates **Bank** with **many Accounts**.

A *structural relation* models a meaningful association between two objects in the system. A *structural link* graphically represents the structural relation. The structural link is an arrow, $\longrightarrow$, with an open head. This is contrasted with the closed triangular arrowheads of the dynamic links (the input and output links and the consumption and result link, $\longrightarrow$, and the effect link, $\longleftrightarrow$). The structural link open arrowhead points from one of the objects to the other. An optional meaningful *tag* is usually recorded along the structural link, allowing for countless variations. The Aggregation-Participation relation is a type of structural relation. Due to its prevalent use, the Aggregation-Participation link is assigned a special symbol. This symbol is the black triangle we have already encountered.

Since each bank holds many accounts, we add the multiplicity constraint, which is denoted by the label "m" next to the **Account** end of the link. Continuing with our problem statement, we next read:

An account is accessible through a cash card. A customer may hold one or two cash cards.

To reflect these facts, in Figure 2.3 we add to the OPD of Figure 2.2 the objects **Customer** and **Cash Card** and the appropriately labeled structural links between pairs of objects. For example, the tag **accesses** from **Cash Card** to **Account** denotes the problem statement sentence "An account is accessible through a cash card." The numbers "**1..2**" next to **Cash Card** are the multiplicity constraint, reflecting the specification that **Customer** can own one or two **Cash Cards**. Tags such as **owns** from **Customer** to **Account** do not have to appear explicitly in the problem statement. They may result from other sources, such as knowledge gained from interviewed domain experts about the problem or familiarity of the system architect with the domain.

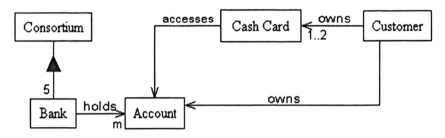

Figure 2.3. Customer and **Cash Card** added to the OPD of Figure 2.2

2.3 Processes and Procedural Links

So far, we have dealt only with objects. It is therefore no wonder that the resulting OPD did not look much different from the static, object-class model of an OO method such as the Unified Modeling Language, or UML for short (Object Management Group, 2000). Reading on the problem statement, we find the following sentence:

> The main process the ATM system is designed to carry out is the execution of cash dispensing and depositing transactions.

We are now ready to introduce the main process of the ATM system, **Transaction Executing**. As in a Data Flow Diagram (DFD), an ellipse denotes a process, with the process name recorded within it. The **Transaction Executing** process is added in the OPD of Figure 2.4. To see what this process does, we continue reading:

> Each transaction refers to a specific account. Executing the transaction will result in one of the following two outcomes: It can yield a (successful) transaction or it can issue a denial notice.

The problem statement indicates that **Transaction Executing** potentially changes, or *affects*, the object **Account**. The effect of **Transaction Executing** is to change the value of the **Balance** attribute of **Account**. However, at this abstract OPD level, **Balance** is not drawn, so the exact nature of the change cannot be reflected.

Transaction Executing is attached to the object **Account** by an *effect link*. The effect link is symbolized graphically as a bidirectional arrowhead that connects the affected object, the **Account,** with the affecting process, **Transaction Executing.**

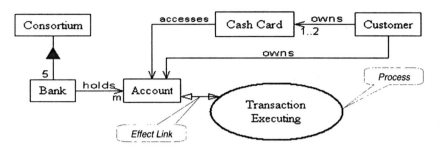

Figure 2.4. The process **Transaction Executing** of Figure 2.3 is attached to the object **Account** via an effect link.

While only the object **Account** is affected by the **Transaction Executing** process, other objects in the OPD are also involved in the process and enable its occurrence, although they are not changed by it. These objects are called *enablers*. Enablers can be of two types: *agents* and *instruments*. The difference between the two is that agents are humans, while instruments are not. Agents require an interface to interact with the system and are capable of making intelligent decisions. Chapter 5 discusses enablers in more detail.

In the OPD shown in Figure 2.5, the **Transaction Executing** process is attached to one agent and three instruments. **Customer** is the user that interacts with the system. In OPM terms, it plays the role of an *agent* in the **Transaction Executing** process. Therefore, customer is linked with this process by an *agent link* – a line ending with a filled circle, —● (sometimes called "black lollipop"). **Cash Card**, **Consortium** and **Bank** play the role of *instruments* in the process. Hence, they are linked to it by an *instrument link* – a line ending with a hollow circle: —○ ("white lollipop").

As noted, both agents and instruments are *enablers* – objects that need to be present in order for the process to occur but do not undergo any change due to the process occurrence. Without the agent **Customer,** or without the instruments **Consortium, Cash Card,** or **Bank, Transaction Executing** could not take place. However, the object **Account** is both required and *affected* by the occurrence of the **Transaction Executing** process: the balance of the account changes as a result of **Transaction Executing.** The effect link, ◁—▷, which results from a superposition of an input link and an output link, denotes the effect of a process on an object.

Recall that procedural links are links that connect an object with a process to show how that process transforms an object, or how an object enables a process. Procedural links are contrasted with structural links, which, as we have seen, express long-term relations among objects without explicitly involving processes. In Chapter 1 we saw three types of such links: input link, output link, and result link. Effect link and the two enabling links, agent link and instrument link, are additional types of procedural links.

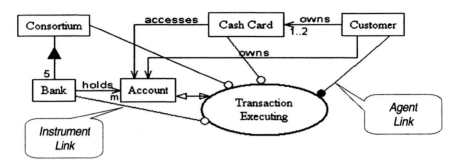

Figure 2.5. Transaction Executing is linked with the top-level objects via enabling links.

2.4 System Diagram: The Top-Level OPD

Every artificial system is designed to carry out a certain *function*. A combination of objects and processes, called the system's *architecture*, enables the execution of this function. In the ATM case study, **Transaction Executing** is the top-level process. It is responsible for interacting with the top-level objects to carry out the main system's function, which is executing monetary transactions. This **Transaction Executing** process, along with the objects involved in its occurrence, is presented in Figure 2.6 as the *system diagram*.[2] The system diagram, labeled **SD**, is the top-level OPD of the ATM system. To complete this **SD**, we look again at the problem statement:

> Each transaction refers to a specific account. Executing the transaction will result in one of the following two outcomes: It can yield a (successful) transaction or it can issue a denial notice.

Based on this statement and problem domain knowledge, we add the objects **Transaction, ATM, Cash,** and **Denial Notice,** and link them to **Transaction Executing,** using the appropriate procedural links. **SD** shows that **Transaction** and **Denial Notice** are two mutually exclusive outcomes of **Transaction Executing.** Upon successful completion, **Transaction Executing** yields the object **Transaction.** To express the generation of this new object, the result link connects the process **Transaction Executing** with the object **Transaction.** As we saw, a hollow-headed arrow originating from the generating process and pointing at the newly generated object denotes the result link. The result link symbolizes the fact that the process yields (or generates, or results in) a new object, which had not existed prior to the process occurrence or execution. If the **Transaction Executing** process fails, it does not generate the object **Transaction.** Rather, it generates the object **Denial**

---

[2] In terms of Data Flow Diagrams (DFDs), the system diagram is analogous to the *context diagram,* the top-level diagram in the process-oriented functional decomposition approach, which uses DFDs.

Notice. A second result link is therefore shown from **Transaction Executing** to the object **Denial Notice**. To denote the fact that the two result links are mutually exclusive, they originate from the same point on the **Transaction Executing** ellipse.

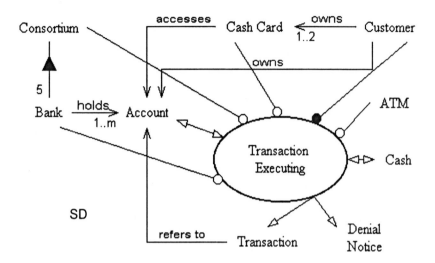

Figure 2.6. System diagram (SD) – the top-level OPD of the ATM System

The **Transaction Executing** process affects the object **Cash**. Depending on the **Type** of the **Transaction**, the effect changes the state of the **Ownership** attribute of **Cash** from **bank** to **customer** or vice versa. However, to avoid clutter, this low-level detail is not shown in **SD**. To denote the fact that an effect exists without specifying what it is, **Transaction Executing** and **Cash** are connected by the effect link. Like **Consortium, Bank,** and **Cash Card**, the object **ATM** is an *instrument* for **Transaction Executing**. **ATM** is required for the process **Transaction Executing** to take place, but this process does not affect **ATM**. An instrument link therefore connects **ATM** with **Transaction Executing**.

We have now finished drawing the system diagram of the ATM system, which is presented in Figure 2.6. It includes details concerning both the structural and the procedural aspects of the system. An example of a structural relation is the one stating that **Consortium** consists of **5 Banks**. In this respect, it resembles an object-class diagram, which is the basic model in object-oriented methods. However, in addition to objects and structural links, **SD** also presents a major stand-alone process, **Transaction Executing**. This process is connected to objects in the diagram via procedural links. For example, **Customer** is connected to **Transaction Executing** via an agent link, **Bank**, via an instrument link, and **Account**, via an effect link. Indeed, a prominent feature of OPM is that it truly and seamlessly integrates function, structure, and behavior in the same model. Until now, this unifica-

tion of the various system aspects was considered impossible, or otherwise impractical. However, as we see in this example, and many others that follow, not only is
this unification possible, but it is also beneficial. Due to the synergy that results
from the concurrent modeling of structure, function and behavior, OPDs explicitly
express these major system aspects in a single model. This is a unique, invaluable
benefit of OPM. Multi-model approaches that advocate the separation of structure,
behavior, function, and other system aspects into different models and diagram
types deny themselves the benefits of this model integration.

2.5 Zooming into the Transaction Executing Process

We proceed by examining the details of the **Transaction Executing** process and
add the following details from the problem statement:[3] To specify the details of the
Transaction Executing process, we will need to draw more elements and links
within this top-level process. However, adding more details to the top-level OPD
would make it too hard to follow, due to excess clutter. Therefore, we apply refinement. When an OPD gets too complicated, loaded or cluttered, we start a new,
descendent OPD, which is consistent with the ancestor OPD.

> Executing a transaction consists of two steps: checking the account and
> processing the transaction. The account checking process establishes the
> approval for processing the transaction. If approval is denied, the ATM
> issues a denial notice, otherwise the transaction is processed, using the
> account number the ATM read from the cash card.

In our case, the ancestor OPD is the top-level one. In the newly created OPD, we
zoom into[4] the details of the **Transaction Executing** process. The result is shown
in Figure 2.7. The new processes inside **Transaction Executing**, namely **Account
Checking**, **Transaction Processing**, and **Notifying**, are lower-level, subprocesses
of **Transaction Executing**. The new objects, **Card Data** and **Approval**, are
attributes of **Transaction Executing**. Focusing on the result of the in-zooming
operation, we have temporarily omitted the objects that are outside the in-zoomed
process. Embedding a detailed process within an ancestor OPD can yield a complex descendant OPD. To avoid this, the OPM scaling principle allows details of
the system to be distributed across various OPDs, as long as they are consistent
with each other. Details that appear in one OPD need not be repeated in any one of
the other OPDs that belong to the OPD set. This provides for reasonably small and
legible diagrams.

---

[3]   To keep the original problem statement in Frame 2 simple, this part did not appear there.
[4]   Note on terminology: we zoom into a process, which is then in-zoomed.

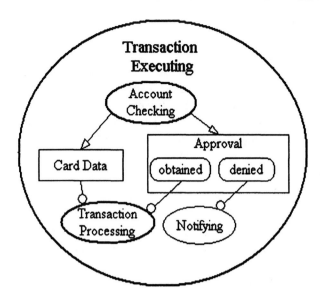

Figure 2.7. The **Transaction Executing** process of the system diagram **SD** (Figure 2.6) is in-zoomed in a new OPD labeled **SD1**. For clarity and simplicity, surrounding objects are temporarily removed.

2.6 The OPD Set

The OPD in Figure 2.7, labeled **SD1**, adds to our knowledge of the system, expressed by the top-level system diagram **SD**. Together, they constitute the first two elements of the system's *OPD set* – the collection of OPDs, which, when complete, provides a full graphic specification of the system. The system diagram, denoted **SD**, is the root node of a directed acyclic graph, i.e., a graph with no cycles or loops, called the *system map* and discussed in Section 2.8. **SD** is considered to be at detail level zero. The idea behind the OPD set concept is to trade off completeness in exchange for legibility of each individual OPD. Specification details of the system can be distributed across various OPDs in the OPD set. While adding new details, the system architect may elect to omit details that are presented in another OPD in order to keep the OPD's appearance readable and its size manageable. Each OPD provides part of the system specification. Like a mosaic, all the OPDs in the OPD set portray the complete image of the system at various levels of detail. To do so, they need to be consistent. No two OPDs contain any details that contradict each other.

An OPD set typically contains several OPDs at each level. Since we define **SD** to be at level 0, **SD1**, **SD2**, and **SD3** are three examples of labels of OPDs at level 1, right beneath the system diagram **SD**. **SD1.1**, **SD1.2**, and **SD3.1** are examples of OPDs at the second level from the top, **SD1.1.1** and **SD2.3.4** are at the third level, etc. The level number is equal to the number of separating periods in the OPD label plus one.

2.7 How to Read an OPD

How do we go about interpreting OPDs and constructing them? When examining an OPD, understanding the *structure* of the system does not mandate a particular reading order of the symbols in the OPD, as time is not an issue insofar as structure is concerned. However, to understand the *dynamics* of the system, one needs to follow a certain path, or thread of execution. The key to interpreting and constructing OPDs is the understanding of three concepts: flow of control, the timeline, and object states and conditions.

2.7.1 Flow of Control

The concept of "flow of control" in computer programs has to do with the way the program "behaves" based on its logic and the data that is input to it. When generalized to systems that are defined in terms of OPM, flow of control is the sequence of process triggering (Peleg and Dori, 1998). It is determined by a series of automated decisions based on the system's own state and the state of the environment.

To understand the flow of control concept we note that every process has two sets of objects: the preprocess object set and the postprocess object set. The *preprocess object set* is a set of objects that are needed for the process to occur. The *postprocess object set* is a set of objects that exist after the process has terminated. Some preprocess objects may be required to be at certain states, and some postprocess can also be output only at certain states. Objects in some preprocess object set can be affected by the process, while other can be generated. These transformed objects are now available and can belong to the preprocess set of the next process.

When possible, a process that occurs before another process should be depicted above it. This convention helps reading the OPD, as it follows the timeline, as discussed below. Each process is checked for determining its preprocess object set by examining all the incoming procedural (effect and enabling) links. If all the required and affected objects exist (in their proper states, if such states are specified), the process occurs and outputs the results, which can be new object(s) and/or change in the state of other object(s) and/or destruction of previously existing objects. We now look at the process that contains as its preprocess object set one or more objects that were transformed by the process that just ended. We repeat the same procedure for this process, until we have traversed all the subprocesses. If two processes require the same preprocess object set, they may be parallel or mutually exclusive. If possible, they should be depicted at about the same height in the OPD.

2.7.2 The Timeline in OPDs

By default, the timeline in an OPD flows from the top to bottom. Hence, the semantics of a process being depicted above another process is that the process on top occurs prior to the one below it, unless flow of control implies a different order. In general, the top-to-bottom process layout in the OPD represents their *default scenario*, i.e., the default order of their execution. This layout helps in reading OPDs and understanding them. While reading an OPD set, we start at the top-most subprocess within a zoomed-in process, as time flows from the top of the diagram to the bottom. Since flow of control allows for such control structures as loops, the order of process execution in an OPD that is determined by flow of control takes precedence over the order implied by the timeline. As noted, processes at the same height in the diagram can happen concurrently or alternatively. Thus, **Notifying** in Figure 2.7 is an alternative process to **Transaction Processing** and is therefore depicted at the same height as **Transaction Processing**. Here too, one can deduce that **Notifying** and **Transaction Processing** are independent (in fact, mutually exclusive) regardless of their spatial arrangement, because, as we shall see next, **Transaction Processing** requires **obtained Approval**, while **Notifying** requires **denied Approval**.

2.7.3 Object States and Conditions

As explained in Chapter 1, an object can exist in one of several states. The state at which the object is can potentially affect the behavior of the system. In the ATM case study, the object **Approval** can be in one of its two possible states: **obtained** or **denied**. If **Approval** is (at the state of being) **obtained**, then the process **Transaction Processing** occurs. If, however, **Approval** is **denied**, then instead of **Transaction Processing** the **Notifying** process takes place. Figure 2.7 expresses this graphically. There are two result links in Figure 2.7, both emanating from the **Account Checking** process. One of them points to the object **Card Data** and the other – to **Approval**.

An instrument link in Figure 2.7 points from the object **Card Data** to the process **Transaction Processing**. This means that this object is required for the occurrence of the process **Transaction Processing**. Another link that looks like an instrument link runs from the state **obtained** of **Approval** to **Transaction Processing**. Such a link, which looks like an instrument link, but goes from a *state* of an object (rather than the object itself) to a process, is a *condition link*. The semantics of the condition link is that the process is enabled if and only if the object is in the state from which the condition link originates. This semantics is similar, but not identical, to that of the instrument link. In order for the process to occur, it requires that the object be in a certain state, so the condition of being in a certain state is the "instrument" that the process requires. The condition link puts a stricter requirement on the occurrence of the process than does the instrument link. The condition link requires not only the presence of the object, as the instrument link does, but also that the required object be at a certain state.

The graphical identity between the instrument and the condition links is a second example of the context-sensitivity of OPM links. OPM links are context-sensitive because their semantics is determined not only by their graphic appearance, but also by the source and destination entities (objects, processes and states) they connect. We have encountered the first example of links' context-sensitivity in Chapter 1. Recall that the three procedural links (the result, input, and output links) we saw were arrows that looked the same, but had three related, although not identical, meanings.

Since the instrument link from **Card Data** to **Transaction Processing** and the condition link from state **obtained** of **Approval** to **Transaction Processing** terminate at two different points along the circumference of the **Transaction Processing** ellipse, they are interpreted as a logical **AND**. In other words, **Transaction Processing** can occur if and only if both **Card Data** exists and **Approval** is **obtained**.

2.8 Completing the In-Zoomed Transaction Executing OPD

Now that we have completed specifying the details of the inner content of the in-zoomed **Transaction Executing** process, we complete the OPD, labeled **SD1**. In Figure 2.8, the objects that surround **Transaction Executing** at the top-level system diagram are brought back into the picture.

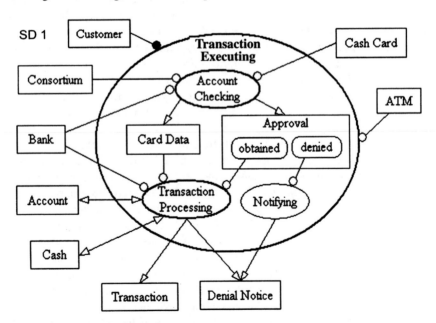

Figure 2.8. The complete **SD1**. Objects surrounding the in-zoomed **Transaction Executing** process are brought back from the top-level system diagram **SD** and linked to specific subprocesses of **Transaction Executing**.

The result was obtained by adding to Figure 2.7 the objects involved in the in-zoomed **Transaction Executing** process and linking them to the appropriate inner subprocesses. Structural relations that are shown in previous OPDs may be repeated in this new OPD, but they may also be omitted. For example, the structural link between **Customer** and **Cash Card**, labeled **owns**, is not shown in this OPD, although both objects are present. Note that the OPDs in the set are nevertheless consistent and do not contradict each other.

At the top-level OPD, shown in Figure 2.6, the object **ATM** played the role of instrument with respect to the process **Transaction Executing**. In the more detailed OPD in Figure 2.8, **ATM** is still attached to the same **Transaction Executing** process. Since **Transaction Executing** is now shown to contain three subprocesses, **ATM** is an instrument for each one of these subprocesses. This is an example of applying the distributive law to procedural links. Other objects are only linked to certain subprocesses. For example, **Bank** is an instrument of **Account Checking** and **Transaction Processing**, but not of **Notifying**.

2.8.1 Logical XOR, AND, and OR Operators

Two logical **XOR** operators exist in **SD1** (Figure 2.8). One is from **Transaction Processing** and the other is to **Denial Notice**. The **XOR** operator for **Transaction Processing** is expressed by the fact that the result links originate from the same point along the **Transaction Processing** process ellipse. The graphic notation of two procedural links originating at the same point indicates that as a result of **Transaction Processing** either **Transaction** or **Denial Notice** can be created, but not both. In a similar manner, the **XOR** operator for **Denial Notice** is expressed by the fact that the two result links, one from **Transaction Processing** and the other from **Notifying,** point at the same point of the **Denial Notice** object rectangle. The graphic notation of two procedural links terminating at the same point tells us that **Denial Notice** can only be created by one of the two processes, either **Transaction Processing** or **Notifying,** but not by both.

In summary, procedural links originating from or pointing at the same point on the circumference of a thing's symbol (rectangle or ellipse) convey the semantics of the logical **XOR** operator. Conversely, procedural links originating from or arriving at different points on the circumference of a thing's symbol convey the semantics of the logical **AND** operator. The number of links is not limited to two. Logical **XOR** and logical **OR** can be denoted by a dotted arc and a double dotted arc, respectively, connecting the incoming or outgoing links. This is useful in case it is graphically inconvenient to draw the links incoming to or outgoing from the same point.

2.8.2 The System Map

The *system map* is a directed acyclic graph,[5] a graph with directed edges that do not produce cycles. Each node in this graph is an OPD and each directed edge is a link from an abstract thing (object or process) to its detailed version in the target OPD (another system map node). Figure 2.9 shows the system map generated by the first two OPDs in the OPD set in our ATM case study. The OPD on the top-left is the root node – the system diagram, labeled **SD**. A thick gray arrow points from the contour of the abstracted **Transaction Executing** process within **SD** to the detailed (in-zoomed) **Transaction Executing** process in the second OPD, which is system diagram at the second level of detail, labeled **SD1**.

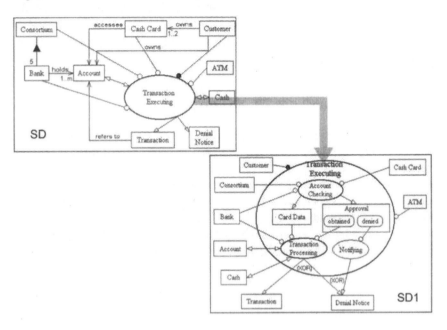

Figure 2.9. The first two nodes in the ATM system map

The OPD labels, such as **SD** and **SD1**, help identify the level of detail of the thing within the OPD. For example, considering **SD** to be at Level 0, **Transaction Executing** in **SD1** is at detail level 1. Examples of lower level OPDs are **SD2.1**, **SD2.1.3**, etc. Each node in the system map below the **SD** root is an OPD that shows some portion of the system. Each thing (object or process) in an OPD is at some level of detail, which may be different from the level of detail of other things in the same OPD. The collection of all the nodes in the system map is the OPD set – the set of all the OPDs, discussed in Section 2.6. Together, they completely specify the system. The system map provides for easy navigability among the entire set of

---

[5]   More accurately, the system map is a hypergraph, because each node is an OPD, which is a graph in its own right.

OPDs, which can vary in size from tens for medium scale projects to the hundreds for larger ones. We will continue to develop the ATM system in Chapter 3. Section 9.5 includes the system map of the complete ATM system with seven nodes, which means that the size of the OPD set of this system is 7.

2.8.3 The Ultimate OPD

A system architect or a project manager may be interested in seeing the entire system at once, in as much detail as possible. Combining all the OPDs in the OPD set can achieve this result. Lower-level OPDs can be embedded within the ones above them in the refinement hierarchy, while duplicates that appear in more than one OPD are removed. To demonstrate this, let us carry out an exercise of generating an ultimate OPD. Suppose that the entire OPD set of our ATM system is of size 2 and that it consists only of the two OPDs labeled **SD** and **SD1.** Embedding **SD1** into **SD** creates the ultimate OPD, presented in Figure 2.10. In this OPD, all the details of **Transaction Executing** from **SD1** have been embedded inside **Transaction Executing** in **SD**.

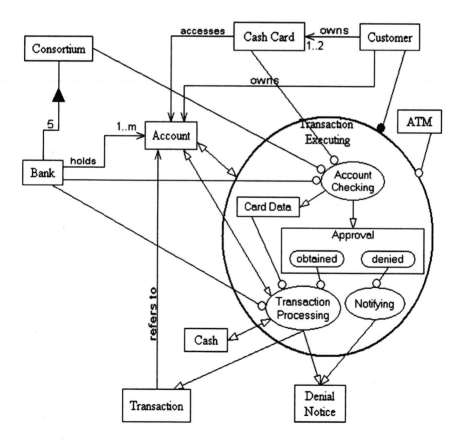

Figure 2.10. The ultimate OPD resulting from an OPD set that consists only of **SD** and **SD1**

An important point to note is how links are handled. Each link that is attached to the **Transaction Executing** ellipse in **SD** can remain as is or migrate to one or more of the subprocesses within **Transaction Executing**. Links that were attached to the **Transaction Executing** ellipse in **SD** but are not applicable to all three subprocesses inside **Transaction Executing** need to migrate from the **Transaction Executing** ellipse to the ellipses of the pertinent subprocesses. For example, the instrument links from **Consortium** and **Cash Card** were moved from the enclosing **Transaction Executing** ellipse to the inner **Account Checking** ellipse. Similarly, the instrument link from **Bank** was moved from the **Transaction Executing** ellipse to the **Account Checking** and the **Transaction Checking** ellipses. However, the instrument link from **ATM** and the agent link from **Customer** remained attached to the **Transaction Executing** ellipse, because these objects are still involved in all the three subprocesses inside **Transaction Executing**.

Examining the resulting ultimate OPD, we see that while merging two OPDs is possible, it already creates an ultimate OPD that is quite complex and not so easy to follow and digest. Imagine how an ultimate OPD would look like, one resulting from the merger of tens of OPDs at various detail levels for medium-size systems, perhaps hundreds of OPDs, in systems that are more complex. While such a diagram might be helpful in simple cases, it is evidently impractical even for a medium-size system due to the sheer number of entities and links that such an OPD would contain. The whole purpose of breaking the system into OPDs, where things are represented at various levels of detail, is to manage the system's complexity by controlling the visibility of details in each OPD. For all practical purposes, the ultimate OPD is a theoretical concept that should not be attempted in practice. In Appendix A, where we provide the OPD set of the ATM system, we do not even try to generate the ultimate OPD of the system, but provide the ATM system map instead.

2.8.4 Zooming Out of Transaction Executing

The consistency of the top-level system diagram with the more detailed OPD can be verified by zooming *out* of the **Transaction Executing** process. Out-zooming is the inverse process of in-zooming. Like an algebraic inverse operator, it acts to nullify the effect of the operator for which it is the inverse. When we apply out-zooming, we achieve *abstraction*, whereas in-zooming results in *refinement* – a more concrete, refined, or detailed view of the system. Zooming out (i.e., applying out-zooming) of the process **Transaction Executing** causes all the things (objects and processes) drawn inside this process to disappear. These things include all the subprocesses and the objects that are internal to the process, along with the links among them.

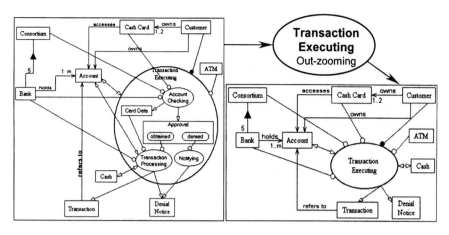

Figure 2.11. Out-zooming of **Transaction Executing** from the ultimate OPD in Figure 2.10 yields back the system diagram of Figure 2.6.

The question that arises is what happens to the links that crossed the border of the out-zoomed ellipse? In our case, these links connected objects outside **Transaction Executing** with the subprocesses recorded within **Transaction Executing**. Upon out-zooming, the edges of these links migrate, such that they are now attached to the contour of the large ellipse representing the entire **Transaction Executing** process. This is the reverse of what we did with these links while constructing the ultimate OPD. We indeed see that what we get in the OPD of Figure 2.8 conforms to the system diagram. There is no point in keeping two separate links of the same type from the same object to the same process. Therefore, two instrument links leading from **Bank** to **Account Checking** and to **Transaction Processing** are merged into one instrument link.

Summary

- Through the ATM case study, we have encountered the basics of Object-Process Diagrams (OPDs).
- An OPD uses three types of entities: objects, processes, and states.
  - Objects, symbolized by rectangles, are things that exist.
  - Processes, symbolized by ellipses, are things that transform objects by changing their states or by generating or consuming them.
  - States, symbolized by rountangles, are situations objects can be at.
- Two types of links connect the entities in an OPD: structural links and procedural links.
  - Structural links connect objects to each other or processes to each other.
    - Aggregation and tagged structural links are examples of structural links.

- Participation constraints can be tucked on structural links next to the object to denote that a quantity other than one of that object participates in the relation.
  - Procedural links connect objects or their states to processes.
    - Input, output, effect and result links are procedural links that involve some transformation of the object.
    - Agent link and instrument link are enabling links, which are required for the process to occur but are not transformed by it.
- In order to manage the complexity of large systems, new OPDs with additional, refined details can be created from existing ones.
- All the OPDs of a system constitute the OPD set.
- The system map is a graph that shows how OPDs relate to each other.

Problems

1. Draw an OPD that contains three objects related to each other by structural relations.
2. Add one or two participation constraints to the OPD of Problem 1.
3. Add a process to the OPD of Problem 1 or draw a new OPD that has one process, one agent, one instrument, and one resulting object.
4. Based on your problem domain knowledge, add the structural relations "is issued by" and "is recognized by" between the appropriate pairs of objects in the OPD presented in Figure 2.3.
5. For each structural link in Figure 2.3 add a link that goes in the reverse direction and write its appropriate tag.
6. What two objects are affected by the process in Figure 2.6? What might be the effect on each one of them?
7. The result links emanating from the process in Figure 2.6 originate from the same point on the process ellipse. Is this a coincidence? If not, explain the semantics and how it would be different if these links were to originate from different points.
8. Complete in **SD1** all the structural links that exist in **SD**. What OPD did you get?
9. What is the difference between the two XORs pictured in Figure 2.8?
10. The door system.
    a. Draw a system diagram (**SD**) of a person opening a locked door of a house from inside the house.
    b. Zoom into the **Door Opening** process in a separate diagram, **SD1**. Specify the Door parts: board, handle, lock, three hinges, and peeping hole.
    c. Add to **SD** the object **Stranger** and the process **Knocking**, and link them appropriately.
    d. Bring **Stranger** and **Knocking** into **SD1**. Link them appropriately.
11. The bathroom system.

a. Name at least three processes a person normally does in the bathroom.
b. What are the instruments for each process? Draw a single OPD including all three processes.
c. What objects are affected by the processes? Add them to the OPD.
d. Determine for which processes the person is an agent and for which the person is affected. Update your OPD.
e. Create a zoomed-in OPD (**SD1**) of **Showering**.

12. The Coffee Challenge.

Each morning, a person drinks coffee he prepares for himself. He first grinds fresh coffee beans, taken from a sealed bag, in a coffee grinder. Then, he moves the ground coffee into a paper bag he had taken from the paper bag box and put in the plastic holder of the percolator. Using the coffeepot, he fills an amount of two cups of hot water from the distilled water tank, which he then pours into the coffee maker. He then turns the coffee maker on. While the water is heated and drips through the ground coffee beans, he shakes milk in a plastic flask to generate foam. He pours the foamy milk into a cup, filling one third of it, and heats it for 20 seconds in a microwave oven. By the time the microwave oven beeps to signal that the milk is heated, the dripping of the coffee stops. He then takes the cup out of the microwave and fills it with the fresh coffee. He stirs the coffee and the foamy milk with a teaspoon and drinks the coffee. When he finishes the first cup, the milk in the flask is still foamy. He repeats whatever needs to be done to drink his second cup. When finished, he turns the coffee maker off, throws the wet ground coffee beans with the paper bag in the trash and rinses the coffeepot. He puts everything back in place so it is ready for tomorrow. Draw a system diagram of the coffee making system described above. You may elect to disregard details that are not as important as others. However, the more details you include in the OPD, the more complete is the specification.

Chapter 3
Object-Process Language

We went next to the School of Languages. ... The first Project was to shorten Discourse by cutting Polysyllables into one, and leaving out Verbs and Participle, because in Reality all things imaginable are but Nouns. ... However, many of the most Learned and Wise adhere to the new Scheme of expressing themselves by Things.

Gulliver's Travels by Jonathan Swift, 1726

In Chapter 2, we started developing the ATM system, and were exposed to OPDs, the graphic facet of OPM. Spoken or written language is the modality of OPM that is complementary to the graphics. Winograd and Flores (1987) noted: "Nothing exists except through language... In saying that some 'thing' exists (or that it has some property) we have brought it into a domain of articulated objects and qualities that exist in language." Indeed, language greatly enhances our ability to understand systems and communicate our understanding to others. In this chapter, we introduce the Object-Process Language (OPL) and show the equivalence between graphic specification through OPDs and natural language specification through OPL sentences and paragraphs. We will add the language element to the set of OPDs we started developing in Chapter 2. We will then proceed with the ATM case study, adding more detailed OPDs and their corresponding OPL paragraphs.

3.1 Motivation for a Language

OPL is the textual counterpart of the graphic OPM system specification. OPL is an automatically generated textual description of the system in a subset of natural English. OPL is extracted from the diagrammatic description in the OPD set. Devoid of the idiosyncrasies and excessive cryptic details that characterize programming languages, OPL sentences are understandable to people without previous programming experience. OPL is a dual-purpose language. First, it serves domain experts and system architects engaged in analyzing and designing a system, such as an electronic commerce system or a Web-based enterprise resource planning system. Second, it provides a firm basis for automatically generating the designed application. In other words, OPL serves two goals. One is to convert the set of Object-Process Diagrams (OPDs) into a natural language text that can be used to express and communicate analysis and design results among the various stakeholders involved in the system under construction. Users include domain experts and their executives on the customer side, who are likely to prefer reading text over examining and interpreting OPDs. These stakeholders do not usually have a command of low-level programming languages, and it is not realistic to expect them to read code. For them, OPL serves the purpose of customer verification and validation.

The second goal of OPL is to provide the infrastructure needed for continuing the application development. The same OPL text file can be the basis for generation of the two important application facets: executable code and database schema. Changes in the analysis, design and specification are almost automatically reflected in the final application. These traits make the combination of the graphic-oriented OPD and its equivalent text-based OPL counterpart an ideal infrastructure for systems specification.

3.1.1 Real-Time Textual Feedback

To enhance OPM's expressive power, we associate with each OPD a collection of sentences in OPL as a textual, natural interpretation of the OPD's graphic representation. Each graphic symbol has a textual OPL equivalent. The syntax of OPL is designed such that the resulting text constitutes plain natural, albeit syntactically restricted, English sentences. Any other natural language could have been selected as the target language, just as well. The identity of OPL and English, the *lingua franca* natural language, makes it readable and understandable to people without the need to learn any programming or pseudo-code-like language. The syntax and semantics of OPL are well defined and unambiguous. This eliminates ambiguity often found in natural languages. The system's OPL specification resulting from an OPD set is thus amenable to being scrutinized, modified, and ultimately confirmed by domain experts, who need not be software experts. The fact that the system's OPL script resulting from an OPD set is amenable to verification by domain experts, who need not be software experts, is of utmost importance.

Using an OPM-supporting software product, such as Systemantica® (Sight Code, 2001), OPL sentences are constructed automatically in real time as a response to laying out the OPD graphic symbols on the screen. Conversely, typing a syntactically correct OPL sentence complements to OPD to reflect the sentence. This capability provides for immediate system interpretation of the human developer's intents. This real-time feedback is indispensable in spotting and correcting errors at an early stage of the system lifecycle, before they had a chance to propagate and cause costly damage. Any correction of the graphic layout changes the OPL script. Changes can be implemented until a satisfactory result is obtained.

3.1.2 Closing the Requirements-Implementation Gap

The capability to directly and precisely translate analysis and design results to a subset of natural language has a tremendous advantage. As noted, prospective users and customers may be more comfortable with reading text than with interpreting OPDs, let alone deciphering program code. The OPL text and its OPD graphic equivalent help close the gap between the original requirement specification expressed as free prose, and the actual OPM system specification resulting from the analysis and design. While producing English sentences, the strict syntax of OPL[1] provides a firm basis for executable code generation and database schema definition. The OPL script, along with the OPD set, closes the gap between the requirement specification, which is usually expressed in free prose, and the actual system specification resulting from the OPM analysis and design. OPL is defined in the following definition frame[2] below.

> *Object-Process Language (OPL) is a subset of English that expresses textually the OPM specification that the OPD set represents graphically.*

In summary, OPL serves two goals. One goal is to convert the set of Object-Process Diagrams (OPDs) into a natural language script for communicating analysis and design results back to the prospective users and customers and getting their feedback and approval. The other goal of OPL, which is attained by its formality, is to provide the infrastructure needed for automated application generation, including code, database scheme, and user interface generation.

3.2 Structural Links and Structure Sentences

As noted, a meaningful tag, which is a phrase expressing the nature of the structural relation, is optionally recorded along a structural link arrow. Each structural link is translated to a sentence or part of a sentence in OPL. As we have seen, certain common structural relations, such as the Aggregation-Participation (whole-part) relation, have dedicated symbols that enable their quick recognition without the need to spell out the text of the relation's name as a tag. For example, the dedicated symbol for the Aggregation-Participation relation is the black triangle, while the one for Generalization-Specialization is the white triangle. Each such symbol is translated into a reserved OPL phrase. The reserved OPL phrase for the Aggregation-Participation relation is **consists of**.

---

[1] The syntax of OPL is defined by a context-free grammar. A set of production rules unambiguously defines how OPL sentences are to be constructed and parsed.

[2] Definition frames, which appear throughout the book, contain important definitions and principles. They are distinguishable by their thick, shaded frame.

3.2.1 The First OPL Sentence

To generate an OPL sentence from two objects that are linked with a tagged structural link, we spell out the textual equivalents of the graphic symbols in the order defined by the direction of the structural relation. We start by recording the name of the object, from which the link originates.

Figure 3.1. The OPD that gives rise to the OPL sentence **"Bank holds many Accounts."**

Concatenating the name of the object from which the link originates with the link tag and then with the name of the object to which the link points yields a meaningful OPL sentence. Applying this to the OPD in Figure 3.1, the OPL structure sentence resulting from the concatenation is:

> **Bank holds Account.**
> *(Tagged structure sentence)*

This is our first OPL sentence. It shows that an OPL structure sentence is constructed by concatenating[3] the phrase that constitutes the name of the source object, **Bank**, with the tag, **holds**, and then with the phrase that constitutes the name of the destination object, **Account**. The parenthesized note in italics below the sentence specifies that this is a structure sentence. Such notes are not part of OPL. Rather, they identify the various types of OPL sentences and appear whenever a new OPL sentence type is introduced. To make the sentences sound more natural in English, an automatic OPL sentence constructor can add the indefinite article "a" or "an" before an object class. The resulting sentence is:

> **A Bank holds an Account.**

This is a legal OPL sentence, which means the same as the sentence **Bank holds Account** and sounds better. The addition of the indefinite article is considered "syntactic sugar," a term borrowed from the domain of computer languages and compiler construction. Syntactic sugar enhances the sentence readability for humans but has no effect on the automated processing of the OPL script. In this book, we usually omit the indefinite article. Due to the multiplicity constraint, which is denoted by the label **m** next to the **Account** end of the link, the correct OPL sentence is:

> **Bank holds many Accounts.**

---

[3] juxtaposing, or arranging into an ordered list

The word many is reserved. The syntax of OPL takes care of generating the plural form (usually by adding the letter **s**) when the word many, or any number greater than one, precede the object.

If no tag is present, the default reserved phrase relates to is used. Hence, if the tag **holds** in Figure 3.2 were not present, the OPL sentence would read:

> **Bank** relates to **Account.**
> *(Relation sentence)*

As the OPL sentences we have seen demonstrate, our convention is to list OPL sentences framed in Arial font. Non-reserved phrases (names of objects, processes, states, tags, etc.) are in **bold Arial**, and reserved phrases (such as many) are in non-bold Arial font. This convention has proven to be instrumental in conveying the semantics of the OPL sentences, as it emphasizes the non-reserved words, which usually are more significant in the sentence than the reserved ones. To make punctuation marks, such as period and comma, noticeable, they are in **bold Arial** as well, although they are considered "reserved."

3.2.2 The First OPL Aggregation Sentence

Looking back at the OPD in Figure 2.1, we see that the object **Consortium** consists of five banks. To generate the OPL sentence that expresses this, we start by listing **Consortium** as the first word in the sentence. To the right of **Consortium** we juxtapose the reserved phrase consists of, which is the textual equivalent of the black triangle link symbolizing Aggregation-Participation. Finally, we add the name of the object **Bank**, which is part of the Aggregation-Participation relation. The resulting OPL sentence is:

> **Consortium** consists of **Bank.**
> *(Aggregation sentence)*

Adding the indefinite article, we get:

> A **Consortium** consists of a **Bank.**

As we have seen in the previous OPL sentence, if a participation constraint exists, it is inserted in front of the multiple object. In our case, there is a participation constraint, which is **5**. As before, we add the letter **s** to convert the single form **Bank** into the plural form **5 Banks**. The resulting OPL sentence, which is equivalent to the OPD of Figure 2.1, is:

> **Consortium** consists of **5 Banks**.

This textual interpretation of the graphic symbols yields a meaningful OPL sentence, which is a syntactically legal and semantically legible.

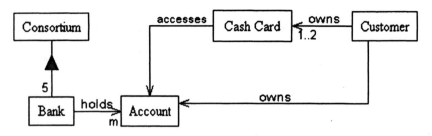

Figure 3.2. Structural relations in the ATM system diagram

3.3 The OPL Paragraph and the Graphics-Text Principle

The OPD in Figure 3.2 contains five structural links: **owns**, which appears twice, **holds, accesses,** and ▲, the Aggregation-Participation relation, which translates to the reserved OPL phrase consists of. Each structural link gives rise to a corresponding structure sentence. Hence, the five OPL sentences in Frame 3 express the same information depicted in the OPD of Figure 3.2.

> **Consortium** consists of **5 Banks**.
> **Bank holds** many **Accounts**.
> **Customer owns Account**.
> **Customer owns 1** to **2 Cash Cards**.
> **Cash Card accesses Account**.

Frame 3. The OPL sentences of the OPD in Figure 3.2 before joining the third and fourth sentences

The two sentences, **Customer owns Account** and **Customer owns 1 to 2 Cash Cards,** use the same relation **owns**. They can therefore be combined into the sentence **Customer owns Account** and **1 to 2 Cash Cards**. Such combination is possible because the same structural relation, **owns,** originates twice from **Customer**. We now have four OPL sentences that together express the content of the OPD in Figure 3.2. In the following definition frame we next define the term OPL paragraph.

> *An **OPL paragraph** of an OPD is a collection of OPL sentences that express the content of the OPD.*

An OPL paragraph that reflects the OPD in Figure 3.2, repeated from Chapter 2, is expressed in Frame 4.

> **Consortium** consists of **5 Banks.**
> **Bank holds** many **Accounts.**
> **Customer owns Account and 1** to **2 Cash Cards.**
> **Cash Card accesses Account.**

Frame 4. The OPL paragraph of the OPD of Figure 3.2

The graphics-text conversion is a two-way street: just as it was possible to extract the OPL paragraph in Frame 4 from the OPD in Figure 3.2, that OPD is reconstructible from its OPL script. The principle in the following definition frame expresses this:

> *The **graphics-text equivalence principle**: The OPD and its OPL paragraph contain the same information and are therefore reconstructible from each other.*

As an exercise, this principle can be verified by reconstructing Figure 3.2 from Frame 4 and vice versa. The graphics-text equivalence principle is true in general. One may wonder why are two modalities needed? If the text and the graphics express the same contents, wouldn't it make more sense to stick to just one of them and leave the other out?

The answer is that this apparent text and graphics equivalence is in fact a major source of OPM's expressive power. Graphics and text trigger different areas in the brain. Individuals have different preferences regarding the way they look to read and write specifications. Usually, technically oriented people prefer diagrams while business-oriented people favor text. Moreover, even for the same individual, the content may sometimes become clearer by looking at one modality while at other times the complementary modality is more helpful.

By providing the ability to switch back and forth between graphics and text, OPM system specification writers are less likely to make costly design errors, while readers of the specification are more likely to fully comprehend the system and detect mistakes or omissions that may have slipped by. The specification reader can fill knowledge gaps in her/his understanding of the system that were formed while examining one modality by looking at the other one. By doing so, the reader reinforces familiarity with the specification and can more easily detect design errors or omissions.

3.3.1 Extending the OPL Paragraph

Having generated an OPL paragraph for the OPD in Figure 3.2, we wish to extend it such that it becomes equivalent to the ATM system diagram, presented in Figure 2.6. Figure 3.3 shows part of the system diagram that we have not yet dealt with from an OPL viewpoint. It contains the process **Transaction Executing** and the set of all the objects that are required in order for this process to occur. This set is the preprocess object set of **Transaction Executing**. It consists of five enablers (agents or instruments) – **Customer, Cash Card, Consortium, Bank,** and **ATM** – and two affectees (affected objects) – **Account** and **Cash**. Of the five enablers, **Customer** is an agent and the rest are instruments.

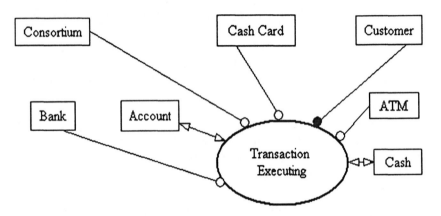

Figure 3.3. The preprocess object set of **Transaction Executing**

3.3.2 Enabling Sentences

Recall that an object is identified as an agent with respect to a process if an agent link leads from the object to the process. Since an agent is a human or an organizational unit with decision making ability, OPL uses the reserved word handles to express the agent link that connects the agent to the process. The agent sentence is constructed by the sequence of words that starts with the name of the object playing the role of the agent (**Customer,** in our case). The agent name is followed by the reserved word handles and then by the name of the process (**Transaction Executing,** in our case). The resulting sentence is:

> **Customer** handles **Transaction Executing.**
> *(Agent sentence)*

We continue with the instruments. An object is an instrument with respect to a process if an instrument link leads from the object to the process. An instrument is an object that the process requires, but it is unaffected by the process occurrence.

The reserved OPL word requires is therefore used to represent an instrument link. The instrument sentence is constructed by the sequence of words that starts with the name of the process (**Transaction Executing**, in our case). The process name is followed by the reserved word requires and then by the name of the object that plays the role of the instrument (**ATM**, for example). The resulting sentence is:

> **Transaction Executing** requires **ATM**.
> *(Instrument sentence)*

Cash Card is another instrument. The instrument sentence that links it to Transaction Executing is:

> **Transaction Executing** requires **Cash Card**.

The two instrument sentences can be combined into one:

> **Transaction Executing** requires **ATM** and **Cash Card**.

The reserved OPL word **and** links two things of the same type, in our example these two things are instruments: **ATM** and **Cash Card**. Not just two, but any number of things of the same type can be enumerated in such a list. When there are more than two elements in the list, each element is separated from the next by a comma. Hence we get the OPL sentence:

> **Transaction Executing** requires **ATM**, **Cash Card**, **Consortium**, and **Bank**.

3.3.3 Transformation Sentences

Account and **Cash** are the two affectees of **Transaction Executing**. Upon completion of the **Transaction Executing** process, the states of both **Account** and **Cash** are different than their states just before the process started. An OPL effect sentence expresses this. Its construction is identical to that of an instrument sentence, except that the reserved OPL word **affects** is used instead of **requires**. The resulting sentence is:

> **Transaction Executing** affects **Account** and **Cash**.
> *(Effect sentence)*

The agent, instrument and effect sentences we generated by now account for all the procedural links in Figure 3.3. In Figure 3.4, the two objects **Transaction** and **Denial Notice** that can result from **Transaction Executing** are added. Suppose that

Transaction is the only object that can result from **Transaction Executing**. The result sentence in this case would be:

> **Transaction Executing** yields **Transaction.**
> *(Result sentence)*

As noted in Chapter 2, the semantics of the result links originating from the same point on the **Transaction Executing** ellipse is that of the logical exclusive OR (XOR) operator: either one object is generated or the other, but not both. The resulting sentence is therefore:

> **Transaction Executing** yields either **Transaction** or **Denial Notice.**
> *(XOR result sentence)*

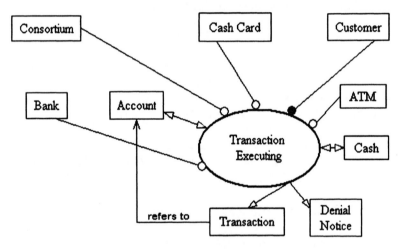

Figure 3.4. Adding the objects that alternatively result from **Transaction Executing**

Recall that the semantics of two procedural links originating from or terminating at different points on the thing's circumference is that of a logical AND operator. Therefore, if the result link from **Transaction Executing** to **Denial Notice** would originate from a point other than the one from which the other result link originates, the result sentence would read:

> **Transaction Executing** yields **Transaction** and **Denial Notice.**

Finally, Figure 3.4 also gives rise to the following structure sentence:

> **Transaction refers to Account.**

3.3.4 The SD Paragraph

Combining the sentences we have extracted from **SD**, we obtain the OPL paragraph of the system diagram, which is listed in Frame 5. An OPL paragraph starts with an OPL paragraph header, which is a phrase consisting of the OPD node name in the system map (**SD** in our case) followed by the reserved word Paragraph, followed by a colon. The header in Frame 5 is **SD Paragraph**. This OPL paragraph contains five structure sentences and four behavior sentences. The reader should be able to verify that **SD** Paragraph contains the same information that the OPD in Figure 3.5 describes.

SD Paragraph:
Consortium consists of **5 Banks**.
Bank holds many **Accounts**.
Customer owns Account and **1** to **2 Cash Cards**.
Cash Card accesses Account.
Customer handles **Transaction Executing**.
Transaction Executing requires **ATM, Cash Card, Consortium**, and **Bank**.
Transaction Executing affects **Account** and **Cash**.
Transaction Executing yields either **Transaction** or **Denial Notice**.
Transaction refers to Account.

Frame 5. The OPL paragraph that reflects the information in the system diagram of Figure 3.5

3.4 More OPL Sentence Types

Zooming into **Transaction Executing**, shown in the OPD of Figure 3.5, we switch from **SD** to **SD1** and encounter a number of new OPL sentence types. Let us see how such a sentence is applied to the zooming into **Transaction Executing** done in Figure 3.5. To express zooming in OPL, we use an *in-zooming sentence*.

Transaction Executing from **SD** zooms in **SD1** into **Account Checking, Transaction Processing**, and **Notifying**, as well as **Approval** and **Card Data**.
(In-zooming sentence)

This in-zooming sentence describes the transition from **SD** to **SD1**. It is therefore positioned between **SD paragraph**, the OPL paragraph that corresponds to **SD**, and **SD1 paragraph**, the OPL paragraph that corresponds to **SD1**. The beginning of the sentence reads:

Transaction Executing from **SD** zooms in **SD1** into...

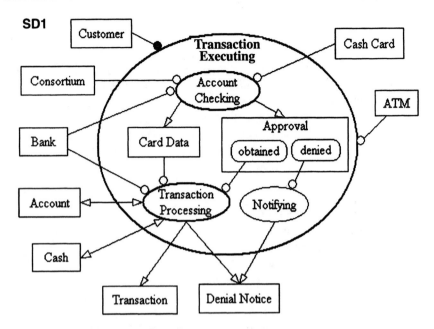

Figure 3.5. The **Transaction Executing** process zoomed-in

This part explains the relationships between the source OPD, which is the system diagram labeled **SD**, and the newly created destination OPD, labeled **SD1** (see Section 2.4 and Section 2.5). Following the reserved word **into** are the names of processes and objects that are depicted inside the in-zoomed process. The processes are listed first, followed by a comma and the reserved phrase **as well as**, followed by a list of objects. We elaborate on in-zooming and out-zooming sentences in Section 3.5.

3.4.1 State Enumeration and Condition Sentences

As Figure 3.5 shows, the object **Approval** can be in one of its two possible states: **obtained** or **denied**. In Chapter 1 we saw a similar example with **Person**, who can be **single** or **married**. The following OPL sentence expresses the two states of **Approval**:

> **Approval** can be **obtained** or **denied.**
> *(State enumeration sentence)*

Likewise, the state enumeration sentence in the wedding example is

> **Person** can be **single** or **married.**

A condition link connects a state of an object to a process. Two condition links exist in Figure 3.5. One goes from the state **obtained** of **Approval** to **Transaction Processing** and the other, from the state **denied** of **Approval** to **Notifying**. A condition link is translated to an OPL condition sentence, which uses the reserved phrase **occurs if** and the reserved word is. The simple OPL condition sentence that corresponds to the first condition link in the OPD of Figure 3.5 is:

> **Transaction Processing** occurs if **Approval** is **obtained.**
> *(Condition sentence)*

A condition sentence is like an "if ... then" statement in a typical procedural programming language. Indeed, the condition sentence can be rephrased as "If **Approval** is obtained then execute **Transaction Processing**." The OPL condition sentence that corresponds to the condition link from the state **denied** of the object **Approval** to the process **Notifying** is:

> **Notifying** occurs if **Approval** is **denied.**

The OPL phrases **Approval** is **obtained** and **Approval** is **denied** are the conditions of the two OPL condition sentences above, respectively. Since **obtained** and **denied** are the only two states of **Approval**, it might have been possible to combine these two sentences into the following compound condition sentence:

> **Transaction Processing** occurs if **Approval** is **obtained**, otherwise **Notifying** occurs.

A compound condition sentence is like an "if ... then ... else ..." statement in many procedural programming languages. It can be rephrased as "If **Approval** is obtained then execute **Transaction Processing** else execute **Notifying**." Reversing the order of the conditioned processes, the following compound condition sentence with the same semantics is obtained:

> **Notifying** occurs if **Approval** is **denied**, otherwise **Transaction Processing** occurs.

The disadvantage of both these compound condition sentences is that neither of them explicitly state the conditions for the occurrence of both **Transaction Processing** and **Notifying**. Therefore, Frame 6, which is the OPL Paragraph that corresponds to the OPD in Figure 3.5, lists the two separate condition sentences rather than the compound one. An automatic OPL generator converts each condition link into a condition sentence, avoiding the generation of compound condition sentences. As we shall see below, compound condition sentences are useful for expressing conditions that result from Boolean objects. Since the states of Boolean objects are simply **yes** or **no**, rather than states like **obtained** or **denied**, no information is lost by joining the sentences.

3.4.2 AND, XOR, and OR Logical Operators

As noted, when two or more links originate from different points along the contour of a thing (object or process), their default semantics is the logical **AND** operator. In Figure 3.5, for example, the two result links from the **Account Checking** process that generate the objects **Card Data** and **Approval** leave from different points along the ellipse of this process. The corresponding result sentence is therefore:

Account Checking yields **Card Data** and **Approval**.

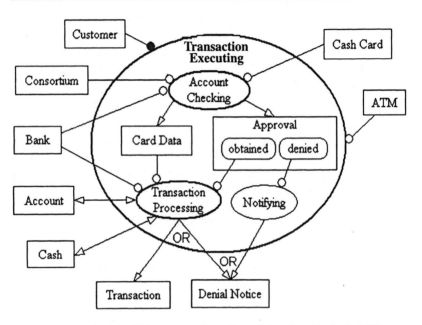

Figure 3.6. The two logical XOR operators from Figure 3.5 replaced by logical OR operators

Conversely, the semantics of links that originate from, or arrive at, the same point along the process ellipse, object box or state oval, is one of logical **XOR**. Note that the two result links in Figure 3.5 originate from the same point along the **Transaction Processing** ellipse. Textually, this is expressed using the reserved phrase **either ... or**. The resulting OPL sentence is:

Transaction Processing yields either **Transaction** or **Denial Notice**.
   (XOR result sentence)

Likewise, the two result links terminating at the same point of the **Denial Notice** object box, give rise to the following OPL sentence:

Either **Notifying** or **Transaction Processing** yields **Denial Notice**.
   (Process XOR result sentence)

To denote the logical **OR** operator, as Figure 3.6 shows, we write **OR** between the pertinent links next to their common point of origin. Another possibility is to join the two links with a small arc near that common point. The reserved phrase is simply **or**. If the logical **OR** operator is used, the XOR result sentence above becomes:

Transaction Processing yields **Transaction** or **Denial Notice.**
(OR result sentence)

The process XOR result sentence above becomes:

Notifying or **Transaction Processing** yields **Denial Notice.**
(Process OR result sentence)

Examining these sentences we see that the use of the logical **OR** operator does not make sense in this case. **Transaction Processing** cannot yield both **Transaction** and **Denial Notice**, neither can **Denial Notice** be generated by both **Notifying** and **Transaction Processing**.

3.4.3 The SD1 Paragraph

Assembling the OPL sentences we have developed so far, Frame 6 lists **SD1 Paragraph,** the OPL paragraph that expresses the information contained in the OPD labeled **SD1** in Figure 3.5.

Transaction Executing from **SD** zooms in **SD1** into **Account Checking, Transaction Processing,** and **Notifying,** as well as **Approval** and **Card Data.**
SD1 Paragraph:
Account Checking yields **Approval** and **Card Data.**
Customer handles **Transaction Executing.**
Transaction Executing requires **ATM.**
Account Checking requires **Consortium, Bank,** and **Cash Card.**
Account Checking yields **Card Data** and **Approval.**
Approval can be **obtained** or **denied.**
Transaction Processing requires **Bank** and **Card Data.**
Transaction Processing occurs if **Approval** is **obtained.**
Notifying occurs if **Approval** is **denied.**
Transaction Processing affects **Account** and **Cash.**
Transaction Processing yields either **Transaction** or **Denial Notice.**
Either **Notifying** or **Transaction Processing** yields **Denial Notice.**

Frame 6. SD1 Paragraph, the OPL script of the OPD in Figure 3.5

3.4.4 In-Zooming and Out-Zooming Sentences

At this point, we have completed writing the OPL paragraphs of SD and SD1 developed in Chapter 2. From now on, as we proceed with modeling the details of the ATM system and encounter new types of OPD and OPL sentences, we develop more detailed OPDs and their OPL paragraphs concurrently. Let us continue reading the problem statement:

> The checking of the account starts as the ATM validates the cash card details with the consortium and the appropriate bank. If the card is valid, the password is checked with the customer, and only then, approval is obtained.

Since the problem statement passage above drills down into the **Account Checking** process, we need to apply another in-zooming sentence to this process in order to expose its three lower-level subprocesses, which we call **Cash Card Validating**, **Password Checking**, and **Approval Denying**. The result is shown in the OPD of Figure 3.7, where these subprocesses are now visible. The OPL in-zooming sentence that expresses the generation of a new OPD labeled **SD1.1** from **SD1** is:

> **Account Checking** from **SD1** zooms in **SD1.1** into **Cash Card Validating**, **Password Checking**, and **Approval Denying**, as well as "**Cash Card** is **valid?**"

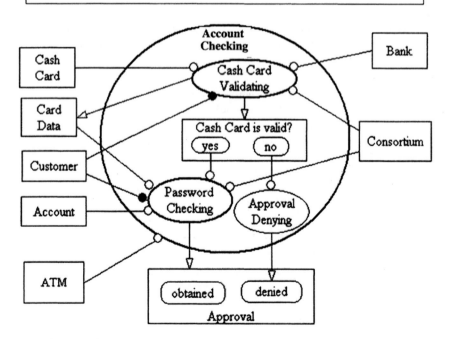

Figure 3.7. Zooming into the **Account Checking** process

Frequently, analysis and design activities of real-life systems do not follow the nice top-down specification, which we have been following up until now, but rather a middle-out fashion. Middle-out analysis and design occur, for example, when the system architect realizes that a set of processes already specified in the OPD can be consolidated into one higher-level process. This abstraction, called *out-zooming*, is the inverse of the in-zooming that we have encountered. The out-zooming sentence from **SD1.1** to **SD1** is:

Cash Card Validating, Password Checking, and **Approval Denying,** as well as **"Cash Card** is **valid?"** from **SD1.1** zoom out as **Account Checking** in **SD1.**

(Out-zooming sentence)

In-zooming is a *refinement* sentence. In-zooming sentences act as the glue among OPD paragraphs, relating them and the OPDs they represent to each other. Chapter 9 discusses refinement and abstraction.

The top subprocess in the OPD of Figure 3.7, which is the first one to occur, is **Cash Card Validating.** This process yields the Boolean object **"Cash Card is valid?"** Returning to the in-zooming sentence from **SD1** to **SD1.1,** we note that the things into which **Account Checking** zooms are listed following the reserved word **into.** The three subprocesses, **Cash Card Validating, Password Checking,** and **Approval Denying** are listed first, followed by a single object called **"Cash Card is valid?"** This is a Boolean object, which is discussed next.

3.5 Boolean Objects and Determination Sentence

In most programming languages, a *Boolean variable* is a logical variable which can be represented by a single bit, since it has exactly two-values, called True and False. Similar to this concept, an OPM *Boolean object* is an object generated by a decision process, which answers a certain "yes or no" question. A process generates a Boolean object just as it generates any other object. A Boolean object ends with a question mark, as in **"Cash Card is valid?"** The question, which is also the name of the Boolean object, is phrased such that it starts with a name of an object (**Cash Card**), followed by the reserved word **is** or the reserved word **are,** followed by a possible state or value of the object (**valid**).

As Figure 3.7 shows, **Cash Card Validating** is the process that generates **"Cash Card is valid?"** Unlike regular objects, however, the generation of a Boolean object is not expressed by a result sentence, but rather by a determination sentence:

Cash Card Validating determines whether **Cash Card** is **valid.**

(Determination sentence)

3.5.1 Boolean Condition Sentences

Each of the Boolean states **yes** and **no** gives rise to another *Boolean condition sentence*. The condition links from **yes** and **no** to the processes **Password Checking** and **Approval Denying** respectively yield the following sentences:

> **Password Checking** occurs if **Cash Card** is **valid.**
> *(Boolean condition sentence)*
> **Approval Denying** occurs if **Cash Card** is not **valid.**
> *(Negative Boolean condition sentence)*

The negative Boolean condition sentence is obtained by plugging the reserved word **not** following the reserved word **is**, which is part of the Boolean object name.

3.5.2 Compound Condition Sentences

A compound condition sentence is useful for expressing conditions that result from a Boolean object, because the Boolean object states are simply **yes** or **no**, so no information is lost. Joining the two Boolean condition sentences above, we get:

> **Password Checking** occurs if **Cash Card** is **valid,** otherwise **Approval Denying** occurs.
> *(Compound condition sentence)*

3.5.3 State-Specified Generation Sentence

Password Checking is the process that generates the object **Approval. Approval** can be generated at one of its two possible states, **obtained** or **denied.** This is expressed by the generation sentence we already know:

> **Password Checking** yields **Approval.**

Approval Denying also generates the object **Approval,** but unlike **Password Checking,** it always generates **Approval** at the state **denied.** The following sentence expresses this:

> **Approval Denying** yields **denied Approval.**
> *(State-specified generation sentence)*

3.5.4 Converting a Dual-State Object into a Boolean Object

A dual-state object, i.e., an object with two states, can be converted into a Boolean object with a name that contains one of its states. For example, the OPD in Figure

3.8 is an alternative OPD to the OPD in Figure 3.5, in which the object **Approval** with its states **obtained** and **denied** is replaced by the Boolean object **"Approval is obtained?"** with its states **yes** and **no**. The two Boolean condition sentences are:

> **Transaction Processing** occurs if **Approval** is **obtained**.
> **Notifying** occurs if **Approval** is not **obtained**.

Since in this OPD the state **denied** is not explicitly specified, no information is lost if we combine these two sentences into the following compound condition sentence:

> **Transaction Processing** occurs if **Approval** is **obtained**, otherwise
>   **Notifying** occurs.

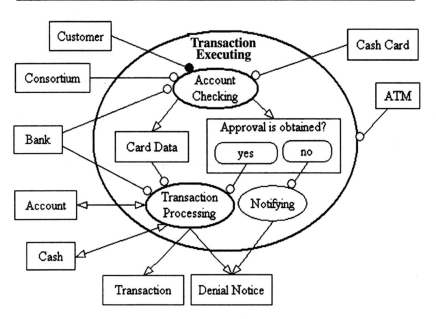

Figure 3.8. An alternative OPD to the original OPD in Figure 3.5, in which the object **Approval** with its states **obtained** and **denied** is replaced by the Boolean object **"Approval is obtained?"**

The complete ATM case study contains four more OPDs in addition to the three we have constructed. Including them all here does not serve the purpose of this chapter. They are therefore provided in Appendix A, where the entire ATM system with seven OPDs and their corresponding OPL paragraphs demonstrate the principles of complexity management. The system map is provided in Appendix A.

3.6 OPD-OPL Item Pairs and Synergy

OPM uses both a graphic modality, Object-Process Diagrams (OPDs) and a textual modality, Object-Process Language (OPL). Each OPL sentence is a natural, albeit constrained, English sentence that describes a portion of the OPD comprising entities (objects, processes, or states) and links among them. The combination of OPD entities and links is called an OPD sentence. Each OPD is a collection of such OPD sentences. It is described textually by a collection of OPL sentences called an OPL paragraph. A set of consistent, interrelated OPDs, called the OPD set, constitutes the system's graphic representation. The OPL script is its equivalent textual system specification. The OPM system specification is the top of a hierarchy of OPM pairs of graphic and textual OPM item pairs, which are defined in Section 9.6.

We noted that many individuals prefer one modality (graphics or text) over the other. Often, designers and engineers prefer the diagrams, while business and law people prefer the text. The coexistence of the two modalities provides for the ability to clarify the model and gain more insight into it. The ATM case study has demonstrated vividly that graphics and text complement each other as they reinforce the understanding of the modeled system. The synergy gained from the graphics-text combination is orthogonal to the synergy obtained from the structure-behavior integration, and further increases the expressive power of OPM. By jointly expressing the three major system aspects in a single, coherent model and in equivalent graphic and textual representations, OPM constitutes a common thread that enables continuous development throughout the system's lifecycle. It facilitates communication among the many types of stakeholders, each having a different background. Customers and their domain experts, who set the requirements, can interpret both representation modalities with relative ease. After their final approval, the resulting OPL script can be ultimately compiled into an executable application.

Summary

- Object-Process Language (OPL) is the textual equivalent of the information presented graphically in an OPD.
- OPL is expressed in sentences that read like natural English and require almost no training.
- Every type of OPD link gives rise to an OPL phrase that is used in some OPL sentence type.
- Each OPL sentence type has a unique template, in which the reserved and nonreserved phrases appear in a particular order that conveys a precise meaning.
- A subset of OPD entities and links that connect them constitute an OPD sentence.
- Every OPD sentence gives rise to an OPL sentence or to a part of an OPL sentence.

- OPL sentences are constructed by combining OPL reserved phrases with non-reserved, system-specific, or user-defined phrases, which are the names of the entities in the system.
- OPM specification is a hierarchy of OPD-OPL pairs, each comprising an OPD item and a corresponding OPL item:
- An OPD item and an OPL item at the same hierarchy level are reciprocally reconstructible from each other.
- Each of the two elements in an OPD-OPL pair is semantically equivalent to its counterpart. Syntactically, though, there may be more than one way to express the OPL interpretation of the OPD
- OPL and OPD complement each other in that one can be consulted to clarify questionable points in the other.
- Domain experts do not need to be OPD experts; they can figure out the meaning of the links form reading the OPL.

Problems

1. Draw an OPD that contains three objects related to each other by structural relations. Write the OPL sentences that relate to them.
2. Add one or two participation constraints to the OPD of Problem 1 and rewrite the OPL sentences.
3. Write an OPL paragraph that has one process, one agent, one instrument, and one resulting object. Draw the corresponding OPD.
4. Referring to Figure 3.7, redraw the relevant part of the OPD where instead of the Boolean object **"Cash Card is valid?"** add the attribute **Validity** with states **valid** and **invalid**. Write the corresponding OPL.
5. Make a list of the various OPL structure sentences you have encountered, i.e., sentences that involve only objects. For each sentence type invent a new sentence of the same type and draw its OPD.
6. Repeat Problem 5 for behavior sentences, i.e., sentences that involve objects and processes.
7. Write an OPL paragraph describing a student brushing his/her teeth. Note enablers and affectees, as well as the structure of the toothbrush. Draw the OPD.
8. Refer to the OPD in Figure 3.9.
   a. Hand-write the OPL paragraph for this OPD.
   b. Suppose that instead of **"Sitting,"** the process in Figure 3.9 is **"Being Seated Process."** How would this affect the link between **Person** and the process? Draw your own version of the OPD with this change.
   c. Add to the OPD in part b the states **standing** and **seated**, and rewrite the pertinent part of the OPL paragraph.

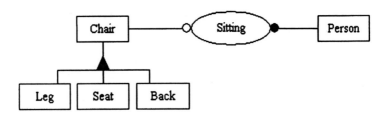

Figure 3.9. OPD for Problem 8

9. Hand-draw the OPD for the following OPL paragraph:

> **Airline Carrier** and many **Customers** handle **Flying.**
> **Customer** exhibits **Location** and **Mileage Record.**
> **Location** can be **origin** or **destination.**
> **Flying** exhibits **Mileage.**
> **Flying** changes **Location** from **origin** to **destination.**
> **Flying** affects **Mileage Record.**

10. Add to the OPD and OPL of Problem 9 the fact that **Airline Carrier** consists of at least one **Airplane**, and at least one **Pilot**. Note the difference between **Pilot** and **Airplane**.
11. The Coffee Challenge continues.
    a. Generate by hand the OPL paragraph of **SD**, the system diagram of the **Coffee Making** system from the Coffee Challenge Problem of Chapter 2. Compare your results to those obtained by Systemantica® (Sight Code, 2001).
    b. Add OPD(s) at level 1, i.e., the level below the system diagram (**SD1, SD2,** etc.) to **SD** of the **Coffee Making** system.
    c. Generate by hand the OPL paragraphs for the OPDs you created in Part b. Compare your results to those obtained by Systemantica®.
    d. Generate the ultimate OPD from the system diagram and the OPDs at level 1.
    e. Generate the OPL script of the system from the ultimate OPD of the **Coffee Making** system.
    f. Use the solution for the previous parts to complete the OPM (OPD and OPL) specification of the coffee making system described in Problem 12 of Chapter 2.

Chapter 4

Objects and Processes

> *Objects in the mirror appear closer than they really are.*
> Oldsmobile Royale 88, left mirror (bottom)
>
> *Do not throw diapers, sanitary napkins and objects into*
> *the toilette.*   Boeing 747, lavatory cover (when raised)

Immanuel Kant said, "objects are our way of knowing." While this is the truth, it is not the whole truth. Objects are our way of knowing the structure of systems. To understand systems' behavior, processes are required. We know of the existence of an object if we can name it and refer to its unconditional, relatively stable existence, but without processes we cannot tell how this object changes over time.

Objects and processes, collectively referred to as "things," are the two types of OPM's universal building blocks.[1] OPM views objects and processes as being on equal footing, so processes are not modeled as "second class citizens" that are subordinate to objects, as is the case with the object-oriented (OO) approach. Major system-level processes can be as important, or even more important than objects in the system model. Hence, they can be modeled independently of a particular object class. This paradigm enables OPM to model real world systems in a single simple model that is faithful to reality and specifies it in graphics and text.[2]

Being able to tell them apart and use them properly is key to mastering OPM. To define these fundamental concepts and to communicate their semantics, we shall first discuss "existence" and "change," laying the foundation for defining objects and processes and distinguishing between them. We will then introduce the "essence of things" and examine the difference between "physical" and "informatical" things. The word "informatical" refers to a generalization of data, information, knowledge, expertise and ingenuity without any reference to their physical manifestation.

4.1 Existence, Things, and Transformations

Webster's New Dictionary (1997) defines *existence* as the noun derived from exist, which is *be, have being, continue to be*. To exist means to stand out, to show itself, and have an identifiable, distinct uniqueness within the physical or mental realm. A

---

[1] Objects, processes, and states are OPM *entities*. States, which are discussed throughout the book, are considered secondary entities, below objects.

[2] This paradigm seems to violate the OO encapsulation principle, thus rendering OPM a "non-pure" object oriented approach. OPM does not present itself as an OO method, but rather as a holistic approach to modeling systems. We elaborate on this issue in Chapter 15.

thing that exists in physical reality has "tangible being" at a particular place and time. Because it stands out and shows itself, we can point to it and say: "Now, there it is."

To stand out means to present a stable form against a background of something else that exists (with which, for the time being, we are not concerned). The notion of "background" is essential, for if there were nothing else that existed, there could not be the contrast of one thing standing out and distinguishing itself from a background of things that exist along with it. The stable form that the existing thing must exhibit is "substantially unchanging" long enough (relative to the typical rate of change of the background) for it to be recognized as "standing out". That which we can never identify, or have its identity be inferred in some way, can have no existence for us. In other words, "to stand out" requires a continuous identifiability over an appropriate duration of time. When we consider existence along the *time* dimension, there are two modes of "standing out," or existence of things. In the first mode, the "standing out" takes place throughout a positive, relatively substantial time period that occurs in such a way that it is basically unchanging, stable, or persistent. We call that which stands out in this mode *object.*

4.1.1 Objects

Webster's Dictionary (1984) provides the following two relevant definitions for the word object:

(1) anything that is visible or tangible and is stable in form.
(2) anything that may be apprehended intellectually.

These two definitions correspond to our notion of *physical* and *informatical* objects. The first definition is the one we normally think of when using the term object in daily usage. The second definition pertains to the informatical, intangible facet of objects. Informatical objects are different from their physical counterparts in that informatical objects have no physical existence and, being intangible, they do not obey the basic laws of physics. However, the *existence* of informatical objects does depend on their being symbolically recorded, inscribed, impressed or engraved on some tangible medium. The medium can be the human brain, paper, some electromagnetic medium, stone, etc.

Qualifying the human brain as a tangible medium that can store intangible things may perhaps seem to some readers cynical or inappropriate. It therefore deserves special discussion. The central nervous system, of which the brain is the major part, is the informatical system in humans and other organisms that controls and regulates the entire organism. The human recollection or the mental record of a thing is a mostly mysterious way that a thing is inscribed in one's mind. Among many other, more elated capabilities of intelligence and emotions, the magnificent capability of the human brain to remember things qualifies it as a superb recording medium. This by no means belittles humans (or maybe other organisms or, in the future, perhaps even some very intelligent machines), since they are, of course, a whole lot more than mere memory machines. A human brain stores vast amounts of data, information, and knowledge of various forms that are the essential basis for intelligence, including inference, prediction, decision-making and behavior. Human memories are not just a series of objects representing facts, images, faces, names, shapes, figures, forms and symbols. They also include structural and behavioral relationships that exist among these objects, and the rules that govern them. Anything that is recorded in the human brain is an informatical object. This informatical object may be the record of some (tangible and/or intangible) set of objects and the processes that the objects in the set undergo.

Since OPM uses objects that are physical or informatical, we define object as something that captures these two facets without committing to either one, while including the element of "existence throughout time."

> An *object* is a thing that has the potential of stable, unconditional physical or mental existence.

This definition is quite remote from the classical definitions of object found in the OO literature, which can be phrased as "An object is an abstraction of attributes and operations that is meaningful to the system."

According to Webster's (1997), an object is "a material thing; that to which feeling or action is directed; end or aim; word dependent on a verb or preposition." To make it possible to refer to objects, natural languages assign names to objects. The name of an object constitutes a primary identifying symbol for that object. The name makes the object amenable to reference and communication among humans. These names are known as *nouns*. However, as part of speech, noun is a syntactic term, while the term object is semantic. While natural language nouns primarily represent objects, not every noun is an object. In fact, as we next show, many of the nouns are processes. This is a source of major confusion. We must not only be aware of this, but also be able to tell apart nouns that are objects from nouns that are processes.

4.1.2 Transformation and Processes

We noted that there are two modes of standing out. The first was in space. The second is in time. In the second mode of "standing out," the standing out occurs "in a changing way" against a background, which is substantially stable. Because that which stands out is changing, it may have different names as it undergoes its transformation. In particular, the name given to "that, which stands out after the change is complete," may be different from the name the thing had before the change occurred. In this case, it is convenient to think of the thing that has brought about a transformation as some carrier that is "responsible" for this transformation. When we are inclined to think in this way, what we really are thinking about, is the patterned changing, the series of transformations that one object or more undergo. For the convenience of language or thinking, we associate this patterned changing with the "carrier" to which we mentally assign the "responsibility." We define *transformation* as a generalization of change, generation and destruction of an object.

> **Transformation** is a generalization of consumption (destruction), change (effect), and generation (construction) of one or more objects.

We call the carrier that causes transformation *process*, and we say that the process is "that which brought about the transformation" of an object. However, that carrier is just a metaphor, as we cannot "hold" or touch a process, although it may be entirely physical. What we may be able to touch, see, or measure, is the object, or one or more of its attributes, as the process is transforming the object. At any given point in time before, during, or after the occurrence of the process, that object can potentially be different from what it is in another point in time. Using our human memory, we get the sense of a process by comparing the present form of the object being transformed, to its past form.[3]

According to Webster's (1997), a process is "a state of going on, series of actions and changes, method of operation, action of law, outgrowth." The American Heritage Dictionary (1996) defines process as "a series of actions, changes, or functions, bringing about a *result*." Focusing on transformation and the result or effect that it induces, we will adopt the following definition.

> A **process** is a pattern of transformation that an object undergoes.

This definition immediately implies that no process exists unless it is associated with at least one object, which is responsible for the transformation. Hence, while we could refer to an object without necessarily using the term "process," the opposite is not possible – the "existence" of a process depends on the existence of at least one object, upon which that process applies some transformation.

When we say that the process brought about the creation (or construction) of an object, we mean that the object, which had not existed prior to the occurrence of the process, has undergone a radical change. This change made it stand out and become identifiable against its background. Analogously, when we say that the process brought about the elimination (or destruction) of an object, we mean that the object, which once stood out, has undergone a radical change, due to which it no longer exists. These radical changes of creation and elimination are extreme versions of transformation. Processes are the only things that cause creation and elimination, of objects as well as changes in the objects' states. Collectively, creation, elimination and change are termed *transformation*.

---

[3]   Hence, we can almost say that a process is in a human's mind, as only through comparing objects at various points in time can we tell that a process took place.

4.2 Processes and Time

There are two perspectives from which a system can be contemplated. One perspective is an instantaneous, snapshot-like, structural one, which views the world as it is in any particular moment of time. This perspective has no time dimension. It represents the objects in the world and the time-independent relationships that may exist among them. A second perspective is one that does include time and represents the time relationships among the things in the world. From this temporal viewpoint, the existence of an object is persistent – the object statically "sits there," waiting to be transformed by a process. As long as no process acts on the object, it remains in its current state.

A process, on the other hand, is typically transient. It is a thing that "happens" or "occurs" to an object rather than something that "exists" in its own right. We cannot think of a process independently of at least one object, the one that the process transforms. By their nature, happenings or occurrences involve time. What actually exists as a process is an informatical object that represents the pattern of behavior, which the objects involved in the interaction exhibit along the time axis.

Newton's first law is often formulated as "Every object persists in a state of rest or uniform motion in a straight line unless compelled by an external force to change that state" (University of Arizona, 2001). This law can be thought of as a special case of the relationships between an object and a process. In Newton's first law, the object is a physical object, the "external force" is the process that acts on the object to change its state, and the state can be either "resting" or "moving at constant speed." As long as no process acts on the object, the object persists in its current state.

4.2.1 Cause and Effect

One insight from investigating the time relationship is *cause and effect*. Certain objects, when brought into the right physical and spatial relationship at the same moment in time, enable a process to take place. When the process is over, at least one of the involved objects is transformed (consumed, generated, or changed). The "cause" is the collection of objects that enabled the process, and the transformation is the "effect."

In feedback, cause and effect are circular: The effect at a given time is the cause for a change later. Another dimension of the time relationship is *history*. Every object has a history. Its history begins at the time when it becomes an identifiable entity and ends at the time when it is no longer the same identifiable entity. Its history includes a time record of the changes it has undergone while it maintained its identity.

Cause and effect are tightly linked with the concept of time. The transformation, which is at the heart of reasoning about what caused what, takes time. To be able to talk explicitly about a change in an object, we assign to it a number of possible, "legal" *states*, or *values*, which are elaborated upon at Chapter 5 and 12. At any time in the life of the object, when no process is acting on it, that object is at one of its states, or values. A transformation of the object amounts to changing its state or value, or to the generation or consumption of that object. Only a process can bring about such a transformation.

During the transformation, i.e., the time at which the process takes place, the object has already left its input state, but has not yet entered its output state. For example, when a white car is painted red, its input state (the value of its color attribute when it enters the body shop for painting) is "white." Its output state (the value of its color attribute when it leaves the body shop) is "red." During a time interval when it is being painted in the body shop, which may be a couple of days, the value of the car's color attribute is not white any more but is not red yet. The duration of the time when the car is neither red nor white is equal to the duration of the painting process. This transformation mechanism, which consumes time, as if time were a commodity, underlines the intimate, inseparable link between objects and processes and the role time plays in relating these concepts to each other.

4.2.2 Syntactic vs. Semantic Sentence Analysis

In natural languages, both objects and processes are syntactically represented as nouns. Thus, for example, both brick and construction are nouns. However, brick is an object, while construction is a process. This can be verified by the fact that the phrase *the construction process* is plausible, while *the brick process* is not. Likewise, the phrase *the object brick* is plausible, while *the object construction* is much less plausible. Paradoxically, the object building (noun), which is the outcome of the constructing process, is spelled and uttered the same as the process of building (verb), adding to the confusion.

These examples demonstrate that we need a semantic, content-based analysis to tell objects from processes. The difficulty we often experience in making the necessary and sufficient distinction between objects and processes is rooted in our education. It is primarily due to the fact that as students in high school we have been trained to think and analyze sentences in syntactic terms of parts of speech: nouns, verbs, adjectives and adverbs, rather than in semantic terms of objects and processes. This is probably true for any natural language we study and use, be it our mother tongue or a foreign language. Only through *semantic sentence analysis* can we overcome superficial differences in expression and get down to the intent of the writer or speaker. Nevertheless, the idea of semantic sentence analysis, in which we seek the deep meaning of a sentence beneath its appearance, is probably a relatively less accepted idea. Sentences in OPL are constructed such that their syntax is expressed using unambiguous semantics.

To apply OPM in a useful manner, one should be able to tell the difference between an object and a process. The *process test*, discussed in the following section, has been devised to help identify nouns that are processes rather than objects. Its importance lies in the fact that it is very instrumental in helping analysts make the essential distinction between objects and processes, a prerequisite for successful system analysis and design.

4.2.3 The Process Test

The object-process distinction problem is simply stated as follows: given a noun, how can we tell if it is an object or a process? Providing a correct answer to this question is crucial and fundamental to the entire object-process paradigm, and is therefore dwelled upon at some length. By default, a noun is an object. To be a process, the noun has to pass the following process test. A noun passes the process test if and only if it meets each one of the four *process criteria*: object dependability, object transformation, association with time, and association with verb.

(1) Object dependability: For a process to occur, it must rely on at least one object for it to occur. There needs to be a set of one or more nouns (which would be objects) that must be present in order for the noun in question to occur, happen, operate, transform, change, alter, or be executed or carried out. If the noun is indeed a process, these objects constitute the *preprocess object set* of that process. Existence of all the objects in this preprocess object set, possibly in specified states, is a *pre-condition set* for the occurrence of the process. Let us consider two examples: **Flight** and **Manufacturing**. For **Manufacturing**, the preprocess object set may consist of **Raw Material, Operator, Machine** and **Model**. In the **Flight** example, **Airplane, Pilot,** and **Runway** may constitute the preprocess object set, since **Flight** cannot exist without them. Moreover, there may be requirements on the state of each of these objects. For **Flight** to take off, it is required that **Airplane** be **operational, Pilot** be **sober,** and **Runway** be **obstacle-free.**

(2) Object transformation: A process must transform at least one of the objects in the preprocess object set. At least one object from the preprocess object set must be *transformed* as a result of the occurrence of the noun in question. If the noun is indeed a process, these transformed objects constitute the *postprocess object-set* of that process. Existence of all the objects in this postprocess object set, possibly in specified states, is a *post-condition set* of the process. **Flight** transforms **Airplane** by changing its **Location** attribute from **origin** to **destination. Manufacturing** transforms two objects: **Raw Material** (by consuming it) and **Product** (by generating it), but only **Raw Material** was a member of the preprocess object set.

(3) Association with time: A process must represent some happening, occurrence, action, procedure, routine, execution, operation, or activity that takes place along the timeline. Both **Flight** and **Manufacturing** start at a certain point in time and take a certain amount of time. Both time and duration are very relevant features of these two nouns in question.

(4) Association with verb: A process must be associated with a verb. Flying is the verb associated with **Flight**. The sentence "The airplane flies" is a short way of expressing the fact that the **Airplane** is engaged in the process of **Flight**. Similarly, to manufacture (yield, make, create, generate) is the verb associated with **Manufacturing**. The sentence "The operator manufactures the product from raw material using a machine and a model" is the natural language way of saying in OPL that **Operator** handles **Manufacturing**. It is not mandatory that the verb be syntactically

from the same root as the process name, as long as the semantics is the same. For example, **Marrying** is a process, which is associated with the verb to marry. To wed is also a legal verb, albeit less frequently used. Alternatively, we could use **Wedding** to fit it to the verb wed. The system in Chapter 1 is called the **Wedding** system, but its main process is **Marrying** (it could have been called the Marriage system).

Many objects, such as **Apple** and **Airplane**, are not associated with any verb, so they do not fulfil this process criterion. It is easy to verify that both **Apple** and **Airplane** do not meet any one of the previous three process criteria either. As noted, however, failure to fulfill even one of the four criteria results in failure of the entire process test.

4.3 Things

We have seen that objects *exist* as static things, while processes *occur* as dynamic things. Objects are relatively *persistent*, while processes are *transient*. Objects cannot be transformed (generated, affected or eliminated) without processes, while processes have no meaning without the objects they transform and the objects that enable their occurrence. Objects and processes are two types of tightly coupled complementary *things*. The extent of this coupling is so intense that if we wish to be able to analyze and design systems in any domain as intuitively and naturally as possible, we must consider objects and processes concurrently. We have to be bale to specify what state the object was at before the process affected it, which objects were consumed and which were generated. At the same time, we need to be able to show how parts, features and specializations of the objects play role in sub-processes of the higher-level process. This is the extent to which objects and processes are interwoven.

As we shall see, objects and processes have a lot in common in terms of relations such as aggregation, generalization and characterization. The need to talk about a generalization of these two concepts necessitates the advent of a yet more abstract term, which generalizes an object and a process. We call this a "thing."

> *Thing is a generalization of object and process.*

The concept of "thing" enables us to think and express ourselves in terms of this abstraction and refer to it without the need to reiterate the words "object or process" repeatedly. Based on the ontology of Bunge (1987; 1989), Wand and Weber (1989; 1993) have used the term *thing* as a synonym to what we refer to as object. Their first premise is that *the world is made of things that have properties*. According to this definition, *thing* seems to be synonymous with *object*. However, during the last two decades, the term *object* has become deeply rooted, at least in the software engineering community. It therefore seems justified to change the word *thing* from its meaning of being an object to the generalization of object and process.

4.3.1 Things and Entities

The word **Thing** may sound too mundane, so one might wish to use the more sophisticated word "entity" instead. However, entity tends to be interpreted more statically than dynamically, i.e., more as an object than as a process, while we wish to use a word that is as neutral and abstract as possible. However, we do use entity when we wish to refer collectively to things and states.

> **State** and **Thing** are **Entities**.
> **Object** and **Process** are **Things**.
> **State** is a situation of **Object**.
> **Object** can be at **State**.
> **Process** changes **State**.

Frame 7. The OPL paragraph of **Figure 4.1**

As we saw, a state is a situation at which an object can be. As such, the semantics it conveys is also static. OPM therefore uses the term **Entity** to generalize a **Thing** and a **State**.[4] The OPD in Figure 4.1 and its corresponding OPL paragraph in Frame 7 show the Generalization-Specialization hierarchy of **Entity, Thing, Object, Process** and **State**, and the tagged structural relations between **Object** and **State**.[5] In the following sections, we discuss four basic attributes of thing: perseverance, essence, origin and complexity.

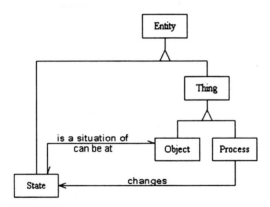

Figure 4.1. The Generalization-Specialization hierarchy of **Entity, Thing, Object, Process** and **State**, and the tagged structural relations between **Object** and **State** and between **Process** and **State**

---

[4] As we shall see, the meanings of **Value** and **State** are very close, but **Value** is more general. In this chapter, we use value and state interchangeably.

[5] Figure 4.1 is a metamodel of OPM, i.e., a model of OPM described in OPM terms. Here, processes and states are referred to as objects. We discuss this subtle point in detail in Chapter 10.

4.3.2 The Perseverance of Things

Since thing is a concept that generalizes an object and a process, we need to define an attribute[6] of thing, the value of which will enable us to differentiate between its two specializations. As Figure 4.2 shows, the name of this attribute is **Perseverance**,[7] and its values are **static** and **dynamic**.

> *Perseverance is an attribute of a thing that pertains to its persistence and determines whether the thing is static – an object, or dynamic – a process.*

The OPL sentences for Figure 4.2 follow.

> **Thing** exhibits **Perseverance.**
>   *(Exhibition sentence)*
> **Perseverance** can be **static** or **dynamic.**
>   *(State enumeration sentence)*

The two sentences can be joined into one:

> **Thing** exhibits **Perseverance,** which can be **static** or **dynamic.**
>   *(State enumerated exhibition sentence)*

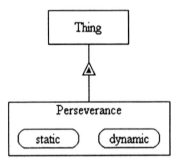

Figure 4.2. The **Perseverance** attribute of **Thing** with its **static** and **dynamic** values

The two possible values of **Perseverance** provide for distinguishing between objects and processes: objects are things with **static Perseverance**, while processes are things with **dynamic Perseverance**. Before continuing with this discussion, let us introduce **Vehicle** as an analogy of **Thing**, as shown in Figure 4.3(a). **Ship** and **Airplane** are specializations of **Vehicle**, which differ in the value of their **Traveling Medium** attribute:

---

[6] An attribute is an object like any other, but it is unique because it describes another object. Attributes are noted by the Exhibition-Characterization link, as in Figure 4.2. This link is elaborated upon in Chapter 7.

[7] Alternative names for the values *static* and *dynamic* of the attribute Perseverance might be *persistent* and *transient*. However, in database terminology, these words are taken. A persistent object is an object that needs to be stored in a database, while a transient object is created at run-time and not maintained in a long-term memory.

> **Vehicle** exhibits **Traveling Medium,** which can be **water** or **air.**
> *(State enumerated exhibition sentence)*
> **Ship** is a **Vehicle,** the **Traveling Medium** of which is **water.**
> **Airplane** is a **Vehicle,** the **Traveling Medium** of which is **air.**
> *(Qualification sentences)*

The two qualification sentences above tell us that both **Ship** and **Airplane** are **Vehicles,** which differ in the value of their **Traveling Medium** attribute: The **Traveling Medium** of **Ship** is **water,** while that of **Airplane** is **air.** In an analogous manner, Figure 4.3(b) shows **Object** and **Process** as specializations of **Thing,** where the distinction is based on the value of the **Perseverance** attribute. The corresponding OPL sentences follow.

> **Object** is a **Thing,** the **Perseverance** of which is **static.**
> **Process** is a **Thing,** the **Perseverance** of which is **dynamic.**

The value **static** of the **Thing**'s **Perseverance** attribute denotes the fact that the thing persists. This thing is an **Object,** which is stable in form and remains at its current state as long as it is not transformed by a process. The value **dynamic** of the **Thing**'s **Perseverance** attribute denotes the temporal nature of the thing. This thing is a **Process.** It represents a pattern of behavior, which starts at some point in time, has a finite (albeit possibly very long) duration, and then ends. The dynamic thing, the process, transforms at least one static thing, an object, by creating or eliminating it, or by affecting it through changing its state.

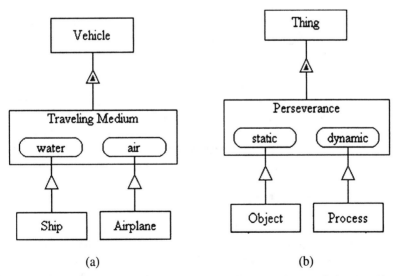

(a)   (b)

Figure 4.3. (a) **Ship** and **Airplane** are specializations of **Vehicle** that are distinguished by the value of their **Traveling Medium** attribute. (b) **Object** and **Process** are specialization of **Thing,** distinguished by the value of their **Perseverance** attribute.

Note that while **Perseverance** is inherited from **Thing** as an attribute of both **Object** and **Process**, the inheritance is *qualified*, because **Object** can only assume the value **static** of **Perseverance**, while **Process** can only assume the value **dynamic** of **Perseverance**. In other words, once we specify an attribute value of **Thing**, we automatically refer to either one as **Object** (if the perseverance value is **static**) or as **Process** (if the value is **dynamic**). Using the term **Object** instead of the much more cumbersome form "**Thing, the Perseverance** of which is **static**" is completely analogous to using the term **Ship** instead of "**Vehicle, the Traveling Medium** of which is **water**." In both cases, to enhance their expressive power, natural languages come up with words that shorten long phrases that are in frequent use.

While objects are persistent (i.e., exhibit static perseverance) and processes are transient (i.e., exhibit dynamic perseverance), boundary examples of persistent processes, as well as transient objects, exist. Persistent processes include such verbs as holding, maintaining, keeping, staying, waiting, prolonging, delaying, occupying, persisting, including, containing, continuing, supporting, and remaining. Rather than induce any real change, the semantics of these verbs is leaving the state of the object as it is, in its status quo for some more time. This is contradictory to the definition of process, which requires that it transforms some object. However, many of these verbs can be considered as working against some "force" which would otherwise change some object. For example, a **Pedestal** supporting a **Statue** works against gravity, so we can think of **Supporting** as a "falling prevention" process, without which the state of the **Statue** would change from **stabilized** to **fallen**. Such state maintaining processes, which keep an object in its current state, are discussed in Chapter 12.

Examples of transient objects are unstable materials, such as an interim short-lived compound in a chemical reaction or an atom in an excited state that spontaneously decays to the ground state by emission of X-rays and fluorescent radiation 9 University of Maryland, 2001). An example of a transient object is a packet in a telecommunication network. Such a packet can reside for a short while at some router on its way and leave no trace of its stay there once the target node has received it. The former example concerns a physical object, while the latter, an informatical one. This distinction leads to the concept of the *essence* of things, discussed below.

4.3.3 The Essence of Things

A thing does not necessarily have to be in the physical world. It can be a *representation* of a thing in the physical world, primarily originating from the human mind, in which case it is an *informatical thing*. Thus, we distinguish between two essentially different things: *physical* things and *informatical* (or *logical*) things. **Essence** is the attribute of **Thing**, the values of which are **physical** or **informatical**.

> *Essence* is an attribute of a thing that determines whether the thing is physical or informatical.

OPM clearly distinguishes between these two types of things. Physical things have materialistic existence. Physical things may carry informatical objects or enable informatical processes. However, the physical aspect of an informatical thing is typically not the focus of discussion or attention at the analysis stage, when the informatical thing is used to describe a physical thing. Informatical things are contrasted with physical ones in that the laws of physics do not apply to them. The physical carrier of the informatical thing becomes significant at the design stage, where implementation issues are of primary consideration.

Having made this distinction between physical and informatical things, we can talk about two related worlds, each residing at a different level. The first is the physical, "real" world, which we sense, and the second is the informational world, which is what we make up in our minds, a mental picture of the physical world. The informational world represents the physical world, as we perceive it. The former is at the materialistic, concrete level, while the latter is at the cognitive, abstract level. When we do wish to refer to an informatical thing as a thing in itself, then we are usually interested in the physical medium that carries it. In the following two subsections we elaborate on physical vs. informatical things.

A *physical object* consists of matter. Therefore, in the broad sense, physical objects are tangible. In order for some tangible, physical object to exist, it must be identifiable directly by observation, or its identity must be inferable from direct observation. Detection is enabled by one or more of our senses, or indirectly, through instrumentation that provides us with sensible information. Sensing can be stimulated by touching or smelling materials, seeing emitted light, hearing vibrating air, or smelling airborne molecules. Due to the limitation of our senses, many physical objects are evident only indirectly. The indirect detection of the existence of physical objects, such as black holes in the universe or some tumors in the human body, is done using some instrument that provides evidence of this existence. For example, a sufficiently small electric charge cannot be seen, heard, or felt, but its existence can be inferred by measuring the potential of its carrier.

An *informatical object* is a piece of informatics.[8] Examples of informatical objects are extremely diverse. They include a database, a recipe, an essay, an algorithm, a book, a childhood memory, a binary digit (bit), a grocery list, a letter, a language, a rule, a constraint, an agreement, an insurance policy, an essay, a law, a bank account, a memorandum, a command, a daughter, an application, an encyclopedia, an information system, a vote, a secret, a password, a test, a grade, a file, a report, a computer program, a geographic map, a thought, an idea or a model.

The record of an intangible, informatical object on some physical medium is what determines whether it exists: if the informatical object is recorded on a physical object, it exists. Thus, if the only copy of some old manuscript disappears, and no living human knows or remembers what it contained, then this informatical object is lost forever.

A *physical process* is a change that a physical object undergoes. Breaking of a cup, falling of a stone, charging of a battery, rusting (oxidation) of iron and solar

---

8   We discuss the informatics hierarchy in Chapter 14.

irradiation are examples of physical processes. In the macro, non-quantum world, all physical objects have mass and occupy coordinates in space and time. Physical processes are abstractions of patterned changes physical objects go through, which yield to such nature laws as mass/energy conservation. All physical objects and processes obey the "laws of nature." These laws are generalizations from observations humans have made throughout the history of science.

An *informatical process* is some transformation (or manipulation) of an informatical object, such as writing, reading, storing, duplicating, learning, memorizing, thinking, transmitting, deducing, dreaming, planning, designing, or analyzing. There are some informatical processes, notably high-level thinking and cognition, that probably only humans can carry out. Others can be done by lower level organisms or by machines that were designed to exhibit a certain degree of intelligence.

4.3.4 Symbolizing Physical Things

A double frame (or shading) distinguishes a physical thing from its informatical counterpart (Figure 4.4). These symbols are applied in Figure 4.5, which is an OPD of a manufacturing system. We use the physical thing symbol when we wish to emphasize the fact that we represent the actual object and/or process, as opposed to its informatical representation in an information system. This distinction is critical when our OPD set serves as a basis for automated code generation, in which case, only informatical things are converted into code. The distinction is even more critical when modeling reactive and real-time embedded systems, where interactions among hardware and software components are significant. However, for initial analysis and design, we will frequently use the default informatical symbols and not bother to denote this distinction in the diagram. We can convert the thing's symbol in the diagram from physical to informatical and back to physical at any time.

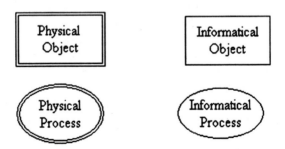

Figure 4.4. The symbols of physical and informatical objects and processes

Since informatical objects and processes are the default, physical object and process sentences are needed to express the fact that the OPD contains physical things. The physical object and process sentences of the OPD in Figure 4.5 are listed in the OPL paragraph in Frame 8. The OPD in Figure 4.5(a) and the OPL paragraph in Frame 8 show that the physical process **Manufacturing** is an operation of the physical object **Factory**. The OPL reserved word physical denotes the fact that **Manufacturing** is a physical process.

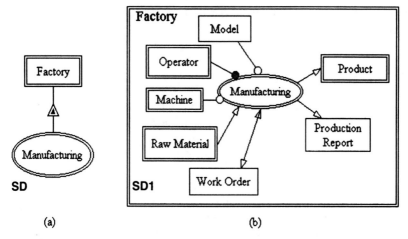

Figure 4.5. Symbols of physical and informatical things applied in an OPD of a **Manufacturing System**. (a) **Manufacturing** as an operation of **Factory**. (b) **Factory** is in-zoomed to expose the **Manufacturing** operation and **Factory**'s lower-level constituent objects.

SD Paragraph:
Factory is a physical object.
   (Physical object sentence)
Manufacturing is a physical process.
   (Physical process sentence)
Factory exhibits **Manufacturing.**

Frame 8. The OPL paragraph of Figure 4.5(a)

Factory from SD zooms in **SD1** into **Operator, Machine, Raw Material, Model, Work Order, Product,** and **Production Report,** as well as **Manufacturing.**
   (In-zooming sentence)
SD1 Paragraph:
Operator, Machine, and **Raw Material** are physical objects.
   (Physical objects sentence)
Manufacturing is a physical process.
   (Physical process sentence)
Operator handles **Manufacturing.**
Manufacturing requires Model and Machine.
Manufacturing affects **Work Order.**
Manufacturing yields **Product and Production Report.**

Frame 9. The OPL paragraph of Figure 4.5(b)

The OPD in Figure 4.5(b) zooms into **Factory**,[9] exposing lower level objects along with the process **Manufacturing**, which **Factory** exhibits. The objects **Machine, Raw Material** and **Product**, as well as the process **Manufacturing**, are physical, while the objects **Model, Work Order**, and **Production Report**, are informatical.

4.3.5 The Origin of Things

The **Origin** attribute of **Thing** relates to the source from which the thing originated. **Origin** can have two possible values: **natural** and **artificial**.

> *Origin is an attribute of a thing that pertains to the source of the thing, that is, whether it is natural or artificial.*

A **natural Thing** is anything that is not artificial, or man-made. Examples of natural objects are planets and organisms, while natural processes are volcano erupting and cell splitting. Examples of artificial objects are product and book, while artificial processes are manufacturing and reading. We will discuss the **Origin** attribute in more detail while dealing with natural and artificial systems in Chapter 10.

4.3.6 The Complexity of Things

The **Complexity** attribute of **Thing** relates to the place the thing occupies along the simplicity-complexity continuum. At this point, we are only interested in two values of the complexity attribute: **simple** and **non-simple**.

> *Complexity is an attribute of a thing that determines whether it is simple or non-simple.*

A **simple Thing** is a thing for which no parts, attributes, or specializations are defined in the context of the system being studied or developed. Any other **Thing** is **non-simple**. We use the tern **non-simple** rather than complex to denote the fact that even things that look "simple," such as an object that has just one attribute, are non-simple according to this definition. We will discuss the **Complexity** attribute while dealing with complexity management of systems in Chapter 9.

4.3.7 Thing Types

The Cartesian product of the four **Thing** attributes, each having two values, yields the following $2^4=16$ combinations, each yielding another **Thing** type. Below we list only eight variations, leaving out the Complexity attribute.

---

[9] Zooming is discussed in Chapter 9.

1. static natural physical things, or natural physical objects;
2. static natural informatical things, or natural informatical objects;
3. static artificial physical things, or artificial physical objects;
4. static artificial informatical things, or artificial informatical objects;
5. dynamic natural physical things, or natural physical processes;
6. dynamic natural informatical things, or natural informatical processes;
7. dynamic artificial physical things, or artificial physical processes; and
8. dynamic artificial informatical things, or artificial informatical processes.

4.3.8 The Relativity of Object and Process Importance

The transformation of an object, discussed above, may be as minor as moving the object from one location to the other, or as drastic as destroying or creating the object. In a theoretic, frozen, static universe at absolute zero, no processes exist and no transformation occurs. However, such a system is of little interest. In a more realistic setting, processes and objects are of comparable importance as building blocks in the description and understanding of systems and the universe as a whole. Without processes, all we can describe are static, persistent structural relations among objects. The question of what came first, an object or a process, is a metaphysical question that is almost analogous to the question of what came first, the chicken or the egg.

Unlike the object-oriented approach, in OPM, the importance or status of the Thing in the system is not based upon the distinction of whether it is an object or a process. Rather, it is based on the level of detail at which the thing first gets to be represented in the OPD. The most important things in the system are shown at the top level OPD, or the system diagram. There, we usually see a small number of high-level, abstract objects and processes that together carry out the function the system, i.e., what it is intended or designed to do. As we drill down, the detail levels using OPM's refinement mechanisms, we find objects that are contained within processes, as well as processes that are contained within objects. In this respect, there is no across-the-board supremacy of one type of thing over the other.

4.3.9 Object and Process Naming

Naming conventions for processes and objects enable both humans and machines to tell them apart. The process naming convention, unless in English it makes no sense, is to name a process by its gerund, i.e., the root of the verb followed by the **ing** suffix, as in **Igniting**.

This naming convention has two advantages. First, it clarifies the dynamic nature of the process as a thing that *happens* rather than a thing that *exists*. To enhance clarity, a process name may include the name of an object before the gerund, as in **Engine Igniting**. If so, the gerund (**Igniting**) transforms the object (**Engine**) that precedes it in the process name. Secondly, the **ing** suffix enables automated detection of processes.

The object name preceding the gerund changes the process name. For example, **Wall Painting** and **Roof Painting** are two different processes. If the gerund form makes no sense in English, or a non-gerund form is strongly preferred, the name of the process should be followed by the OPL reserved word **Process**. For example, if we prefer to call a process **Addition** rather than **Adding**, to avoid ambiguity we should call it **Addition Process**.

For object names, a problem arises when using the same English noun that ends with **ing**, be it a suffix or not. If an object name ends with **ing**, it must be followed by the OPL reserved word **Object**. A common example is **Ceiling**. While it is obviously an object, an automated system would interpret it as a process. To solve this problem, we would write **Ceiling Object**. A more confusing example is the word **Casting**, which is defined in The American Heritage Dictionary (1996) as having two noun meanings: (1) the act or process of making casts or molds, and (2) something cast in a mold. In this case we call the second meaning (2) **Casting Object** to distinguish it from **Casting**, which is the process of pouring molten material into the mold, as in meaning (1) above.

4.4 Informatical Objects

Informatical objects share the fact that they convey a piece of informatics, such as data, information, knowledge, or expertise. The informatical object can relate to a physical thing or to an informatic one, in which case it is often referred to as a meta-informatical object. The intangible semantics of the informatical object, or the message it conveys, rather than the medium on which it is inscribed, is of interest. As noted, unlike physical objects, informatical objects do not yield to the laws of physics. They are not made of matter and they do not occupy space, they do not wear out.

4.4.1 Telling Informatical and Physical Objects Apart

As noted, one feature that is common to all informatical objects is that they do not obey basic laws of nature, to which physical objects yield. Most notably, informatical objects consistently violate the law of matter and energy conservation: If I give you an idea and you give me an idea, we each have two ideas, not one! Unlike matter, an informatical object cannot be characterized by mass, energy, shape, weight, color, electric charge, or any other feature that may characterize a physical object.

An informatical object may be created in a person's brain. Its only real existence may be as an idea in humans' minds, where it is inscribed by some complex manner within the neural network of the brain. An informatical object can be broadcast by word of mouth, communicated among humans. It may be as simple as the number three or as abstract and high-level a concept as money, mortgage, freedom, or justice. Informatical objects and processes, such as oral traditions, facts, events, beliefs, and transactions, reside in the human brain. However, if they are to be recorded outside human cognition they must be inscribed on some physical medium. The medium may be as ancient as stone or as modern as an optical disk, a biologically based artificial memory, or any other existing or futuristic kind of persistent memory. Some kind of medium is and will always be required to preserve informatical objects. In the past, this matter was stone, papyrus, or clay tablets. Traditionally it has been paper. Increasingly, the medium is shifting from an analog to a digital one.

4.4.2 Systems and Information Systems

An *information system* is a system that uses informatical objects and processes to reflect the reality of another system and potentially provides means to manage the modeled system. Consider, for example, the physical object **Automobile Repair Facility**, which operates a **Stockroom**. The **Stockroom** stores **Tools** and spare automobile **Parts**. **Mechanics** borrow **Tools** and use **Parts** for repairing cars in the **Automobile Repair Facility**. The **Parts** and **Tools** are physical objects. The process of **Tool Borrowing** is a physical process: it causes a **Tool** to be moved from the **Stockroom** to the **User** who borrowed it. In OPM terms, the **Tool Borrowing** process changes the state of the attribute **Keeper** of the **Tool** from **stockroom** to the **Name** of the specific **Mechanic** who borrowed it. The information system of the stockroom models the physical entities, which are the **Tools** and **Parts**, where they are located in the **Stockroom**, which **Mechanic** took what **Tool**, at what **Date**, for which **Project**, etc. This provides for management operations like **Stock Monitoring**, **Order Initiating** and **Part Dispensing**. Without the information system, the stockroom would be much more difficult to manage. This information system is an informatical object. It models what exists and what happens in the physical world. In the pre-computer era, or for a small enough stockroom, a deck of note cards, or even the phenomenal memory of the stockroom manager might have been sufficient. Normally, though, developing and maintaining an information system is made possible through computers.

Ideally, while analyzing the objects represented by the information system of the stockroom and the processes they undergo, one should not be concerned about the medium on which the information is recorded. It can be a remote computer disk, an index-card-based system or, if it is simple enough, the excellent memory of the stockroom manager. Once we start discussing *how* that information will be stored and processed, we move from analysis into the design of the information system. Consequently, the carrier of the information becomes relevant, and the discussion is about the world of physical things that carry and process the information. The design should specify how the physical materialization of the information system integrates into the physical system that had originally been analyzed and modeled.

4.4.3 Translation of Informatical Objects

Syntax is the form, arrangement, or layout of symbols that the informatical object conveys on some medium. Semantics pertains to the deep meaning an informatical object conveys. Hence, two informatical objects are *semantically identical* if they contain the same informatic contents, i.e., they *mean the same* for the same person. However, if these two objects are *syntactically different*, there needs to be a process that translates one representation into the other. This process establishes isomorphism from the pattern (informatics content, or semantics) of one to the pattern of the other.[10] We use the term *translation* rather than transformation, because we have reserved the term transformation to denote what a process does to an object in general. Translation is a specialization of transformation, in which an informatical object assumes a new form that is semantically identical to the old one.

The simplest, most trivial case is the null translation. Two instances of the informatical object 5, namely 5 and 5, are the same, and only the null translation is needed to assert this. A slightly more involved translation is required when we want to check if 5 and the Roman numeral V are the same. Indeed, these are two different symbols for the same idea, and can be translated from one to the other. One can easily think of successively more complicated translations from 5 to mathematical expressions that evaluate to 5. We deem the "5 to V" example trivial, because we, as humans, do an excellent job of character recognition with almost no effort. For machines, however, perfect Optical Character Recognition (OCR), in particular if the symbols are noisy and/or handwritten, is still a challenge. Moreover, depending on the context, V can also be the alphabetic character or the symbol for victory.

In general, the translation between two informatical objects can be complex. It may require that the different matter components on which one informatical object is recorded, be extracted, or segmented out from their background and associated with a symbol. The pattern itself in that segmentation of the informatical object has then to be correlated with the symbol representing the pattern of the other informatical object in order to establish their semantic identity.

We can think of semantic identity at increasing levels of abstraction. For example, are two sentences that express the same idea differently, or in two different natural languages, the same informatical object? If not, how close are they? To find out a series of two or more iterations of translation may be required, yielding intermediate informatical objects.

4.4.4 Toward "Pure" Informatical Objects

The price per bit of stored data for electronic medium has been decreasing significantly. Already, the price of the storage hardware and the effort to make the physi-

---

[10] Even two syntactically identical informatical objects may differ in their level of noise.

cal record is negligible compared to the value of information it stores. Hence, a book or software that took many man-years to develop, and in which huge resources and efforts have been invested, is easily mounted on a single CD-ROM, the raw price of which is close to zero. Moreover, with the emerging Web technology, informatical objects may reside anywhere in the world and transferred in seconds to any other place. We have access to them without the need for being aware of or concerned about their physical location, as we access them only by their logical address. With current technology, once an informatical object has been generated, it can be perfectly duplicated any number of times and distributed at a speed that is close to the speed of light to any place at negligible cost. Thus, as information technology advances, it becomes more justifiable to disconnect the links that traditionally existed between informatical objects and the media upon which they are inscribed.

4.5 Object Identity

The identity of objects is an important, yet elusive problem. Physical objects must be treated differently than informatical objects. Since matter is a basic feature of a physical object, two instances of a physical object are identical if and only if they occupy the same space at the same time. This is possible if and only if the two are actually the same object, implying that no two distinctly identifiable instances of a physical object are the same. Thus, two new identical cars of Model X that just emerged from the assembly line are different instances of the same object class.

The situation with object identity is different with respect to informatical objects, since here the essential feature is the idea, concept, pattern, or symbol, rather than physical matter or energy. From the physical medium point of view, each informatical object instance is distinct, but from the informatical point of view, all copies represent or refer to the same informatical object. Two or more copies of a file system directory, a file on a diskette, or of the same book, are identical insofar as their informatical content (semantics) is considered. Two identical (paper or computer) files containing blueprints and manufacturing instructions for cars of Model X are, from the informatical viewpoint, the same object, because the informatical content they convey is identical. Physically, the pieces of media, on which these informatical objects are recorded, are different, since they are physical matter that obeys the laws of nature. Thus, two copies of the same book are printed on separate pages and bound as two distinct physical object instances. Likewise, two copies of the same file are physically different, as they occupy different address spaces in the computer's primary or secondary memory, or even on different machines, diskettes, tapes, etc. However, when viewed as informatical objects, they are identical.

4.5.1 Change of State or Change of Identity?

During their life, physical objects as well as informatical ones can undergo a host of transformations. We have already mentioned the lifecycle of a system, which is a kind of object. As noted, transformation of an object can only take place when a process acts upon it. This transformation generates, affects or eliminates the object. The extent of the change can vary from very small to very large. If the change is small, we tend to say that the object was altered from one of its states to another while keeping its identity. As the extent of the effect grows, so does the difference between the object before the process started and after it ended. At some point, the two are so different, that the human modeler is inclined to think of the object resulting from the process as a newly created object. The object that had existed before the process took place may have been eliminated or at least changed radically. As we show below, the issue of whether an object changed only its state or its entire identity is similar in natural and artificial systems.

In nature, living organisms undergo a striking variety of transformations. Some of the transformations are deemed as just a change in state while others are a change in object identity. The transformation from a cub to a grown-up lion is considered a change in the state of a lion from young to adult. Similarly, growing of a baby into an adult is considered a change in the person's state. The silkworm, on the other hand, has four distinct forms of existence. It transforms from egg to larva (worm, or caterpillar) to pupa, The larva undergoes complete transformation within a protective cocoon or hardened case, to butterfly, which, in turn, lays the eggs of the next silkworm generation. Each transformation yields an object that is very distinct from its predecessor in shape and function. The difference is so profound that each such transformation is called metamorphosis. We are inclined to view each reincarnation as a separate object rather than a mere change of the same object's state. A frog, like other amphibians, transforms from egg to tadpole to mature frog, providing an example similar to the silkworm. The frog and silkworm examples are conveniently thought of as changes of object, because the various incarnations of these creatures are profoundly different from each other in appearance and behavior. The human and lion examples, on the other hand, are more naturally modeled as a change of the object's state.

The situation with transformations of artificial objects is similar to natural ones: If the change is profound, objects change identity. If it is not – the same object just alters its state. What is "profound" is subjective and context-sensitive. Consider, for example, two processes from a manufacturing realm: **Molding** and **Testing**. **Molding** acts on the object **Raw Material** (e.g., plastic) to convert it to another object, that we call **Product**. The identity of **Raw Material** changed as a result of the **Molding** process to the extent that we need to refer to the process outcome by a different name. Hence, the object **Raw Material** has been eliminated or consumed (or, in object-oriented programming terms, destroyed), while a new object, called **Product**, has been created (or constructed or generated).

Suppose **Product** now undergoes the process of **Testing**, in which its shear strength is measured. If the test succeeds, **Product** is **approved** for **Marketing**, otherwise it is rejected. Unlike the **Molding** process, which altered the identity of the processed object from **Raw Material** to **Product**, **Testing** does not change **Product** to the extent that we would be inclined to say that it lost its identity. Instead, the only effect of the **Testing** process is to alter the state of **Product** (from **untested** to **tested**). While there is a difference between **Product** before and after the **Testing**, (since after the test we have information about the product's strength, which we did not have before), this difference is not profound enough to justify change of identity. However, it does cause a change in state. Hence, transformation can be thought of as a general term that encompasses creation, effect (change of state) and elimination of an object. We will elaborate on this when discussing system dynamics.

Generalizing the natural and artificial examples, when the change is not drastic, we are inclined to think that the object only alters its state while retaining its identity. When the transformation is extreme, a change in object identity takes place. As is the case in similar situations, the borderline between "drastic" and "non-drastic" is not well defined. When in doubt, appeal to common sense. Analyzing the same system, different modelers may provide different viewpoints on whether or not particular objects should lose their identity and become new objects. Indeed, we will see instances where it makes sense to model changes in objects either as a change in their state or as a change in their identity, and both versions would be acceptable.

4.5.2 Classes and Instances of Objects and Processes

A class is a collection of things that share a common set of features, attributes and operations, which are discussed in detail in the Chapter 7. Each member of the class is referred to as an instance. An object class and a process class are two distinct types of classes. Object classes and object instances are similar to their object-oriented counterparts. Processes provide for a powerful abstraction of occurrences of the same pattern of happening. In OPM, the object class and instance concepts are extended to processes. A single actual process occurrence is a *process instance*, while the pattern of happening is a *process class*. The power of the process class abstraction is that it enables one to talk about the process as a *template* or a *protocol* for a transformation sequence that a class of objects undergoes. It does so without specifying the temporal framework or the particular set of object instances that is associated with it. The process class **Flight Process**, for example, brings to mind a generic movement through the air of some **Aircraft** (a typical member, or *instance*, of the class **Aircraft**) which took off at some **Start Time** and landed at **End Time**. The **Aircraft** that executed the **Flight** departed from some **Origin** and arrived at some **Destination**. A process instance is a particular occurrence of a process class to which that instance belongs. Thus, the process instance Flight 123 of Carrier Q brings to mind a specific occurrence of the process class **Flight Process**. In flight instance, an instance of the object class **Aircraft**, having Carrier Q Serial Number 5678234, flew at a specific time from a particular origin to a particular destination.

Summary

- An object is a thing that has the potential of stable, unconditional physical or mental existence.
- Transformation is a generalization of consumption (destruction), change (effect), and generation (construction) of one or more objects.
- A process is a pattern of transformation that an object undergoes.
- Thing is a generalization of object and process.
- Semantic sentence analysis seeks the deep meaning of a sentence beneath its appearance (syntax).
- A noun passes the process test if and only if it meets four criteria: object dependability, object transformation, association with time, and association with verb.
- Entity is a generalization of thing and state.
- Things exhibit four attributes:
  - Perseverance pertains to the thing's persistence and determines whether the thing is static, an object, or dynamic, a process.
  - Essence pertains to the thing's nature and determines whether the thing is physical or informatical.
  - Origin pertains to the thing's source and determines whether it is natural or artificial.
  - Complexity pertains to the thing's level of complexity and determines whether it is simple or non-simple.
- The highest (most abstract) level of OPD in which objects and processes appear determines their relative importance. There is no across-the-board supremacy of objects over processes.
- Rules for naming things are simple:
  - A process name should end with "ing," and if it does not, the reserved word Process should be appended to it.
  - An object name should not end with "ing," and if it does, the reserved word Object should be appended to it.
- Informatical objects convey a piece of informatics, such as data, information or knowledge.
- Each physical object has a unique identity, even it has identical copies. Identical informatical objects are not necessarily distinct.
- Change of identity is more dramatic than change of state, in which the identity of the object is maintained. The decision as to when the change is of identity or only of stare is not always clear-cut.
- In OPM not only object instances belong to classes; process instances belong to process classes as well.

Problems

1. When a white car is painted red, its input state (the value of its color attribute when it enters the body shop for painting) is "white". Its output state (the value

of its color attribute when it leaves the body shop) is "red". Draw an OPD and write the corresponding OPL script.

2. Write six pairs of semantically equivalent sentences. The first sentence in each pair should contain at least a noun and a verb while the second should contain two nouns, one object and one process (with the possible addition of a verb). Draw an OPD for the second sentence and write down the OPL script for the second sentence. Example: (1) The dog barks at a stranger. (2) The dog engages in barking at a stranger (see Figure 4.6).

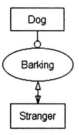

Figure 4.6. Example for Problem 2

3. Apply the process test to six nouns to determine whether these nouns are objects or processes. Find at least three processes. For each process, draw an OPD and list its preprocess and postprocess object-sets.
4. Use OPM to specify the process test as a system.
5. Apply OPM to model the stockroom system described in Section 4.4.
6. Draw an OPD and the OPL script for one natural and one artificial system.
7. Write the OPL paragraph for the OPD in Figure 4.7.
8. Compare and contrast physical and informatical objects.
9. Give two examples for each one of the eight types of things listed in Section 4.3.
10. Give two examples of informatical objects. For each of these objects list three different physical objects that could serve as media to keep record of the informatical object.
11. What processes led to the creation of the clothes in your closet? Classify each process as informatical or physical, natural or artificial.
12. Apply the process test to three processes in the ATM case study.

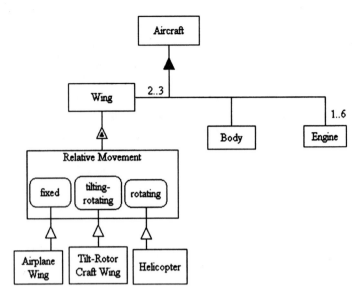

Figure 4.7. OPD for Problem 7

Part II

Concepts of OPM Systems Modeling

Chapter 5

Dynamics

> *Every day we are confronted with systems that have an inherent tendency to change. The weather, the stock market, or the economic situation are examples.*
>
> Hans Meinhardt (1995)

System dynamics deals with system changes over time. The dynamic aspect of a system is the complement to the static. It specifies how the system operates to attain its function. Time is an essential concept in analyzing the dynamic aspect of systems. An important motivation in the development of OPM has been to strike a needed balance in systems modeling between the structural and procedural aspects of a system. OPM is designed to incorporate structure and behavior in one coherent frame of reference. This chapter addresses the issue of system dynamics; in other words, the system's behavior and the changes it undergoes over time.

5.1 States

In Chapter 4, we defined transformation as a generalization of change, generation and destruction of an object. Process was defined as the pattern of transformation applied to one or more objects in the system. We did not have the tools to specify what exactly happens to the object that undergoes transformation. Building upon the definitions of process and transformation, we discuss the terms *state* and *status* of an object, and argue that when transformation occurs, the state of the object changes.

5.1.1 Object States and Status

To be able to talk explicitly about a change in an object, we assign to it a number of mutually exclusive values, situations, or positions, which we frequently refer to as *states*.

> *State is a situation or position at which the object can exist for a period of time.*

Examples of object states abound. Let us see a few examples. States that a person can be in are "awake" and "asleep." A machine can be in states "operational" or "broken." The weather can be sunny, cloudy, rainy or snowy. An organism can be alive or dead. One distinguished attribute, which is of utmost importance for describing how the object changes over time, is status. **Status** is the name of the attribute. the values of which are states.

> *Status is an attribute of an object whose values are the object's states.*

OPM reserves the word **Status** to refer to the name of the attribute that stores the various states, such that *each state is a different value, or "slot", of the* **Status** *attribute.* In everyday language, however, this fine distinction is blurred. The confusion arises because the term "state" is frequently used both implicitly, as the name of the attribute (which, according to the above definition, should be "**Status**"), and explicitly, as the names of the various values of that attribute. To be precise, we should be careful in the use of these two related, yet different, concepts. We will use the word **Status** to refer to the attribute name, while states will be the values of **Status**.

5.1.2 Change and Effect

Processes and system dynamics are closely associated with the notion of *change*. Change is such a basic concept, that defining it seems both difficult and unnecessary. However, when we talk about a change in OPM, we need to be specific about what a change means.

> *A **change** of an object is an alteration in the state of that object.*

More specifically, a change of an object is reflected in replacing the currently valid state (i.e., the current value of its **Status** attribute) by another value. The only thing that can cause this change is a process. The process causes the change by taking an object at some state, and transforming it to another state. Hence, a change of an object means a *change in the state* at which the object exists. As noted, only a process can carry out such a change. This change mechanism underlines the intimate, inseparable link between objects and processes. We call this change in state the *effect* of the process on the object.

> *The **effect** of a process on an object is the change in the object's state that the process causes.*

While the terms "change" and "effect" are almost synonymous, there is a subtle difference in their usage. We use *effect* when we refer to the *process*, and *change* when we refer to the *object*. The effect is what the process does to the object, while the change of state is what happens to the object due to the process occurring. Later in this section we refine the above definition of effect with the notions of input and output links.

5.1.3 Explicit and Implicit Status Representations

Graphically, as **Lamp** and **Status** in the OPDs of Figure 5.1 show, a state is represented by a "routangle" (rounded-corner rectangle), with the state's name recorded inside it. A state symbol must be drawn within the object that "owns" it. Since a state is a value of an attribute of an object, it cannot be detached from the object and has no meaning unless it is drawn inside an object.

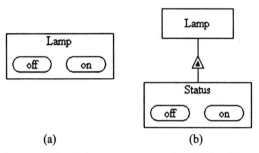

        (a)   (b)

Figure 5.1. States of **Lamp**. (a) Implicit status representation. (b) Explicit status representation.

The OPD of Figure 5.1(a) is the implicit status representation, while that of Figure 5.1(b) is the explicit one. In the *implicit* status representation, the states are shown directly inside the **Lamp** rectangular box symbol, whereas in the *explicit* status representation, **Lamp** explicitly exhibits the **Status** attribute, which "hosts" the two **Lamp** states. In this case, **Lamp** has an attribute called **Status**, which has two possible values: **off** and **on**. Generalizing from this example, the status attribute can be represented either explicitly or implicitly. In *implicit status representation*, exemplified in Figure 5.1(a), the status attribute is not specified explicitly. Instead, the different states of the object are drawn within the box of the object itself. In *explicit status representation*, exemplified in Figure 5.1(b), the different routangles that represent states of the object are drawn within the object box of the **Status** attribute of that object. The OPL script that is equivalent to the OPD Figure 5.1(a) is:

> **Lamp** can be **off** or **on**.
>   *(State enumeration sentence)*

The reserved phrase "can be" identifies the *state enumeration* sentence. It is followed by a list of states ending with the reserved phrase "or" between the last and second to last state names. An object cannot be at more than one state at a time. Therefore, the semantics of the state enumeration sentence is that of the logical exclusive OR, called XOR for short. Hence, a more accurate phrasing of the previous state enumeration sentence might be "**Lamp** can be either **off** or **on**." However, this is a more verbose phrasing that does not sound well when more than two states are possible. The OPL script equivalent to the OPD Figure 5.1(b) is:

> **Lamp** exhibits **Status**.
>   *(Exhibition sentence)*
> **Status** can be **off** or **on**.
>   *(State enumeration sentence)*

The two OPL sentences above can be replaced by one:

> **Lamp** exhibits **Status**, which can be **off** or **on**.
>   *(State enumerated exhibition sentence)*

As the example shows, the implicit representation "**Lamp** can be **off** or **on**" is more succinct and intuitive than the explicit representation "**Lamp** exhibits **Status**, which can be **off** or **on**." It is used whenever we can avoid the explicit denotation of **Status** without compromising the OPD's understandability. If more than two states are expressed in the sentence, a comma follows each state except for the last one. Thus, if **standby** is a third state, the sentence "**Lamp** can be **off** or **on**." would read:

> **Lamp** can be **off**, **standby**, or **on**.

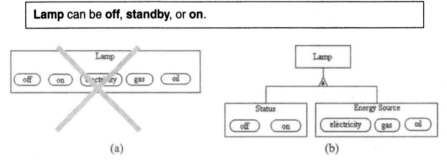

(a)   (b)

Figure 5.2. Incorrect (a) and correct (b) addition of the states **electricity, gas,** and **oil** to **Lamp**

We cannot add states or values that do not belong to the same attribute. Suppose, for example, that our lamp is versatile and can operate with three different energy sources: electricity, gas, or oil, and it is possible to switch among these three sources. To model this fact, we wish to add **electricity, gas,** and **oil** as states of **Lamp**. A wrong way of doing so is depicted in Figure 5.2(a). It mixes apples and oranges, so to speak: **on** and **off** are states of the **Status** attribute, while **electricity, gas,** and **oil** are states of another attribute, **Energy Source**. Since an object can only be at one state at a time, putting all five states together does not make sense. In Figure 5.2(b), **Energy Source** is correctly added as an attribute of **Lamp** beside **Status**, with **electricity, gas,** and **oil** being the states[1] of **Energy Source**.

---

[1]   As we will see in Chapter 12, unless the **Lamp** can be easily switched between **electricity, gas,** and **oil**, these energy sources are better modeled as *values* of **Energy Source**.

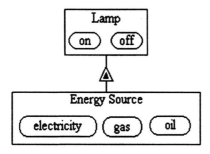

Figure 5.3. Energy Source is added as an attribute to the implicit status representation of **Lamp**.

In Figure 5.3, **Energy Source** was added as an explicit attribute to the implicit status representation of **Lamp**. The corresponding OPL paragraph is:

> **Lamp** can be **off** or **on**.
> **Lamp** exhibits **Energy Source**, which can be **electricity, gas**, or **oil**.

As this example shows, both implicit and explicit status representations provide for adding attributes to the object. The implicit status representation, demonstrated in Figure 5.3, is shorter and usually more intuitive. Hence, unless it poses a problem, it should be preferred.

5.1.4 The Input, Output, and Effect Links

Figure 5.4 shows the OPDs of Figure 5.1 with the addition of the **Lighting** process, which transforms **Lamp** from its **off** state to its **on** state in both the implicit (a) and explicit (b) status representations. The object being transformed has at least two states. One is the *input state*, which is the state that the object is in *before* the occurrence of the process. The other is the *output state*, which is the state the object is in *after* the occurrence of the process.

> *Input state* is the state of the object before the occurrence of the process.
> *Output state* is the state of the object after the occurrence of the process.

We can now refine our definition of *effect* as follows:

> *The **effect** of a process on an object is the change in state that the object undergoes from its input state to its output state as a result of the process occurrence.*

The effect of the process **Lighting** on the object **Lamp** is to change **Lamp** from its input state **off** to its output state **on**. Figure 5.4 expresses this using two identically looking arrows with elongated triangular arrowheads. One arrow, the input link, originates from the input state and terminates at the affecting process. The other, the output link, originates from the affecting process and terminates at the output state.

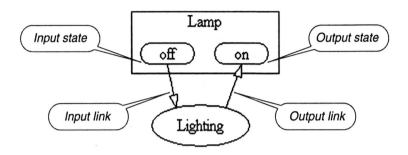

Figure 5.4. The input and output links, the input and output states, and the effect of the process **Lighting** on the status of **Lamp**

> *Input link is a link from the input state of the affected object to the process that changes that object.*
> *Output link is a link from the process that changes the object to the output state of that object.*

The OPD of Figure 5.4 shows an input link from **Lamp**'s input state **off** to the process **Lighting** and an output link from Lighting to **Lamp**'s output state **on**. The OPL script that is equivalent to the implicit state representation in the OPD of Figure 5.4 is:

> **Lamp** can be **on** or **off**.
> **Lighting** changes **Lamp** from **off** to **on**.

5.1.5 State Suppression and the Effect Link

Explicitly expressing the states of an object in the diagram often yields an OPD that is too crowded or busy, making it hard to read. This is an instance of the comprehensiveness-clarity tradeoff, which is discussed in detail in Chapter 9. To avoid this, OPM provides for the option of suppressing the appearance of states within an object.

Figure 5.5(a) shows the same OPD as Figure 5.4(a). Figure 5.5(b) is an illegal, in-between representation, in which the states are suppressed and not shown. Consequently, the input and output links that were previously attached to the state rountangles migrate to the boundary of the object box. They now link the process and the object directly, going from and to the object itself rather than from and to its states. This interim representation is not valid in OPM. Figure 5.5(c) carries the

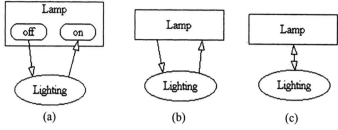

Figure 5.5. State suppression and the effect link. (a The OPD of Figure 5.4 with states expressed. (b) States of **Lamp** are suppressed, and the input and output links move from the suppressed states to the contour of the **Lamp** object box. (c) The input and output links are superimposed to yield the effect link.

same information as Figure 5.5(b). However, to reduce the graphic clutter, the input and output links, denoted by two opposite unidirectional arrows, have been superimposed by joining them into one bidirectional arrow, yielding the symbol of the *effect link*.

> An *effect link* is a transformation link, connecting a process and an object that the process changes.

The effect link is symbolized by a bidirectional arrow connecting the affecting process with the affected object, in which the input and output states are suppressed. The OPL sentence that corresponds to Figure 5.5(a) is:

> **Lighting** changes **Lamp** from **off** to **on.**
>   *(Change sentence)*

The OPL sentence that corresponds to Figure 5.5(c) is:

> **Lighting** affects **Lamp.**
>   *(Effect sentence)*

The elimination of the state symbols from the object is termed *state suppression*. State suppression is one of several abstracting options. Chapter 9 is devoted to abstraction and refinement. For now, we note that abstracting is a means to simplify the OPD at the cost of hiding details related to things in the OPD. Expectedly, as both the OPDs and in their equivalent OPL sentences demonstrate, state suppression eliminates the information about how exactly the process affects the object. This information can be provided in lower-level OPDs.

5.1.6 State Expression

We have seen how state suppression transforms the OPD portion from being detailed to being more abstract. The reverse of state suppression is *state expression*; refining the OPD by adding the information on what the possible states are. Whereas suppression caused the input and output links to be joined to an effect link, expression splits the effect link to its input link and output link components. To see this, consider the following example.

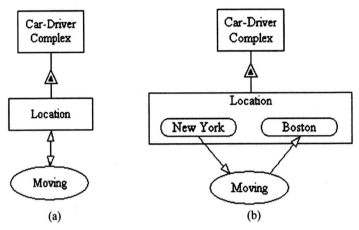

Figure 5.6. The effect of **Moving** on **Location** of **Car-Driver Complex**: (a) States suppressed. (b) States expressed.

Car cannot (yet) move sensibly without a human driver. Therefore, the car is just part of a higher level system, which we call **Car-Driver Complex**. The main service the **Car-Driver Complex** provides is the process **Moving**. The main service this system provides is the process **Moving**. The OPL script equivalent to the OPD of Figure 5.6(a) is:

> **Car-Driver Complex** exhibits **Location**.
> **Moving** affects **Location**.

The OPL script of the OPD in Figure 5.6 (b) is:

> **Car-Driver Complex** exhibits **Location**.
> **Moving** changes **Location** from **New York** to **Boston**.

Evidently, the OPD in Figure 5.6 (b), in which the states of the **Location** of **Car-Driver Complex** are expressed, is more informative than the OPD in Figure 5.6 (a), in which they are suppressed. More specifically, the process **Moving** affects the **Location** of **Car-Driver Complex** by changing the value of its **Location** attribute from one place (**New York**) to another (**Boston**).

5.2 Existence and Transformation

Change is one possibility of what can happen to an object when a process acts on it. A process affects an object to change it, but it can also do things that are more drastic: it can generate an object or consume it. The term *transformation* covers the three possible modes by which a process can act on an object – construction, effect, and consumption.

> **Transformation** *is the generalization of construction, effect, and consumption, which a process can do when it acts on an object.*

Construction is synonymous with creation, generation, or yielding. *Effect* is synonymous with change or switch, and *consumption* is synonymous with elimination, termination, annihilation, or destruction. We have seen that the effect of a process on an object is to change that object from one of its states to another, but the object still exists and maintains the identity it had before the process occurred. Construction and consumption change the very existence of the object and are therefore more profound transformations than effect.

Recall our discussion of existence in Chapter 4. When we say that a process constructs (yields, generates, creates, or results in) an object, we mean that the object, which did not previously exist, has undergone a radical transformation. This transformation made it stand out and become identifiable and meaningful in the system. It now deserves treatment and reference as a new, separate entity. When we say that a process consumes (eliminates or destroys) an object, we mean that the object, which previously existed, and was identifiable and meaningful in the system, has undergone a radical transformation. Consequently, the object no longer exists in the system and is no longer identifiable.

To explain construction and consumption within the framework of state changes, in Figure 5.7 we assign two states to an object: **existent** and **nonexistent**. Using implicit state representation, the **nonexistent** and **existent** states are drawn directly inside the **Object** box. In Figure 5.7(a), **Constructing** affects **Object** by changing its state from **nonexistent** to **existent**. In Figure 5.7(b), **Consuming** affects **Object** by changing its state from **existent** to **nonexistent**. The OPL paragraph follows.

Object can be **nonexistent** or **existent**.
Constructing changes **Object** from **nonexistent** to **existent**.
Consuming changes **Object** from **existent** to **nonexistent**.

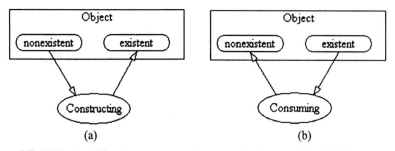

(a)   (b)

Figure 5.7. OPDs describing the states **nonexistent** and **existent** of the implicit **Existence** attribute of **Object** and the effect of (a) **Construction** and (b) **Consumption** on these two **Existence** states

5.2.1 Result and Consumption Links

Since the **nonexistent** state of **Existence** is literally nonexistent, we can think of it as such and eliminate it from the two OPDs in Figure 5.8. We also omit both the input link from the Object's **nonexistent** state to the **Constructing** process and the outgoing effect link from **Consuming** to the same **nonexistent** state. This leaves only the output link from **Constructing** to the **existent** state and the input link from the **nonexistent** state to **Consuming**.

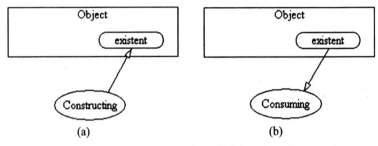

Figure 5.8. (a) The **nonexistent** state and the input link from it to **Construction** are removed. (b) The **nonexistent** state and the output link to it from **Construction** are removed.

Finally, since any object that exists is always at the state **existent**, we can eliminate this state just as we did with **nonexistent**, without losing information. Suppressing the **nonexistent** states of **Object** yields the *result link* in Figure 5.9(a) and the *consumption link* in Figure 5.9(b) as specializations of the output and input links, respectively. The result link shows the creation of an object; the consumption link shows its destruction.

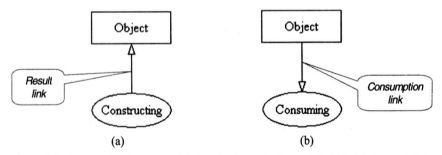

Figure 5.9. Suppressing the states of **Object** in the OPD Figure 5.8 yields (a) the result link and (b) the consumption link.

5.2.2 Procedural Links, Enablers, and Transformees

The definition of OPM process requires that the process transform at least one object. However, the process can have one or more objects that enable that process, but are not transformed by it. Hence, from the viewpoint of a given process, OPM distinguishes between two types of objects: *enablers* – objects that enable the process and *transformees* – objects that the process transforms. In this section, we discuss these two object types and elaborate on the procedural links that tie objects to processes.

Procedural links provide the "glue" that binds and relates objects to processes and vice versa, thereby providing for integration of the system's structure and behavior within a single model. This structure-behavior integration is one of the most important features of OPM. The structure-behavior integration is contrasted with the multiple models advocated by object-oriented approaches. As discussed, this integration within a single model avoids the inherent model multiplicity problem that object-oriented methods introduce. Procedural links are of utmost importance. They are indispensable in such elementary modeling needs as showing how processes transform objects and expressing cause and effect relations.

> **Procedural link** *is a connection between a process and an object or a state of an object.*

To denote the role of each type of object in the set of objects that are involved in a process, a special symbol is assigned to each one of the links connecting that object to the pertinent process. Accordingly, there are two types of procedural links: enabling links and transforming links.

5.2.3 Enablers

Suppose you wish to move from your place to an apartment in another city. To do this, you need a moving truck, which you rent from a moving truck rental company. You return the truck to the same place where you took it and with the same amount of gasoline as you took it. Hence, ignoring the amortization of the truck, nothing in it has changed. However, you would not be able to carry out the moving with out it. We say that the **Truck** is an *enabler* of the **Moving** process.

An enabler of a process is an object that enables the process execution, but is not transformed by it. In other words, an enabler of a process is an object that must be present in order for the process to occur. However, the existence and state of that object after the process is completed remains the same as it was just before the process started.

> *Enabler* of a process is an object that must be present in order for that process to occur, but is not transformed as a result of the occurrence of the process.

An enabler can be one of two types: an agent or an instrument, as defined below.

5.2.4 Agents

The term *agent* is reserved for the human user or user, who interacts with the system or parts thereof.

> *Agent* is an intelligent enabler, which can control the process it enables by exercising common sense or goal-oriented considerations.

This definition of agent as an intelligent enabler, which can control the process it enables by exercising common sense or goal-oriented considerations, implies that it must consist of one or more humans. Usually, it is a single person – the system's user. It can also be an organization, or a unit within a man-made organization, such as department, city council, government, group, team, etc. The notion of agent is important because it provides for modeling the "human in the loop", i.e., how people interact with the system. This is a clear indication to the system designer of points of interaction with the system where human interface needs to be developed. Moreover, the hierarchy of processes that the agent is involved in provides an excellent guideline for the arrangement of a friendly graphic user interface, which can even be automated to some extent.

The agent link is somewhat analogous to the "stick figure" used in UML's use-case model. In OPM, however, no separate model is needed as modeling the user is incorporated into the single OPM model. As shown in Chapter 15, use cases in UML notation can automatically be extracted from the OPM model, as can other UML models, such as the object model and state transition diagrams.

Not any human or organization is necessarily an agent. For example, if a **Student** is engaged in the process of **Studying**, his or her knowledge level attribute can change, say from **shallow** to **deep**, in which case **Student** is a transformee (defined below) rather than an agent. Likewise, if a department is undergoing "Business Process Reorganization", its structure or behavior changes as a result of this process, so it is also a transformee, not an agent.

5.2.5 Instruments

An instrument of a process is any non-human, physical or informatical object, which does not change as a result of the execution of the process.

Instrument is a non-human physical or informatical enabler.

Examples of instruments include machines, tools, computers, robots, controllers, hardware, software, documents, orders, recipes, algorithms, prescriptions, files, commands, information, and data. In practice, physical instruments usually change because of their participation in enabling a process. For example, they can wear out as they are being used as process enablers. However, it is often the case that from the point of view of the system under consideration, these changes are either not significant enough to be accounted for, or they are out of the system's scope.

For example, in developing a manufacturing system, a system architect may be required to account for maintaining a machine tool that wears out due to the metal cutting process it enables. In this case, the machine should not be assigned the role of an instrument. Rather, it will be modeled as an affected object, which is defined below. The attribute of the machine that changes as a result of its operation can be, for example, its amortization level, or hours of operation since the last overhaul. We will have to take this tool wear in account if our system encompasses the maintenance aspect of the machine.

Algorithms and recipes are prime examples of informatical instruments that can be used repeatedly, ideally without wearing out (in practice we may witness "software amortization" as well…). A file is an instrument if it is used as "read only", so that its content is not altered at all. Otherwise, the file is a transformee, as discussed below. The distinction among the two types of enablers in an OPD is done through the different enabler links, explained next.

5.2.6 Enabling Links

Enables are linked to processes through enabling links.

Enabling link is a procedural link that connects a process with an enabler of that process.

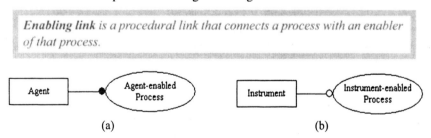

(a)   (b)

Figure 5.10. The enabling links: (a) The agent link. (b) The instrument link.

Graphically, as Figure 5.10 shows, an enabling link is a "lollipop", a line leading from the enabler to the process it enables, which ends with a circle touching the process side. If the enabler is an agent, the enabling link is an agent link, denoted as a "black lollipop", i.e., its ending circle is filled (black). If the enabler is an instrument, the enabling link is a "white lollipop", i.e., its ending circle is blank (white).

5.2.7 Transformees

A transformee of a process is an object that undergoes a transformation as a result of the occurrence of the process. The transformation can be construction, effect (change of state) or consumption.

> *Transformee* *of a process is an object that is transformed by the occurrence of that process.*

Like an enabler, a transformee is a *role* that an object assumes *with respect to a particular process*. Hence, an object can be an enabler with respect to one process and a transformee with respect to another process. A transformee can be one of three types: Affectee, resultee or consumee. The distinction among these transformee types in an OPD is done through the various transformation links, as explained below.

> *Affectee* *of a process is a transformee that was affected by the occurrence of that process.*
> *Consumee* *of a process is a transformee that is consumed and eliminated as a result of the occurrence of that process.*
> *Resultee* *of a process is a transformee that is constructed as a result of the occurrence of that process.*

As noted, a file is an instrument if it is used as "read only" so that its content is not touched. If any data in the file is added, changed, or removed, then it is an affectee. If the process erases the file after it uses it, then the file is a consumee. A new file, which is created as a result of a process, is a resultee.

The transforming link is a unidirectional or bidirectional arrow, connecting the transformee to the transforming process.

> *Transformation* *link is a procedural link that connects a process with a transformee of that process.*

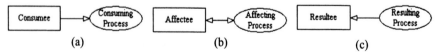

(a)   (b)   (c)

Figure 5.11. Transformation links: (a) Consumption link. (b) Effect link. (c) Result link.

As shown in Figure 5.11(a), the arrow in a consumption link is unidirectional, with the arrowhead pointing from the **Consumee** to the **Consuming Process**.

> *Consumption link* is a transformation link that connects a consuming process with a consumee of that process.

In Figure 5.4, we have shown how a pair of incoming and outgoing effect links is joined to produce the bidirectional effect link, which is shown in Figure 5.11(b).

> *Effect link* is a transformation link that connects an affecting process with an affectee of that process.

In a result link, the arrow is unidirectional, with the arrowhead pointing from the **Resulting Process** towards the **Resultee**, as shown in Figure 5.11(c).

> *Result link* is a transformation link that connects a resulting process with a resultee of that process.

An object and a process can be connected by one procedural link only. This raises interesting questions regarding the precedence of the procedural links, which we address in Chapter 9.

5.2.8 Odd Man Out: The Invocation Link

By definition, any process must transform some object. Sometimes, however, the transformation is not significant in the system and may as well be ignored, but the change in the skipped object invokes another process. Thus, while we are not concerned with the transformation the invoking process has brought about, we do wish to model the invocation of the next process in the series of process executions. This is the purpose of the invocation link.

> *Invocation link* is a direct link between an invoking process and an invoked one.

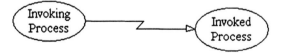

Figure 5.12. The invocation link

The invocation link, shown in Figure 5.12,[2] can be thought of as an event that marks the end of the process from which the link emanates, and at the same time,

---

[2]   In Systemantica®, the lightening symbol is replaced by a double-headed arrow: ⎯⊳⊳.

invokes the process to which it points. A complementary way of looking at an invocation link is as a shortcut in the model that symbolizes a transformed object between two processes that the invocation link connects, which, for the sake of brevity was skipped. This is demonstrated in Figure 5.13 and the corresponding OPL script in Frame 10. Figure 5.13(a) shows an invocation link that "short-circuits" the path between **P1** and **P2**.

Figure 5.13(b) shows the "complete" version, where the interim object **B** is generated by **P1** and immediately consumed by **P2**. When the process **P** is out-zoomed, the two OPDs look the same: **P** consumes the object **B1** and yields **B2**. When **P** is in-zoomed, both OPDs show that **P** zooms into (and consists of) **P1** and **P2**.

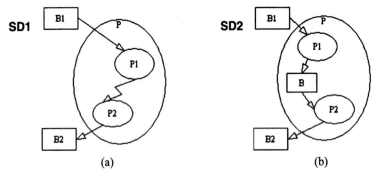

(a)   (b)

Figure 5.13. The invocation link: (a) An invocation link from P1 to P2. (b) A similar OPD, in which the intermediate object B was not skipped.

SD1 Paragraph:	SD2 Paragraph:
P zooms into **P1** and **P2**.	P zooms into **P1** and **P2**, as well as **B**.
P1 consumes **B1**.	P1 consumes **B1**.
P1 invokes **P2**.	P1 yields **B**.
P2 yields **B2**.	P2 consumes **B**.
	P2 yields **B2**.

Frame 10. The OPL paragraphs of SD1 and SD2 in Figure 5.13.

5.3 Object Roles with Respect to a Process

Objects can take on more than one role with respect to a process. In this section we discuss and exemplify the various roles of objects as well as sets of objects with respect to a particular process.

5.3.1 Enablers and Affectees

Enablers and affectees are possible roles that an object assumes with respect to a particular process. This means that the same object can be an enabler for one process but not for another, or it can be an enabler for one process and an affectee for another.

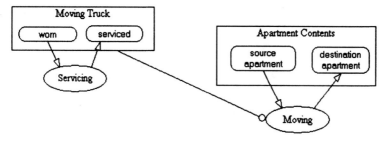

Figure 5.14. Apartment Content is an affectee of **Moving**; **Moving Truck** is an instrument of **Moving** and an affectee of **Servicing**.

Examining Figure 5.14, we see that the **Servicing** process, which the moving company applies periodically to the **Moving Truck**, changes its state from **worn** to **serviced**, which makes **Moving Truck** an affectee of **Servicing**. With respect to the process **Moving**, however, **Moving Truck** is an enabler, while **Apartment Content** is an affectee, as **Moving** changes its state from **source apartment** to **destination apartment**.

5.3.2 The Involved, Preprocess, and Postprocess Object Sets

At least one object must be transformed by an occurrence of the process. More formally, for a process to take place, it requires a non-empty set of objects, called the *involved object set*.

> The **involved object set** *of a process is the set of objects that are transformed by the process or enable it.*

The involved object set is a union of two sets: the preprocess object set and the postprocess object set.

> The **preprocess object** *set of a process is the set of objects that are required for the process to start its execution, enablers as well as transformees.* The **postprocess object set** *of a process is the set of objects that are generated or affected by the execution of the process.*

The involved object set of **Moving** in Figure 5.14 is **Moving Truck** and **Apartment Content.** The preprocess object set of **Moving** is identical with its involved object set. The preprocess object set of **Moving** is **Apartment Content,** since it is the only one that is affected by **Moving.**

The involved object-set is the union of the preprocess and postprocess object sets, which are not necessarily disjoint. If the involved object-set contains affectees, they are common to the preprocess and postprocess object sets. Enablers and consumees belong only to the preprocess object set, while resultees, only to the postprocess object set.

5.3.3 Condition and Agent Condition Links

We have seen that a condition link looks graphically like an instrument link, but its semantics is different from that of the instrument link. Condition links are state-specified enablers: rather than going from the instrument to the process, the condition link goes from a specific state or value of the instrument to the process. The link from state **denied** of **Approval** to **Notifying** in is an example. Similarly, an *agent condition link* looks graphically like an agent link. However, like the condition link, rather than going from the agent to the process, it goes from a state or value of the agent to the process. The link from state **authorized** of **User** to **Accessing** in Figure 5.15 is an example.

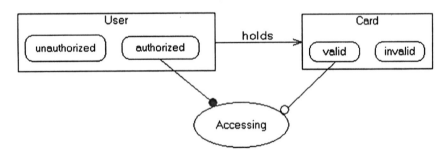

Figure 5.15. An example of a condition link and an agent condition link

The OPL of this OPD is:

> **User holds Card.**
> **User** can be **unauthorized** or **authorized.**
> **Card** can be **valid** or **invalid.**
> **User** must be **authorized** for **Accessing** to occur.
> > *(Agent condition sentence)*
> **Accessing** occurs if **Card** is valid.
> > *(Condition sentence)*

The agent condition sentence follows the style of the agent sentence, which is agent-centric. In an agent sentence we start with the agent name, such as "**User handles Accessing.**" Similarly, an agent condition sentence also starts with the agent name, as in "**User** must be **authorized** for **Accessing** to **occur.**" Agent condition links are less prevalent that condition links, because agents are normally not required to be at a certain state for a process to take place.

5.3.4 Operator, Operand, and Transform?

Before concluding this chapter on the dynamics of systems, it may be interesting to compare the OPM ontology to the definitions of Ashby (2001) regarding operand, operator and transform:

> *Consider the simple example in which, under the influence of sunshine, pale skin changes to dark skin. Something, "the pale skin", is acted on by a factor, "the sunshine", and is changed to dark skin. That which is acted on, the pale skin, will be called the OPERAND, the [causing] factor will be called the OPERATOR, and that what the operand has changed to, will be called the TRANSFORM.*

In our terminology, **Skin** is an object, while **dark** and **pale** are states of an attribute of the object **Skin** called **Color.** **Tanning** is a process, and **Sun** is an instrument that enables the **Tanning** process, the effect of which is to change the **Color** of the **Skin** from **pale** to **dark.** This terminology seems more intuitive and appropriate for non-mathematical systems than the operand, operator and transform ontology. The "sunshine factor" is a bit problematic to describe. It is not clear whether it refers to the shining process of the sun or to the object that aggregates the photons of energy radiated by the sun, which the skin absorbs.

In OPM, we would model **Radiating** as a first subprocess of **Tanning.** **Radiating** requires (in other words, is enabled by the instrument) **Sun** and produces the object **Solar Energy**, which is absorbed by the **Skin** via the second subprocess, **Absorbing.** A third subprocess, **Pigmenting**, is the one that finally changes the **Color** of **Skin** from **pale** to **dark.** In summary, the operator is the process (**Tanning**). The operand is the affectee in its state before the process occurred (**Skin** in its **pale Color** state), while the transform is its state after the process occurred (**Skin** in its **dark Color** state).

Summary

The various procedural links are summarized in Table 1.

Table 1. Procedural links, their semantics and various attributes

Type	Name	Semantics	Symbol	Source	Destination
Transforming	Consumption	The process consumes the object.		Object	Process
	Result	The process generates the object.	→▷	Process	Object
	Input[a]	The process changes from an input state.		state	Process
	Output[a]	The process changes to an output state.		Process	state
	Effect	The process changes the object.	◁—▷	Object	Process
				(Both are source and destination.)	
Enabling	Agent	The human agent enables the process.	—●	Object	Process
	Instrument	The process requires the instrument.	—○	Object	Process
Condition	Agent condition	The agent at state enables the process.	—●	state	Process
	Condition	The instrument at state enables the process.	—○	state	Process

a. Input and output links only appear in pairs: a change from the input state to the output state. Single "Input" or "Output" links are state-specified consumption and result. They have different semantics, as defined in Chapter 12.

Problems

1. For each of the following statements, draw an OPD and write a corresponding OPL sentence:
   a. States that a person can be in are "awake" and "asleep".
   b. A machine can be in states "operational" or "broken".
   c. The weather can be sunny, cloudy, rainy or snowy.
   d. An organism can be alive or dead.
2. Write the OPL sentences for the OPDs in Figure 5.2 and in Figure 5.3. Explain which sentences make or do not make sense and why.

3. Provide three examples of processes. For each one of them, show in an OPD the affected object and the input and output states and write the corresponding OPL script.
4. Using OPDs, show two examples where implicit status representation is better than explicit one and two examples where the opposite is true.
5. Compare the sentence

> **Water** can be **solid, liquid** or **gas.**

with the two sentences

> **Water** exhibits **State of Matter.**
> **State of Matter** can be **solid, liquid** or **gas.**

   a. What is **State of Matter** in OPM terms?
   b. Draw the corresponding OPDs.
   c. Add the processes that transform **Water** from any one of its state to another.
6. Show two cases in which:
   a. A set of situations is best modeled as states of an object.
   b. A set of situations is best modeled as a set of attribute values.
   c. Neither states nor values are better in modeling a set of situations.
7. Show and explain how the three transformation links are related to the input and output links. Compare the OPDs and their corresponding OPL sentences.
8. Why is it important to distinguish between enablers and transformees? Show an example.
9. Why is it important to distinguish between the enablers agent and instrument? Show an example.
10. An agent can be just involved in a process or be in charge of it. Propose an extension to OPM to make this distinction in both OPDs and OPL.
11. Explain and exemplify how the input and output links are related to state-specified consumption and result links. See Section 12.1 for more information. Compare the OPDs and their corresponding OPL sentences.
12. Draw an OPD and explain the relationships between the involved object set and the preprocess and postprocess object sets discussed in Section 5.3.
13. Draw the OPD and derive the OPL paragraph of the tanning system in Ashby's example of operator, operand, and transform in Section 5.3.4.

Chapter 6
Structure

The Piglet lived in a very grand house in the middle of a beech-tree, and the beech-tree was in the middle of the Forest, and the Piglet lived in the middle of the house. Next to his house was a piece of broken board which had "TRESPASSERS W" on it.

Winnie-The-Pooh by A. A. Milne

Structure pertains to the relatively fixed, non-transient, long-term relationships that exist among components or parts of the system. Alternatively, structure can be viewed as a snapshot – an account of the system, or part of it, at some point in time. The snapshot captures the system at some state, at which specific relationships between objects hold. Structure is contrasted with the complementary dynamic aspect of the system, or its behavior, discussed in Chapter 5, which has to do with the changes the system undergoes over time, along with the causes for and effects of these changes. In other words, structure is about the static aspect of the system, while behavior is about its dynamic aspect. This chapter is devoted to discussing the structure of systems and expressing it through OPM.

6.1 Structural Relations

A basic concept that is needed in order to discuss structure is a structural relation, so to kick off our discussion, we start with the following definition.

> A **structural relation** is a connection or an association between things that holds irrespective of time.

A structural relation holds in the system in general, not contingent upon conditions that are time-dependent. A structural relation is usually meaningful among objects and less so among processes, or between objects and processes. However, we will see examples of processes that exhibit structural relations. UML uses the term *association* to denote a structural relation. By its nature, a structural relation is multilateral, because every thing that participates in the association has some relation with the rest of the things. The number of things involved in the structural relation determines the *"arity"* of the relation: a relation of a thing to itself is a *unary* structural relation, a relation between two things is a *binary* structural relation, a relation between three things is a *ternary* structural relation, and so forth.

The most common, important and prevalent structural relation is the binary relation. A unary relation is rare and can be considered as a special case of the binary relation. An n-ary relation with $n \geq 3$ can be analyzed as a set of n binary relations that are held between each one of the n things and the rest of the $n - 1$ things. In view of this observation, we focus our attention on binary structural relations.

A binary structural relation is bidirectional: if thing T_1 relates to thing T_2 through the relation $\Re$, then it is also true that T_2 relates to T_1 through another relation $\Re'$, and vice versa. Symbolically, $T_1 \Re T_2 \Leftrightarrow T_2 \Re' T_1$. $\Re$ is a symbol that stands for the *forward structural relation*, that is, the structural relation as T_1 views it while referring to T_2. Conversely, $\Re'$ is the *backward structural relation*, i.e., the relation as T_2 views it while referring to T_1. $\Re$ and $\Re'$ constitute a *structural relation pair*.

$\Re$ and $\Re'$ may be the inverse of each other. Symbolically, $(\Re')' = \Re$. For example, if $\Re = $ "**is parent of** " then $\Re' = $ "**is child of** " and so if "**A is parent of B**" then "**B is child of A**". These are examples of pairs of mutually anti-symmetric relations: existence of a relation in one direction mandates the existence of the opposite relation in the reverse direction. Examples of non-mutually anti-symmetric relation pairs are $\Re = $ "**pushes**" and $\Re' = $ "**supports**", as well as $\Re = $ "**loves**" and $\Re' = $ "**is indifferent to**". If **A loves B** it is not necessarily true that **B loves A**.

The *name* of the structural relation is a *phrase* – a combination of one or more natural language words (but not a complete sentence), separated by spaces. The name expresses the nature, meaning, or content of the structural relation between the two things that participate in the relation. Examples of phrases that express structural relations are **owns, is next to, resides, borders, represents, is equivalent to**, and **contains**.

6.1.1 Structural Links

Recall that an Object-Process Diagram (OPD) is the visual formalism of OPM that describes graphically the structure and/or behavior of the system or part thereof. An OPD can contain objects, processes and links. Links can be structural or procedural. Since in this chapter we discuss the structure of systems, we focus on structural links. While a structural relation models an association between two things (usually objects) that is meaningful in the system, a *structural link* graphically represents the structural relation.

> A **structural link** is the graphic symbol that represents a binary structural relation between two objects in an OPD.

A structural link is symbolized by an arrow with an open head: $\longrightarrow$. This is contrasted with the closed triangular arrowheads of the dynamic links (the input link and output link, $\longrightarrow\!\triangleright$, and the effect link, $\triangleleft\!\!-\!\!\triangleright$). The structural link arrow points from one object to another. A meaningful label or *tag* is usually recorded along the structural link, which is consequently dubbed as a *tagged structural link.*

Figure 6.1. An OPD with two objects and a unidirectional binary structural link between them

Figure 6.1 is an example of an OPD with two objects – **Highway** and **City**, and a unidirectional binary structural link, labeled **surrounds**, that links these two objects. The corresponding equivalent OPL sentence is recorded below.

Highway surrounds City.

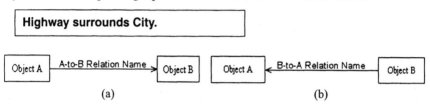

Figure 6.2. A generic OPD showing two objects and two unidirectional anti-symmetric binary structural links between them

Figure 6.2 is a generic OPD that contains two pairs of objects related by two oppositely directed structural links. The tag of the forward link in Figure 6.2(a) is **A-to-B Relation Name**, while the tag of the backward link in Figure 6.2(b) is **B-to-A Relation Name**. The choice of "forward" and "backward" is arbitrary.

Figure 6.3. A The two anti-symmetric binary structural links of the OPD in Figure 6.2 are joined to yield a single bidirectional structural link.

In Figure 6.3, the two anti-symmetric unidirectional binary structural links of the OPD in Figure 6.2 are joined to yield a single bidirectional binary structural link. The structural link is depicted as a bidirectional, harpoon-shaped arrow, which links the two things. The names of ℜ, the forward (A-to-B) relation, and ℜ', the backward (B-to-A) relation, are recorded such that the sticking out of the arrowheads unambiguously determine the direction in which each relation holds.

Structural relations are held between things, processes as well as objects. In particular, the Aggregation-Participation relation is applicable to processes just as well as it is to objects. However, in general, structural relations are much more relevant to objects than they are to processes. Therefore, we focus in this chapter on binary structural relations between objects only.

6.1.2 Structural Relation Directions

Each structural relation has a *source object* – the origin object, from which the arrow denoting the structural link emanates, and a *destination object* – the object to which the arrow points. The two directions of the structural link are called the *forward direction*, which is from the source object to the destination object, and the *backward direction* from the destination to the source object. The decision as to which object is the source (and hence which direction of the structural link is the forward direction) is arbitrary. Following are two examples of OPDs containing binary structural links and their corresponding OPL script.

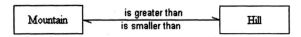

Figure 6.4. A bidirectional structural link example

The structural relation between the two objects **Mountain** and **Hill**, when viewed from the direction from **Mountain** to **Hill**, is that **"Mountain is greater than Hill."** In the opposite direction it is **"Hill is smaller than Mountain."** The OPD in Figure 6.4 is a formal graphic expression of these facts, and the following OPL script is their corresponding textual representation, the *OPL paragraph*.

> **Mountain is greater than Hill.**
> **Hill is smaller than Mountain.**

The OPL sentence of a structural relation is constructed by simply concatenating (chaining text pieces in some order) the name of the source object, the name of the structural link, the name of the destination object, and a period. For the OPL (and the OPD) to make sense, the choice of the names for the forward and backward structural relations should be made so as to enable the construction of two mutually complementary, anti-symmetric sentences that make sense.

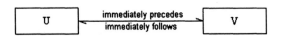

Figure 6.5. The forward and backward structural relations between **U** and **V**

As another example, the two objects **U** and **V** are the two alphabet letters, between which it holds that **"U immediately precedes V."** and **"V immediately follows U."** Here, the name of the forward relation is $\Re$ = **immediately precedes** and the name of the backward relation is $\Re'$ = **immediately follows**. The OPD of Figure 6.5 and OPL paragraph below are the corresponding formal graphic and textual expressions of these facts.

> **U immediately precedes V.**
> **V immediately follows U.**

Often, the reverse structural relation is just the passive voice of the forward relation or is otherwise insignificant. In cases like this, we avoid specifying the reverse direction and the structural link becomes unidirectional.

6.1.3 Unidirectional Structural Link

To see an example of a unidirectional structural link, consider the OPD in Figure 6.6(a), which expresses the OPL sentence **Diskette stores File**. Since the reverse sentence, **File is stored on Diskette**, does not increase our understanding of the system, we do not need to specify it, so the simpler OPD in Figure 6.6(b) is the one that will be employed.

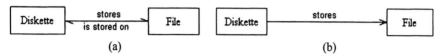

(a)   (b)

Figure 6.6. From a bidirectional to a unidirectional structural link: (a) The bidirectional link is redundant since the reverse structural relation is the passive of the forward relation. (b) Instead, a more economic unidirectional link is employed.

There may be cases other than the backward direction being the passive of the forward direction, is otherwise meaningless or unimportant, in which the analyst can decide not to specify the backward direction. For example, we could have designed the OPDs in Figure 6.4 and in Figure 6.5 with just one of their structural relations, since the other one can be deduced unambiguously. However, we should keep in mind that binary structural relations always come in pairs. One relation is the relation as it is seen from the viewpoint of one of the objects linked by the relation, and the other – as seen from the other object's viewpoint. Specifying only one direction is a shorthand notation.

6.1.4 OPD Sentences

In the simple OPDs that we have seen in this chapter so far, the entire OPD is translated to one or two OPL sentences. A collection of symbols in an OPD that give rise to an OPL sentence is an OPD sentence, also called graphic sentence.

> An **OPD sentence** is a set of OPD symbols from which an OPL sentence is constructed.

A unidirectional structural link with the two objects attached to its two edges, as the OPD in Figure 6.6(b), is an instance of a graphic sentence. A bidirectional structural link is a superimposition of two anti-symmetric unidirectional links, so it produces two OPL sentences. Just as it is possible to extract an OPL sentence from a graphic sentence, it should be possible to construct a graphic sentence from a given OPL sentence. The OPL sentence **"Diskette stores File"** can be generated automatically by a CASE tool by concatenating the source object with the relation name followed by the destination object name, while following the direction of the unidirectional structural link.

Generating a graphic representation from text poses algorithmic challenges that are not present when we generate text from graphics. This is because while text is one dimensional – a string of characters with a single, predefined reading order, graphics are two-dimensional and have no definite reading order. Automatically laying out the various OPD symbols such that they do not overlap, make sense, and are comprehensible and pleasing to the human eye, is not trivial.

6.1.5 The Reciprocity of a Structural Relation

In the examples we have seen, the anti-symmetric forward and backward relations between the source and destination objects differ in name and have opposite semantics. However, the forward and backward relations may be identical. Figure 6.7(a) is an example of a structural relation in which the relation with the same tag, which is the non-reserved OPL phrase **"is married to"**, appears in both directions. The OPL paragraph of Figure 6.7(a) is:

> **Jacob is married to Rachel.**
> **Rachel is married to Jacob.**
> *(Structure sentences)*

To avoid repeating the same relation name in the OPD and to avoid having to write two separate sentences in the OPL paragraph, we change the name of the relation such that a meaningful sentence will result in from **"Jacob and Rachel ..."** Thus, in the OPD of Figure 6.7(b) we use the single reciprocal structural relation **are married** that is semantically equivalent to Figure 6.7(a). The OPL sentence is:

> **Jacob** and **Rachel are married.**
> *(Reciprocal structure sentence)*

In general, a reciprocal structure sentence is constructed by first listing the two objects, joined by the reserved word and, followed by the name of the reciprocal structural relation.

> A *reciprocal structural relation* is a structural relation that replaces a pair of identical forward and backward structural relations.

Conversely, the examples we have seen in the three previous OPDs are of *non-reciprocal* structural relations. Additional examples of reciprocal structural

(a)   (b)

Figure 6.7. Two identical structural relations (a) are substituted by one reciprocal structural relation (b).

relations include: **touch each other, are siblings, are linked, are semantically equivalent, are next to each other, relate to each other,** and **agree.** The name of the reciprocal relation often starts with the word **are,** but as the examples show, this is not always the case. When we will study about the Generalization-Specialization structural relation in Chapter 8, we will see that the word are is also a reserved phrase that is used in specialization sentences, such as "**Jacob** and **Rachel** are **Persons.**" However, when the word **are** is used as part of a reciprocal tag, it is not reserved. At first glance, there seems to be a conflict between the above type of sentence and a reciprocal structure sentence, such as "**Jacob** and **Rachel** are **married.**" The problem seems to be that just as a machine could deduce that "**Jacob** is a **Person**" and "**Rachel** is a **Person**," it could infer the nonsense sentences "**Jacob** is a **married**" and "**Rachel** is a **married**." However, a closer look reveals that this is not a problem, because objects (such as **Person**) are capitalized, while tags (such as **are married**) are non-capitalized.

The distinction between the two structural relation types is made by the **Reciprocity** attribute of the structural relation.

> *Reciprocity is an attribute of a structural relation that denotes whether the relation is reciprocal or not.*

The two values of the **Reciprocity** attribute of a **Structural Relation** are **Reciprocal** and **Non-reciprocal.** A **Structural Relation** is **Non-reciprocal** if the forward and backward relations are different, in which case there are two relation names, as in Figure 6.5. A **Structural Relation** is **Reciprocal** if the structural relation name is recorded only once along the bidirectional link, as is the case in Figure 6.7(b). Since non-reciprocal structural relations are more prevalent, the default **Reciprocity** value is **Non-reciprocal.**

6.1.6 Null Tags and Their Default OPL Reserved Phrases

The tag in both the unidirectional and the bidirectional tagged structural links may be the *null tag*, i.e., an empty tag, as shown in **SD2** and **SD4** in Figure 6.8. These null tags have default OPL phrases, which are shown in **SD1** and **SD3** in Figure 6.8.

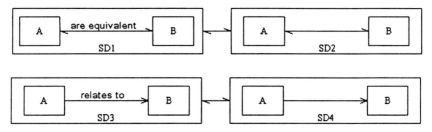

Figure 6.8. The null tagged structural links and their default OPL phrases

The OPL paragraph of both **SD1** and **SD2** is:

A and **B** are equivalent.
   (Equivalence sentence)

The OPL paragraph of both **SD3** and **SD4** is:

A relates to **B**.
   (Relation sentence)

Using the null bidirectional tagged structural link, Figure 6.8 also shows that "**SD1** and **SD2** are equivalent" and "**SD3** and **SD4** are equivalent."

6.1.7 Structural Relations as Static Verbs

Many of the tags of the structural relations are some verb forms. Few examples are **surrounds, precedes, follows, contains, holds, maintains, remains, supports, fastens** and **comprises**. What, then, is the difference between these verbs and processes? The difference lies in the message of continuity, stability, detachment from time, or steady state, that these verbs convey. Recall that the process test, discussed in Chapter 4, stipulates four conditions for a thing to qualify as a process:

(1) Object dependability: A process must rely on at least one object for it to happen.
(2) Object transformation: A process must transform at least one of the objects in the preprocess object set.
(3) Association with time: A process must represent some happening, occurrence, action, procedure, or activity that takes place along the timeline.
(4) Association with verb: A process must be associated with a verb.

Of these four conditions, structural relations do not fulfill the second and third conditions. The verb in the tag of the structural relation does not carry out any object transformation, and therefore no happening, occurrence, action, procedure, or activity takes place along the timeline. Rather, a structural relation has the notion of static, steady state that is true as long as no process acts upon any of the objects involved. Telling the difference between a verb that really expresses a process and one that is actually a structural relation requires deep understanding of the situation at hand. For example, in the OPL sentence **Highway surrounds City**, the word **surrounds** is a structural relation. Once the highway has been built, for as long as it exists, it is static and keeps surrounding the city regardless of the time element. However, in the sentence "Police surrounds House," **Surrounding** is a process that changes the object House from **non-surrounded** to **surrounded**. This is a process that takes time to complete and requires continuous activity.

6.2 Participation Constraints and Cardinality

In all the examples and discussions so far we have tacitly assumed that each thing, be it object or process, participates singly in the relation. Indeed, the convention in OPDs is that when no quantity is explicitly recorded by the side of a structural link, it is taken to be 1, which is the default value. However, in general we may wish to specify a certain number or a range of numbers of instances of the same class of things that participate in the relation. Participation constraints are designed to take care of this.

6.2.1 Participation Constraints

When more than one object is involved in a structural relation at one of the sides of that relation, a participation constraint needs to be specified to denote this.

Participation constraint is a number or symbol recorded along the structural link next to an object, which denotes the number of instances of that object which participate in link.

By default, the implicit participation constraint is 1. Thus, if exactly one thing participates in the relation, no participation constraint needs to be specified. When the participation constraint on the destination side of the structural link is different than 1, it has to be specified explicitly, as shown in Figure 6.9. The OPD in Figure 6.9 is an example of a tagged structural link, for which the participation constraint on the target object is expressed as a specific number.

The participation constraint can be specified either as a certain number or symbol, or as a range whose limits are numbers and/or symbols (lower-case letters) representing numbers.

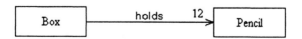

Figure 6.9. A one-sided cardinality with a participation constraint on the destination object

The corresponding OPL sentence is:

> **Box holds 12 Pencils.**

The destination object **Pencil** in the OPD of Figure 6.9 has the participation constraint 12, while the object **Box** has the implicit default participation constraint, which is 1. If the participation constraint is explicit, as it is for **Pencil** in the OPL sentence "**Box holds 12 Pencils**," it means that the participation constraint is greater than 1. In this case, while generating the OPL sentence from the OPD, the numeric or symbolic value of the participation constraint is put before the object name and the object name becomes plural. Usually that means concatenating the letter **s**, but a program that generates OPL sentences from OPDs should also account for exceptions of converting a noun from singular to plural.

Figure 6.10. A one-sided cardinality with a single number participation constraint on the source object

Ignoring the participation constrain in Figure 6.10, the OPL sentence is "**Bolt Fastens Flange**." Since the source object **Bolt** has the participation constraint 6, while the destination object **Flange** has the implicit default participation constraint, which is 1, the corresponding OPL sentence is:

> **6 Bolts fasten Flange.**

Note that to maintain a correct English syntax, the suffix **s** for singular third body present tense was removed from **fastens**, while to **Bolt** the letter **s** was added to make it plural. As we see later in this chapter, when a multiplicity constraint larger than 1 exists on both sides of the link, two OPL sentences are needed.

6.2.2 Parameterized Participation Constraints

The participation constraints need not be a particular number. It can be a parameter, symbolized by a lower-case letter.

> *A **parameterized participation constraint** is a participation constraint which is a symbol, denoting a parameter that stands for some numeric value.*

Figure 6.11 and its OPL sentence below it provide an example of a parameterized participation constraint. Here, the designer wishes to emphasize that the number of cylinders is even.

Figure 6.11. A parameterized participation constraint

> **Engine comprises 2n Cylinders.**

6.2.3 Range Participation Constraints

The participation constraint is not always a particular number. It can be a range of possible numbers.

> *A **range participation constraint** is a participation constraint that allows for a number of possible objects within a specified range to take part in the relation.*

A range participation constraint, which is exemplified in Figure 6.12, is denoted as "$q_{min} .. q_{max}$", where $q_{min} = 3$ is the lower participation constraint value and $q_{max} = 5$ is the maximal range quantity. The two quantities are separated by two consecutive dots.

Figure 6.12. A one-sided cardinality with a range participation constraint of 3..5

> **Machine Center controls 3 to 5 Machines.**

A single-valued participation constraint, i.e., a participation constraint denoted by one number or symbol, as in Figure 6.9 and in Figure 6.10, is a special case of a range participation constraint, in which $q_{min} = q_{max}$. For the sake of brevity, only one of them is written. Often, q_{min} is a small number, such as 0, 1, or 2, while q_{max} is the symbol m, which stands for many. The letter m is a "reserved symbol" in participation constraint, meaning that the exact value of "many" is not fixed as in an algebraic equation. This means that it may be different for different participation constraints even in the same diagram although all bear the same letter m. Any lower-case letter other than m stands for a particular, yet unspecified number (as in algebra).

6.2.4 Shorthand Notations and Reserved Phrases

The reserved phrase "q_{min} to q_{max}" can be used for any of the participation constraints, where both q_{min} and q_{max} can be any whole number. However, it frequently makes more sense to use different phrases that express the participation constraint more naturally.

Table 2. The six abbreviated participation constraint symbols and their reserved phrases

Symbol	?	*	(none)	1	+	m
qmin .. qmax	0..1	0..m	1..1	1..1	1..m	m..n
OPL phrase	An optional	optional (+ plural)	(none)	a / an	at least one	many

Table 2 shows the six abbreviated participation constraint symbols, which are ? for 0..1, * for 0..m, 1 for 1..1, + for 1..m, and m for m..n. Each such abbreviation has a corresponding OPL reserved phrase, which is listed below the symbol.

Table 3. Sample OPDs and OPL sentences using the participation constraint symbols

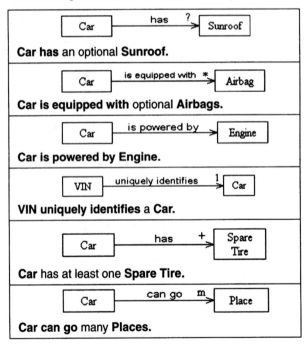

Table 3 shows sample applications of these participation constraint symbols. The participation constraint symbol 1 is the default and therefore needs not be written explicitly. Its reserved phrase is nothing, as in

> **Car is powered by Engine.**

Alternatively, the indefinite article a or an can be used, depending on whether the name of the target object starts with a vowel, as in

> **Car is powered** by **Engine.**

This option is demonstrated in Table 3 by the OPL sentence "**VIN uniquely identifies a Car.**" The reason for allowing these two alternative forms, one with and the other without the indefinite article, is to provide humans with the ability to express their ideas as naturally as possible. The system should interpret the two sentences as having the same semantics. In programming languages, such provisions are known as "syntactic sugar."

6.2.5 Cardinality

Each edge, or side, of the structural relation can have a participation constraint that is in general independent of the participation constraint on the other side. The combination of the two participation constraints is the structural link's *cardinality*.

> *Cardinality* is an attribute of a structural relation whose value depends on the combination of the participation constraints on both sides of the structural link.

We denote the cardinality as $[(q_{min} .. q_{max}), (q'_{min} .. q'_{max})]$, or, more succinctly without the parentheses as $[q_{min} .. q_{max}, q'_{min} .. q'_{max}]$, where q_{min} and q_{max} are the lower and upper bounds of the participation constraint on the source side of the structural link, while q'_{min} and q'_{max} are the corresponding parameters on the destination side.

Multiplicity is an important factor in database schema design, which takes place during the design phase of information systems development. The various participation constraints on the two structural link edges give rise to a number of combinations. Traditionally, these combinations were though of as yielding three possible cardinality values: one-to-one, one-to-many, and many-to-many.

As Figure 6.13 shows, a one-to-one cardinality exists when no participation constraint is recorded on either side of the structural link, in which case the default value 1 is assigned to both sides. A one-to-many cardinality exists when there is an explicit participation constraint with $q_{min} > 0$ and $q_{max} > 1$ on exactly one side of the structural link and 1 on the other. Finally, a many-to-many cardinality exists when the participation constraints on both sides of the structural link are explicit, and in both $q_{max} > 1$.

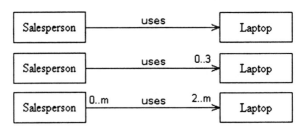

Figure 6.13. Three OPDs with cardinalities of one-to-one (top), one-to-many (middle), and many-to-many (bottom)

The three OPL sentences that correspond to the three OPDs in Figure 6.13 are:

> **Salesperson uses Laptop.**
> **Salesperson uses** zero to **3 Laptops.**
> **Salesperson uses 2** to many **Laptops.**

However, for the third OPD, since the cardinality is many-to-many, we need to add a sentence in the reverse direction to express the participation constraint on the **Salesperson** side of the structural link. The sentence is:

> **Laptop is used by** zero to many **Salespersons.**

Combining pairs of the symbols "?", "*", "1", "+" and "m", we get the 25 cardinality types that are listed in the 5×5 array in Table 4. The four customary cardinalities, [1, 1], which is "one-to-one," [1, m], which is "one-to-many," [m, 1], which is "many-to-one," and [m, m], which is "many-to-many" appear in the highlighted boxes in the table. These cardinalities are the ones recognized in entity relationship diagrams (ERDs), proposed by Chen (1976), which are used to design databases. Here we see that they comprise less than a sixth of the possible combinations.

Combining particular values is also allowed. For example, the participation constraint "?, 3..m" is legal and is translated in OPL as "optional or at least **3**". Another thing that should be reiterated is that while all the examples so far referred to objects, they can be applied to processes as well.

Table 4. The 25 cardinality types obtained by possible pair combinations of the shorthand participation constraint symbols

Destination symbol / Source symbol	'?' = (0..1)	'*' = (0..m)	'1' = (1..1)	'+' = (1..m)	'm' = (m..m)
'?' = (0..1)	[? , ?]	[? , *]	[? , 1]	[? , +]	[? , m]
'*' = (0..m)	[* , ?]	[* , *]	[* , 1]	[* , +]	[* , m]
'1' = (1..1)	[1 , ?]	[1 , *]	[1 , 1]	[1 , +]	[1 , m]
'+' = (1..m)	[+ , ?]	[+ , *]	[+ , 1]	[+ , +]	[+ , m]
'm' = (m..m)	[m , ?]	[m , *]	[m , 1]	[m , +]	[m , m]

6.2.6 Participation Constraints in Procedural Relations

Participation constraints are not restricted to structural links. Procedural links can be assigned to participation constraints as well.

Figure 6.14. Use of participation constraint in an agent link

Figure 6.14 shows a simple use of participation constraint in an agent link. The corresponding OPL sentence is "**Three Mechanics** handle **Repairing**."[1]

Another example is given in Figure 6.15 followed by the corresponding OPL sentences.

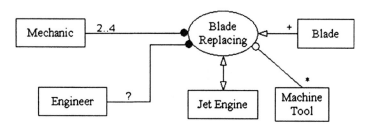

Figure 6.15. Participation constraints in various procedural links

---

[1] Note that as an alternative to digits, numerals up to ten in an OPL sentence can be written in words, as proper English style requires. This reads better than "**3 Mechanics** handle **Repairing**."

> **2 to 4 Mechanics** and an optional **Engineer** handle **Blade Replacing.**
>   *(Agent sentence)*
> **Blade Replacing** affects **Jet Engine.**
>   *(Effect sentence)*
> **Blade Replacing** consumes at least one **Blade.**
>   *(Consumption sentence)*
> **Blade Replacing** uses optional **Machine Tools.**
>   *(Optional use sentence)*

Note that due to the **optional** participation constraint on the **Machine Tool** side of the instrument link, the last sentence is an *optional use sentence* rather than an instrument sentence. It employs the reserved word **uses** rather than **requires**, since the combination "**requires optional**," which would result from using an instrument sentence in this case, is an oxymoron.

6.3 The Distributive Law and Forks

Structural relations are distributive in a sense analogous to the distributive law in algebra.

> **The distributive law of structural relations:**
> *If A, B, and C are objects, and $\Re$ is a structural relation, then*
> $A\Re(B, C) \Leftrightarrow A\Re B, \ A\Re C.$

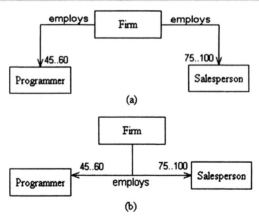

Figure 6.16. The distributive law of structural relations applied in OPDs: (a) disjoint links. (b) joint links

This is not just a law in mathematics and in OPM, but, as we see next, also in natural languages. The two OPDs in Figure 6.16 provide an example of the graphical application of the distributive law of structural relations. In the OPD of Figure

6.16(a), there are two disjoint tagged structural links, both bearing the same tag **employs**. One **employs** tag is recorded along the link from **Firm** to **Programmer** and the other along the link from **Firm** to **Salesperson**. The relation called **employs** is the relation $\Re$. Hence, here A = **Firm**, B = **Programmer**, C = **Salesperson** and $\Re$ = **employs**. This OPD has exactly two graphic sentences, each giving rise to one OPL sentence. One graphic sentence consists of the object **Firm**, the structural relation **employs** and the object **Programmer**. The other graphic sentence consists of the object **Firm**, the object **Programmer** and the structural relation **employs**. The corresponding OPL sentences are:

> **Firm employs 45 to 60 Programmers.**
> **Firm employs 75 to 100 Salespersons.**
> *(Tagged structure sentences)*

In the OPD of Figure 6.16(b), the two structural links of Figure 6.16(a) are joined at their origin and diverge somewhere along the link. Since now only one structural link emanates from the source object, the two OPL sentences become one:

> **Firm employs 45 to 60 Programmers and 75 to 100 Salespersons.**
> *(Fork tagged structure sentence)*

Note that the relation name **employs** appears in this OPL sentence just once. This goes hand in hand with the fact that in Figure 6.16(b) the two separate links labeled **employs** were joined, such that there was no need to write the label twice. The OPL reserved word and in this context has the same function as the comma in the mathematical expression $A\Re(B, C) \Leftrightarrow A\Re B, A\Re C$. Graphically, joining the origin of two structural links having the same tag, such as **employs** in Figure 6.16(b), has the same function as the comma and the OPL reserved word and. Such joining gives rise to forks, which are discussed next.

6.3.1 Forks

In algebra, the distributive law $A\Re(B, C) \Leftrightarrow A\Re B, A\Re C$ is extensible to any number n of elements within the parentheses. Thus, $A\Re(B_1, B_2, \ldots B_n) \Leftrightarrow A\Re B_1, A\Re B_2, \ldots A\Re B_n$. The same is true for OPM and natural languages. To express this in OPM we define fork below.

> A **fork** is an aggregation of two or more structural links with the same tag, which has a joint edge on one side of the link, which splits into two or more edges on the other side.
> The joint edge of the link, called **handle**, emanates from the source object. Each one of the other edges, called **tooth**, diverges from the handle and is connected to a different destination object.

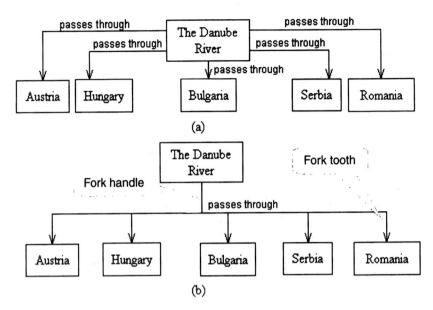

Figure 6.17. Fork, handle, and tooth demonstrated

The OPD in Figure 6.17(b) exemplifies a fork with its handle and teeth, and its OPL script equivalent. OPD (a) contains five separate structural links, all labeled with **passes through**. It is therefore equivalent to the following OPL script that comprises five OPL sentences:

> **The Danube River passes through Austria.**
> **The Danube River passes through Hungary.**
> **The Danube River passes through Bulgaria.**
> **The Danube River passes through Serbia.**
> **The Danube River passes through Romania.**
> *(Tagged structure sentences)*

This set of OPL sentences, though correct, is mechanical, repetitive, and dull. The OPL script reflects the redundancy of links in its corresponding OPD. The application of the distributive law provides for aggregating the five links into a fork. Using the expression $A\Re(B_1, B_2, \ldots B_n) \Leftrightarrow A\Re B_1, A\Re B_2, \ldots A\Re B_n$, and substituting $A =$ **The Danube River**, $B_1 =$ **Austria**, B_2 **Hungary**, B_3 **Bulgaria**, B_4 **Serbia** and B_5 **Romania**. The result is presented in OPD (b), where a fork is shown with one callout pointing to its handle and another pointing to one of its teeth. Only one structural link, labeled **passes through** emanates from **The Danube River**, and diverges into five teeth. The OPL script of this OPD is one compact OPL sentence, which is also a perfect English sentence:

> **The Danube River passes through Austria, Hungary, Bulgaria, Serbia, and Romania.**

The participation constraints of the handle and any tooth of the fork may be different from the default, 1. However, like the handle itself, the participation constraint of the handle is common to all the teeth. If, for some link, a different participation constraint is required on the handle side, then this link needs to be separated from the fork.

6.3.2 Fork Degree

The degree of the fork is the number of teeth that emanate from the fork's handle.

> *Fork degree is the number of teeth emanating from the handle of the fork.*

The degree of the fork in the OPD of Figure 6.16 is 2. The fork's handle object is **Firm** and its two fork objects are **Programmer** and **Salesperson**. The degree of the fork in Figure 6.17 is 5.

> *The **teeth set** of a fork is the set of all the target objects that the fork links.*

The teeth set of the fork labeled **employs** in the OPD of Figure 6.16(b) is {**Programmer, Salesperson**}. The teeth set of the fork labeled **passes through** in the OPD Figure 6.17(b) is {**Austria, Hungary, Bulgaria, Serbia, Romania**}.
If the teeth set is of a size greater than 2, it may sometimes be convenient to omit some of the objects in the teeth set that are not relevant for what that particular OPD is designed to convey. Hence, not each OPD in a system's OPD set must contain all the objects in the teeth set. One OPD may contain one subset of the teeth set, while another OPD may contain another subset.

> *The **maximal teeth set** of a fork is the union of the teeth sets emanating from the same object and having the same tag in all the OPDs in the OPD set.*

The maximal teeth set of the fork labeled **employs** in the OPD of Figure 6.16(b) includes all the pertinent types of occupations employed by **Firm**. The size (number of elements) of the maximal teeth set is equal to the maximal fork degree. For example, say another OPD in the OPD set to which Figure 6.16(b) belongs has the teeth set {**Programmer, Software Engineer, Project Leader**}. Suppose also that these are the only two OPDs in the OPD set of that system, in which the object **Firm** appears with the structural link labeled **employs**. The maximal teeth set of the fork labeled **employs** would then be:

Teeth-set$_{max}$(**Firm employs**) = {**Programmer, Salesperson**}∪{**Programmer, Software Engineer, Project Leader**} = {**Programmer, Salesperson, Software Engineer, Project Leader**}.

In a similar manner, the teeth set of the fork labeled **passes through** in the OPD Figure 6.17(b) includes all the countries through which the **Danube River** passes.

6.3.3 Fork Comprehensiveness

Frequently, showing all the fork objects overloads the OPD both graphically and mentally. Therefore, things that are not relevant in a particular OPD are not drawn, while others may be omitted in other OPDs that belong to the same OPD set. This omission of irrelevant fork objects helps eliminate the excess clutter frequently caused in OPDs of real life systems. Omitting for objects may, however, mislead the reader of an individual OPD to think that the teeth objects presented in that particular OPD are all the teeth objects that can be linked to the handle object. To avoid such confusion, it may be important to indicate whether all the teeth objects that can be linked to the handle indeed appear in the fork. To this end, we define fork's *comprehensiveness* attribute value as follows.

> *A fork link is* **comprehensive** *if and only if it links the handle object with each one of the objects in the teeth set.*

Using the fork's *comprehensiveness* attribute, one can indicate whether the structure implied by the fork is comprehensive or non-comprehensive, as defined below.

> **Comprehensiveness** *is an attribute of a fork that indicates whether the fork is comprehensive.*

The importance of fork comprehensiveness is that it tells the diagram reader whether all the teeth objects that can potentially be linked to the handle object are indeed linked. The default state of the fork's **Comprehensiveness** attribute is **comprehensive**, meaning that all the fork objects that can be attached to the handle are indeed attached. The other state of **Comprehensiveness** is **non-comprehensive**. A non-comprehensive fork can be made comprehensive by completing the missing objects in the forks' teeth set.

Examining the OPD in Figure 6.18, we see that although the **Danube River** passes through **Germany**, it is not depicted in the diagram. The non-comprehensive symbol expresses the fact that not all the countries through which the **Danube River** passes are represented in the OPD. As Figure 6.18 shows, the non-comprehensive symbol is a short segment crossing the fork's handle near the handle object. The OPL reserved phrase that expresses the fact that the fork is non-comprehensive is "and more," which is appended at the end of the list of fork objects. The following OPL sentence, which corresponds to the OPD in Figure 6.18, demonstrates this.

> **The Danube River passes through Austria, Hungary, Bulgaria, Serbia, Romania,** and more.

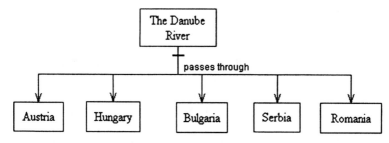

Figure 6.18. A non-comprehensive fork

6.4 The Transitivity of Structural Relations

Transitivity in structural relations has the same following definition and semantics as in algebra:

> A **transitive** structural relation $\Re$ is a structural relation for which it holds that if $A \Re B$ and $B \Re C$ then $A \Re C$.
> **Transitivity** is an attribute of a structural relation, which determines whether the structural relation is transitive.

```
Database --contains--> Folder --contains--> File
```

Figure 6.19. A transitive structural relation

Figure 6.19 is an example of a transitive structural relation, whose OPL paragraph is:

> **Database contains Folder.**
> **Folder contains File.**

Given that the structural relation **contains** is transitive, it can be deduced from the OPD in Figure 6.19 that **Database contains File**. Transitive structural relations yield hierarchies. Figure 6.19 is an example of a containment hierarchy. We can extend this hierarchy by specifying, for example, that **File contains Record, Record contains Field, Field contains Character,** etc. We cannot, however, construct a similar hierarchy for the structural relation **passes through** in Figure 6.17, because the structural relation **passes through** is not transitive.

Some structural relations may be transitive or not, depending on the objects they relate. Hence, a transitive structural relation can be *positive, neutral* or *negative*. When we say "transitive relation" without specifying its value, we mean "positive transitive relation". The relation **contains** in Figure 6.19 is an example of a positive transitive structural relation: it is always true that if **A contains B** and **B contains C** then **A contains C**. Since the structural relation **contains** is transitive, we say that the value of the transitivity attribute of the structural relation **contains** is positive.

From an OPM perspective, we say that **Transitivity** is an attribute of a **Structural Relation**, which exhibits three values: **Positive, Neutral**, and **Negative**. The default value of **Transitivity** is **Neutral**. A neutral transitive structural relation is a structural relation that may or may not be transitive. Consider the relation **is friend of** as an example of a neutral transitive structural relation. It may be true that if **Al is friend of Ben** and **Ben is friend of Chen**, then it may be the case that **Al is friend of Chen**, but this is not guaranteed. Likewise, if **A, B** and **C** are closed shapes in a plane, **A touches B** and **B touches C** may imply that **A touches C**, but this is not guaranteed. Hence, the transitivity value of both **is friend of** and **touches** is neutral.

An example of a negative structural relation is as follows. Suppose **Jack is father of Jim** and **Jim is father of Jill**, then **Jack is not father of Jill**. As an example from plane geometry, consider three points **A, B** and **C** along a straight line, such that **AB, BC**, and **CD** are three line segments. If **AB touches BC** and **BC touches CD**, then the sentence **AB touches CD** is always false. Hence, in this system, the transitivity of the structural relation **touches** between two line segments on a line is negative.

6.5 The Four Fundamental Structural Relations

Four structural relations are most prevalent and play an especially important role in specifying and understanding systems. These relations, termed the *fundamental structural relations*, are:

(1) Aggregation-Participation;
(2) Exhibition-Characterization;
(3) Generalization-Specialization; and
(4) Classification-Instantiation.

Table 5. The fundamental structural relation names, OPD symbols, and OPL sentences

Structural Relation Name		OPD Symbol	OPL Sentence	
Forward	Backward		Forward	Backward
Aggregation	Participation	A▲B	A consists of B.	B is part of A.
Exhibition	Characterization	A⊿B	A exhibits B.	B characterizes A.
Generalization	Specialization	A△B	A generalizes B.	B is an A.
Classification	Instantiation	A⧊B	A classifies B.	B is an instance of A.

Table 5 lists the four fundamental structural relations and their respective triangle symbols. The name of each such relation consists of a pair of dash-separated words. The first word in each such pair is the forward relation name, i.e., the name of the relation as seen from the viewpoint of the thing up in the hierarchy. The second word is the backward (or reverse) relation name, i.e., the name of the relation as seen from the viewpoint of the thing down in the hierarchy of that relation.

Aggregation-Participation denotes the relation between a whole thing and its parts. Exhibition-Characterization denotes the relation between a thing and its features (attributes and operations). Generalization-Specialization denotes the relation between a general thing and its specializations. Finally, Classification-Instantiation denotes the relation between a class of things and the instances of that class. Due to their importance and prevalence, the four fundamental structural relations are assigned triangular symbols, which are shown in Table 5, along with their names and sentences.

The OPL sentences are provided in both the forward and backward directions. Each fundamental structural relation has a default, preferred direction, which was determined by how natural the sentence sounds. The preferred shorthand name for each relation is underlined. Likewise, underlined sentences are the default sentences. They are more useful, as they sound more natural.

Table 6. The fundamental structural relations with one, two, and three or more descendants

Structural Relation name	Number of Descendants					
	One		Two		Three or more	
Aggregation-Participation		A consists of B.		A consists of B and C.		A consists of B, C, and D.
Exhibition-Characterization		A exhibits B.		A exhibits B and C.		A exhibits B, C, and D.
Generalization-Specialization		B is an A.		B and C are As.		B, C, and D are As.
Classification-Instantiation		B is an instance of A.		B and C are instances of A.		B, C, and D are instances of A.

The sentences in Table 5 are shown for the case of one descendant. For two descendants or more, we create a fork for these relations, and the sentences are slightly different. The three different sentence versions are shown in Table 6 for the default sentences. We should note that the special graphic symbols assigned to the fundamental structural relations, shown in Table 5, do not make them special amongst the rest of the structural relations. Diagramming convenience and ease of diagram reading alone have motivated the introduction of these symbols. Following this idea of denoting a frequently used relation by a special symbol, it is possible to add a symbol for one or more structural relations that are widely used within a specialized domain. Consider an example from the domain of chemical laboratory testing of industrial lots. In this domain, the reserved phrase is a sample of is a prevalent and useful structural relation between a sample and the lot from which it was taken. A dedicated graphic symbol (e.g., a piece cut out of a cake) can be introduced in this domain to enable quicker and easier modeling.

These four fundamental structural relations are so important, that the next two chapters are devoted to discussing these relations. Chapter 7 discusses Aggregation-Participation and Exhibition-Characterization, while Chapter 8 examines Generalization-Specialization and Classification-Instantiation.

Summary

- A structural relation is a long-term association from a source object to a destination object. Such relations are persistent in the system and do not change as time progresses.
  - Structural relations are mostly applicable between objects and sometimes between processes, but never between an object and a process.
  - A tag can be placed along the structural link to specify the nature of the structural relation, yielding a tagged structural link.
  - Every relation has a reverse relation, which is the relation between the two objects in the backward direction.
  - Tagged links can be unidirectional or bidirectional. A tagged link is unidirectional if the backward tag is the passive voice of the forward tag or is otherwise meaningless or unimportant.
- Each link can have a participation constraint on either the source or destination side of the link, designating how many copies, or instances, of the adjacent object participate in the relation.
  - The constraint can be:
    - a range of numbers $q_{min} \cdot\cdot q_{max}$;
    - an integer;
    - the letter m, indicating many; and
    - a shorthand symbol: ?, +, or *.
  - Cardinality is the combination of the two participating constraints on either side of the link.
- Some structural relations are transitive, while others are not. The transitivity of any relation can be positive, neutral, or negative.
- Distribution (in the sense of the distributive law in mathematics) is applicable to all OPM relations.
  - Distribution gives rise to forks, where the relation from the object attached to the fork's handle is distributed to any number of objects attached to the fork's teeth.
  - The teeth set is the set of all the objects that can be attached to teeth of the fork.
  - The fork's degree is the size of its teeth set.
  - The fork's comprehensiveness is an attribute of the fork, which specifies whether all the objects in the teeth set are specified in the OPD.
    - A fork is comprehensive if all the objects in the teeth set are specified in the OPD, otherwise it is non-comprehensive.
    - The non-comprehensive symbol is a short bar crossing the fork's handle.
- Four structural relations are fundamental and therefore are assigned graphic symbols. These are:
  - Aggregation-Participation;
  - Exhibition-Characterization;
  - Generalization-Specialization; and
  - Classification-Instantiation.

Problems

1. Write six English sentences that express binary structural relations. For each sentence draw the corresponding OPD and write the OPL sentences.
2. Draw three OPDs that express unidirectional binary structural relations and three OPDs that express bidirectional binary structural relations. For each OPD write the corresponding OPL sentence(s).
3. Add participation constraints in at least six of the 12 OPDs you invented in response to problems 1 and 2, and write the corresponding OPL sentences.
4. Pick up a book, a newspaper or a magazine you have handy and open it at a random page.
   a. Find 7-10 structural relations expressed in this page.
   b. Write their OPL sentences.
   c. Draw their OPDs.
5. List as many verbs as you can that are more naturally used in structural relations than in processes or procedural relations.
6. Use the pattern of Table 3 to invent your own OPDs and OPL sentences.
7. Draw an OPD of a real product in your house or one that you are familiar with that has at least three levels of aggregation and at least two cardinalities. Use parameters if required. Write the corresponding OPL paragraph (the OPL sentences that are equivalent to the OPD).
8. Select five cardinality types from the 25 in Table 4, for each one of them draw an OPD and the corresponding OPL sentence(s).
9. Use OPD and OPL to specify at least 6 participation constraints in procedural relations.
10. Pick at least two domains. For each domain suggest a prevalent structural relation that "deserves" to become fundamental. Invent a symbol and a reserved phrase of that relation and show how it is applied in an OPD and OPL.
11. Find three examples of fork relations. For each example, draw two OPDs: one without fork and one with fork. For each OPD write the OPL paragraph.
12. Find examples of three relations, one with a positive transitivity, one with a neutral and one with a negative transitivity. Draw their OPDs and write their OPL sentences.
13. Don's computer is in his room. When Don turns his computer on, using the power button on the computer's tower box, the monitor turns on as well. The modem, embedded in the tower box, connects to the Internet. His Internet browser is set to open with the URL of his favorite comics site, which he reads on a daily basis. A button near the top of his browser opens his email account. After reading both, he clicks the modem icon at the bottom right corner of the screen, which disconnects the modem. He then shuts down the computer, which automatically turns the power off.

a. Apply OPM to model this description in three different ways:
   (1) Using only objects and structural relations;
   (2) Using as many processes and states as possible, and only procedural relations; and
   (3) Using a combination of the two ways above.
b. Reflecting on the three models you built, answer the following questions.
   (1) Which model most faithfully described the system?
   (2) Which model conveys the most details?
   (3) Which model captures best what Don was doing? Why?
   (4) Which model do you prefer? Explain and point to likes and dislikes in each.

Chapter 7

Aggregation and Exhibition

This large four-wheel chariot is one of the striking finds of the Pazyryk burial mounds. It consists of a number of parts joined together by leather straps and wooden nails. The trunk is made of two frames joined by means of short carved poles and leather straps. The frames constitute the basis for the canopy. Each of the four large wheels has 34 spokes.

Description of a wood and leather Chariot, Eastern Altai, Russia (The State Hermitage Museum, 2001)

I must be able to attribute properties to the objects.

Immanuel Kant (1787)

Any interesting system can be decomposed into parts. The system and its parts can be described using natural language modifiers: adjectives and adverbs. Without the ability to mentally take things apart and examine their features, our ability to study systems would be greatly hindered. Consider, for example, a table. Without describing it, we cannot determine its function: is it for dining, studying, or displaying artifacts? We cannot tell what material it is made of: is it plastic, metal, or wood? What are its dimensions? Is the tabletop round, square or perhaps some other shape? How many legs does the table have, and how long is each?

This chapter discusses two fundamental structural relations, possibly the two most important ones: Aggregation-Participation, the relationship between the whole and its parts, and Exhibition-Characterization, the relationship between a thing and its features. Aggregation-Participation is also known as whole-part (Coad and Yourdon, 1991), composition (Kilov and Simmonds, 1996), or the part-of relationship (Fowler, 1996). Exhibition-Characterization is the fundamental structural relation that binds a thing (object or process) with another thing called *feature*, which characterizes it.

7.1 Aggregation-Participation: Underlying Concepts

Aggregation-Participation is a fundamental structural relation which denotes the fact that a (relatively high-level) thing aggregates (i.e., consists of, or comprises) one or more (lower-level) things, each of which is referred to as a *part*. The higher-level thing is then called the *whole*, or *aggregate*, while the lower-level things that comprise it are the *parts*. The particular context in which the terms "whole" and "part" relate needs to be considered (Latimer and Stevens, 1997). This relationship is very central to OPM, yet since we know the idea of whole-part from our everyday life, it should not be too complicated to comprehend.

Aggregation-Participation is a means to describe the composition of every nontrivial thing by enumerating its parts and specifying their place in the whole-part hierarchy. Naturally, most things can be decomposed further than the deepest decomposition specified in the analysis and design of the system. In particular, physical objects can be decomposed all the way down to the molecular and atomic or even sub-atomic levels. However, the specification of yet deeper participation hierarchy levels always stops at a point that is deemed sufficient by the system analyst.

In psychology, Rescher and Oppenheim (1995) have provided a conceptual framework for the precise explication of the gestalt[1] concept of "whole" and summarize the intuitive requirements or conditions of talking about a whole and its parts:

The whole must possess some attribute in virtue of its status as a whole, an attribute peculiar to it and characteristic of it as a whole. The parts of the whole must stand in some special and characteristic relation of dependence with one another; they must satisfy some special condition in virtue of their status as parts of a whole. The whole must possess some kind of structure in virtue of which certain specifically structural characteristics pertain to it.

These conditions are perfectly fit for objects and systems in general. To specify the concept of part, it is necessary at the very outset to state the conditions under which some object is to be considered part of another whole object. The specification of a particular part-whole relation thus determines for a given object, the whole, just which objects are its parts. Only when the context makes clear which specific part relation is intended, may we even speak of parts and wholes. Until definitions of "whole" and "part" are offered, discussion of these terms cannot proceed (Latimer and Stevens, 1997).

Many OO methods call the Aggregation-Participation "has-a" (as opposed to "is-a" for the Generalization-Specialization relation). It may indeed seem natural to use some form of the verb "to have" to denote the relation between the whole and its parts, as in "A car has body, engine and wheels." However, we avoid the use of this verb to denote aggregation because it is overloaded and may have[2] various interpretations. To see this, it suffices to look at the sentence examples "Dave has a step mother", "Jack has a yellow motorcycle", "We are having a discussion", and "The baby has pneumonia." OPM's choice of reserved words for denoting the Aggregation-Participation relation is explained below.

---

[1]   A physical, biological, psychological, symbolic configuration or pattern of elements so unified as a whole that its properties cannot be derived from a simple summation of its parts.

[2]   Already in this sentence, we *have* a built-in example that shows the multiple uses of "have".

7.1.1 Aggregation-Participation as a Tagged Structural Relation

Like all structural relations, Aggregation-Participation is a pair of forward and backward relations. Aggregation is the forward structural relation – the relation as seen from the aspect of the aggregate, or the ancestor, when it refers to its parts, the descendants. The backward structural relation, i.e., the relation as seen from the aspect of each part, is participation. Aggregation and participation are inverse relations. Aggregating can be thought of as the process of taking parts and making a whole. Participating is the inverse process: taking the aggregate and breaking it into the parts that comprise it. However, as we have noted in the discussion on structural relations (p. 112), to *aggregate* or to *consist of* has a static notion, with time having no relevance to this relation.

The forward (or downward) direction of the Aggregation-Participation relation, from the whole to its parts, is the *aggregation* direction. The reserved phrase used to express the forward direction of the relation is "consists of". The reverse (or upward) direction, from each part to the whole, is the *participation* direction. The reserved phrase used to express the forward direction of the relation is "is part of".

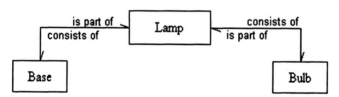

Figure 7.1. Aggregation-Participation expressed by a labeled structural relation

The OPD of Figure 7.1 is an example of an Aggregation-Participation relation expressed using the tagged structural relation notation we have been using so far. The pair of Aggregation-Participation relations is tagged **consists of** for the forward direction, with the source object **Lamp** being the whole and the destination objects **Base** and **Bulb** being the parts. The same link in the backward direction is tagged **is part of** (with the source and destination roles reversed) for the backward direction. The two OPL sentence pairs that are equivalent to the OPD of Figure 7.1 are:

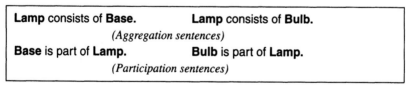

Since both sentence types sound natural, the decision of which of the two forms to use may be up to the system architect. By default, OPM uses the forward relation, i.e., the aggregation sentence, in which the whole consists of the parts.

7.1.2 The Aggregation-Participation Symbol

The OPD of Figure 7.2 has the same semantics as Figure 7.1, but here the Aggregation-Participation relation is expressed by the specific symbol designated for the Aggregation-Participation relation. As we saw in Table 5, this symbol is a solid black triangle, ▲.[3] The triangle is linked with the whole and the parts through orthonormal polylines.[4] The tip of the triangle is linked through an orthonormal polyline to the aggregate, or whole (**Lamp**, in our example). The triangle's base is linked through other orthonormal polylines to each one of the things that are parts of the aggregate (**Base** and **Bulb**, in our example). The triangle simply replaces the pair of forward and backward textual tags of the Aggregation-Participation relation, but this is no more than a matter of a shorthand graphic notation convention for this important and widely used structural relation.

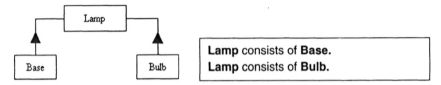

Figure 7.2. The Aggregation-Participation relations of Figure 7.1, expressed by their designated solid triangle symbol.

Figure 7.3. The two Aggregation-Participation relations of Figure 7.2 represented as a fork

As a structural relation, the Aggregation-Participation relation abides by the distributive law. Therefore, two or more structural links can be represented as a fork. Applying the distributive law to the OPD and the corresponding two OPL aggregation sentences in Figure 7.2 yields the OPD and the OPL sentence in Figure 7.3.

The use of participation constraints in Aggregation-Participation relation is the same as in tagged structural relations in general. A participation constraint can be attached to the whole or to any one of its parts. The implicit default is 1, and a participation constraint other than 1 is recorded next to the line linking the solid triangle's base to the part. The OPD in Figure 7.4 and the OPL that follows exemplify this.

---

[3] UML has adopted the diamond symbol for this relation. In OPM, all four fundamental structural relations are triangles.

[4] polylines whose components are parallel to either one of the diagram axes

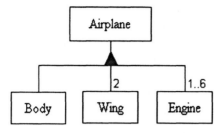

Figure 7.4. Participation constraints applied to Aggregation-Participation links

> **Airplane** consists of **Body, 2 Wings** and **1 to 6 Engines.**

Showing the parts of a whole is called unfolding. The reverse operation, folding, hides shown parts. Folding and unfolding is a pair of abstracting-refining operations, which are discussed in Chapter 9.

7.1.3 Sets and Order

Sets are abstract collections of things that consist of *elements* or *members*. A set may therefore be thought of as an aggregate (whole) and its elements – as parts. Each element in the set is unique.[5] The elements of the set may be *ordered* or *unordered*. The attribute of the Aggregation-Participation relation, which denotes whether or not a set is ordered, is called **Orderability**. The two values of **Orderability** are ordered and unordered, which are reserved OPL words. The default value of this attribute is unordered.

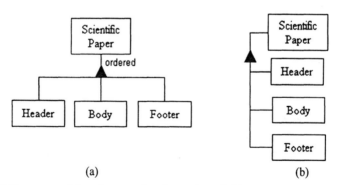

Figure 7.5. Two ways of denoting an ordered set: (a) explicitly. (b) graphically, implicitly

> **Scientific Paper** consists of **Header, Body,** and **Footer**, in this order.

---

[5]   A bag is a collection of elements that may not be unique.

The order in the two OPDs in Figure 7.5 reflects the *order* of the scientific paper, i.e., the order in which the parts of the paper should be read. Each one of these OPDs shows another way of denoting the fact that a set is ordered and specifying the order of its members. The first way, shown in Figure 7.5(a), is by writing ordered next to the tip of the aggregation triangle, and drawing the elements in order from left to right. The second way, shown in Figure 7.5(b), is graphic: the ordered objects are drawn vertically, with the first on top and the last in the bottom.

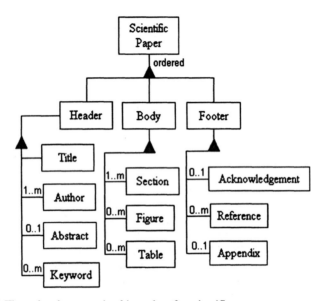

Figure 7.6. The ordered aggregation hierarchy of a scientific paper

For extra clarity, we can always add the reserved word ordered even for a vertically aligned set of ordered objects. If the graphic layout makes it easy to lay out the members of an unordered set vertically, then the OPL reserved word unordered needs to be recorded next to the tip of the aggregation triangle. When the members are not drawn in a clear horizontal or vertical manner and no orderability value is explicitly specified, the default value, unordered, is assumed. Figure 7.6 gives a more detailed view of the OPD in Figure 7.5, which shows both explicit and implicit orderability notations. When dealing with processes (see page 139), orderability is intimately related to the timeline.

7.1.4 Aggregate Naming

Frequently during the analysis, we encounter situations in which we need to name an aggregate, which has no single word in natural language. As we will see, these situations are not unique to aggregates; they are also encountered in a variety of other circumstances, such as naming an attribute when only the names of its values or states are explicit.[6] In cases like these, we must exercise our creativity to generate an appropriate phrase that best captures the essence of what we wish to express. The capability of inventing meaningful names, or generating expressive phrases, is a very important component of the analysis process. It provides us with the power to abstract into a whole a collection of things that would otherwise be very difficult to think about and relate to as a unity.

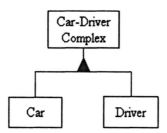

Figure 7.7. Naming an aggregate which has no single word in natural language

Car-Driver Complex consists of **Car** and **Driver**.

To illustrate the point of aggregate naming and the importance of appropriate phrase generation, consider a systems development team, whose assignment is to improve the traffic system in City X. After some thought and discussion, the team agrees that an essential object in the system is the composition of a car and the person that drives it in the city streets. This object is much more central to the system than a car alone or a driver alone. The role a car without a driver plays is restricted to parking issues, while the driver without the car should be considered a pedestrian. Nonetheless, having agreed that the car along with its driver is a major object that needs to be accounted for in the system, our team still lacks an elegant way of referring to it. Since there is no single English word for this object, the team has come up with the name **Car-Driver Complex**, as illustrated in Figure 7.7.

7.1.5 Aggregating Processes

When we consider an OPM whole-part relation, we normally focus on objects as the aggregate and the parts. Objects can consist only of objects, while processes can consist only of processes. Nevertheless, the whole-part relation is meaningful among processes as well.

To see an example, consider the two alternative Aggregation-Participation representations in Figure 7.8 and their corresponding OPL paragraphs in Frame 11. Figure 7.8(a) applies unfolding, which uses the whole-part black triangle symbol.

---

[6] For example, what is the name of the attribute the values of which are **wide** and **narrow**? **Width**? **Narrowness**? Something in-between? Such a neutral word does not exist. This is discussed in more detail in Section 7.7.1.

The result is a tree-like aggregation hierarchy similar to that of objects. Figure 7.8(b) applies in-zooming, where processes are recursively zoomed into, revealing lower-level processes that are contained within their immediate ancestors. Semantically, the two graphic representations are very close. The aggregation hierarchy emphasizes the whole-relation, while in-zooming emphasizes the containment relation. The process of **Population Growing** of a specific place (be it a city, state, country, or continent) consists of two separate subprocesses: **Natural Growing** and **Population Exchanging**. Each of those processes can be further decomposed into two yet lower-level processes. **Natural Growing** is decomposed into **Birth Process** and **Death Process**, while **Population Exchanging** is comprised of **Immigrating** and **Emigrating**, i.e., outflow and influx of population, respectively. The OPL paragraphs listed in Frame 11 differ in that the one on the left, which corresponds to Figure 7.8(a) uses the aggregation reserved phrase consists of, while the one on the right uses the in-zooming reserved phrase zooms into.

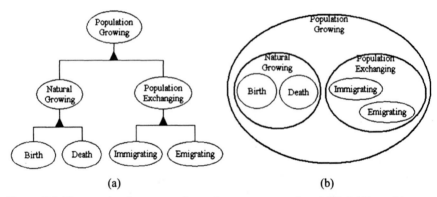

(a)   (b)

Figure 7.8. The two visual representations of process aggregation. (a) Unfolding: a hierarchical tree structure through the Aggregation-Participation symbol. (b) In-zooming: containment through enclosing things within the boundaries of their immediate ancestors

Population Growing consists of **Natural Growing** and **Population Exchanging**. **Natural Growing** consists of **Birth Process** and **Death Process**. **Population Exchanging** consists of **Immigrating** and **Emigrating**. *(Structure sentences)*	**Population Growing** zooms into **Natural Growing** and **Population Exchanging**. **Natural Growing** zooms into **Birth Process** and **Death Process**. **Population Exchanging** zooms into **Immigrating** and **Emigrating**. *(In-zooming sentences)*

Frame 11. The two OPL paragraphs that correspond to the two OPDs in Figure 7.8.

Like processes, objects can be described using either one of the two visual representations. However, the preferred way of specifying each one of the thing types is different. The default way of refining objects is unfolding, while the default way of refining processes is in-zooming. We elaborate on refinement and abstraction in Chapter 9.

7.2 Aggregation Hierarchy and Comprehensiveness

Aggregation-Participation is a structural relation whose transitivity is positive. This is so because if **A** consists of **B** and **B** consists of **C**, then obviously **A** consists of **C**, albeit indirectly. Since transitive structural relations give rise to hierarchy, the Aggregation-Participation relation provides for the construction of an *aggregation hierarchy* – a hierarchy of Aggregation-Participation relations within an OPD, as shown in Figure 7.6.

7.2.1 Aggregation Hierarchy

Figure 7.9 is an OPD of the object **Road Vehicle**, showing a five-level aggregation hierarchy. Assuming the OPD is correct from an automobile mechanics perspective, the OPL script, which is equivalent to this OPD, is shown in Frame 12.

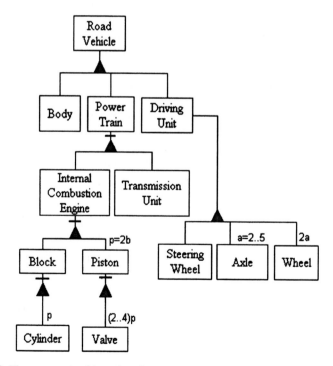

Figure 7.9. The aggregation hierarchy of a **Road Vehicle** with non-comprehensive aggregation and inter-related participation constraints

---

Road Vehicle consists of **Body, Power Train,** and **Driving Unit.**
Power Train consists of **Internal Combustion Engine, Transmission Unit,** and additional parts.
Driving Unit consists of **Steering Wheel, a=2..5 Axles,** and **2a Wheels.**
Internal Combustion Engine consists of **Block, p=2b Pistons** and additional parts.
Block consists of **p Cylinders** and additional parts.
Piston consists of **(2..4)p Valves** and additional parts.

Frame 12. The OPL script equivalent to the OPD of Figure 7.9.

7.2.2 Aggregation Comprehensiveness

Besides being transitive, aggregation is also distributive, giving rise to a fork. Like any fork relation, it is characterized by the **Comprehensiveness** attribute, presented in page 124. The default **Comprehensiveness** value of Aggregation-Participation is **comprehensive,** as is the case with any fork relation.

Four of the OPL sentences in Frame 12 are non-comprehensive aggregation sentences, as the reserved phrase "and additional parts" indicates. For example, in addition to **Block** and **Piston, Internal Combustion Engine** consists of other parts, such as starter and distributor, which do not appear in this OPD and in its corresponding OPL sentence. These parts are not included in the OPD because they are not in the focus of interest of this particular OPD. However, from the viewpoint of the entire system they are significant and therefore they do appear in other OPDs of the OPD set. To denote this fact, we use the OPD symbol for non-comprehensiveness, which is a short vertical bar near the whole end of the aggregation link. Applied to Aggregation-Participation, this symbol denotes that the descendants are not *all* the parts. The corresponding OPL equivalent phrase "and additional parts" follows the list of parts. If, however, there are other parts that are not relevant to the system (and therefore do not appear in any OPD), there is no need for the non-comprehensiveness symbol.

7.2.3 Parameterized Participation Constraints

Parameterized participation constraints quantify the ratio between the whole and parts of the whole that are of the same type or among parts of various types. Figure 7.9 and its corresponding OPL script in Frame 12 demonstrate this. For example, suppose we wish to denote the fact that the number **p** of pistons in an internal combustion engine is even. To express this, we denote the participation constraint of the whole-part relation between **Internal Combustion Engine** and **Piston** by **p=2b.** Since **b** is an integral number, **p** is even. The participation constraint of **Cylinders** is also **p,** because each **Piston** is in exactly one **Cylinder,** so the number of cylinders and pistons in an engine is equal. There are 2 to 4 valves for each **Cylin-**

der. The total number of valves is therefore the number of valves per cylinder times the number of cylinders. Hence, the participation constraint of **Valve** is **(2..4)p**. Since **p=2b**, the total number of valves in an **Internal Combustion Engine** is **(2..4)p = (2..4)2b = 2(2..4)b**, where **b** is any (small) integer number.

Parameterized participation constraints can also constrain the number of participating objects to be above or below a certain threshold. For example, if a requirement exists that a **Company** consists of more than 50 **Employees,** we can specify **n>50** where **n** is the number of **Employees** in the participation constraint.[7]

7.2.4 Participation Level and Aggregational Complexity

We have noted that most things can be decomposed further than the deepest decomposition specified in the analysis and design of the system. Every thing in the system has its own appropriate *participation level* – the level of part details, which is just enough to specify that thing in the context of the system under study or development.

The participation level is a level of part details, beyond which any further decomposition is unnecessary and may potentially damage the analysis and design due to an excessive load of redundant part details. For example, we may or may not wish to specify further that the object **Valve** in Figure 7.9 consists of a **Cap**, a **Ring**, a **Lever**, four **Bolts** and four **Nuts**. The decision will be based on the purpose of the system. Suppose the analysis aims at understanding how a land vehicle operates and be able to communicate and explain our understanding of the system to others. If the purpose and function of a valve are known, we need not introduce any further details about the parts comprising the valve. In this case, the object **Valve** is *atomic*. An atomic object is simple and non-decomposable into parts from the viewpoint of the system being studied or developed.

If, on the other hand, the analysis is supposed to be a full-blown account of all the parts comprising a land vehicle, then we should go ahead and account for the parts of the valve. Such detail level is required, for example, for constructing a comprehensive BOM[8] of this product. To this end, we draw an additional OPD, in which we express the appropriate Aggregation-Participation relations, with **Valve** being either at the top or at some intermediate level of the hierarchy. In this case, **Valve** would be an *aggregationally compound* object, namely, an object that consists of two or more parts.

---

[7] Such expressions solve at least some of the problems for which Object Constraint Language (OCL) was incorporated into UML.

[8] An OPD set can be used as a tool for constructing a complete Bill-of-Material (BOM), which is a detailed description of the "product tree". Indeed, a BOM can be viewed as an aggregation hierarchy with specifications of participation constraints. While traversing the aggregation hierarchy levels, we must account for participation constraints as we move down from one level to the other and carry out the multiplication properly, as we did in this example.

We generalize this example to define the aggregation complexity attribute. The aggregation complexity attribute has two values: *atomic* and *compound*.

> **Aggregation complexity** *is an attribute of a thing that specifies whether it consists of parts that are needed to fully specify the system.*
> *A thing that aggregates other things in the system is* **compound**.
> *A thing that does not aggregate any other thing is* **atomic**.

A thing's aggregation complexity determines how complex a thing is from the viewpoint of the Aggregation-Participation relation. Hence, more formally (but less succinctly), *atomic* is *aggregationally simple,* while *compound* is *aggregationally complex.*

Graphically, an atomic thing has no decomposition into parts in any one of the OPDs in the OPD set specifying the system, since that would contradict its definition as being atomic. A compound thing might be shown as consisting of parts in one or more OPDs. It can also appear with no parts in other OPDs. Put differently, a thing is atomic if in no OPD in the system's OPD set the thing is connected to a handle of an aggregation fork link. Whenever an atomic thing appears in an OPD it must be a *leaf* of the aggregation hierarchy tree. It would occupy the hierarchy's bottom-most level.

7.3 Exhibition-Characterization: Underlying Concepts

To define and describe things in the world, natural languages use adjectives and adverbs. Without these types of words, neither objects nor processes can be adequately distinguished and understood. Exhibition-Characterization is the fundamental structural relation that binds a thing (object or process) with another thing, called *feature*, that characterizes it. An OPM *feature* is a thing that describes a thing. Exhibition-Characterization is a fundamental structural relation. Like any binary structural relation, it involves two things: the exhibitor and the feature.

> **Exhibition-Characterization** *is a fundamental structural relation, which denotes the fact that a thing exhibits, or is characterized by, another thing.*

To be consistent with the names of the other fundamental structural relations, we need the first word in the relation pair to describe the forward direction of the relation. Since Exhibition-Characterization is transitive, it yields a hierarchy. This forward direction, then, is also the downward direction: from a thing higher in the hierarchy (the exhibitor) to one or more things lower in the hierarchy (the features).

The forward (downward) direction of the Exhibition-Characterization relation, from the exhibitor to its features, is the *exhibition* direction, while the reverse (upward) direction, from each feature to the exhibitor, is the *characterization* direction. The above definition assumes the forward direction of the Exhibition-Characterization relation. Viewed in the backward direction, the feature is said to *characterize* the exhibitor.

7.3.1 The Name Exhibition-Characterization

The word *characterization* is much more commonly used in the context of the relation than its forward counterpart, *exhibition*. *Characterization* is therefore the short name of the relation. Based on this, we may occasionally drop the "exhibition" part of the name of this fundamental structural relation and abbreviate it to *characterization*, bearing in mind that this is the direction up the hierarchy level.

This concession exemplifies the kind of design decisions that need to be made while conceiving OPM names. On one hand, consistency and orderliness are imperative, but on the other hand, clarity and expressive power are enhanced when the language is as natural and as terse as possible. Since English, like all natural languages, has its idiosyncrasies, compromises such as this must often be made after weighing the pros and cons of each alternative. It may also be somewhat odd that we spend so much intellectual effort in choosing good names for abstract ideas. However, a meaningful name can make the great difference between a well-understood and appropriately used concept, and one that misses the point due to a term that while being formally correct, it is poorly understood.

7.3.2 The Exhibition-Characterization Symbol

In order to be able to express a thing graphically without needing to repeat it for an object and a process separately, we introduce the symbol of thing, which symbolizes the generalization of object and process. The **Thing** symbol, ⬭, is a superposition of a process on an object, and is used only in meta-OPDs.[9] Using this symbol, Figure 7.10(a) presents the two unidirectional relations, which, in OPL explicitly read as follows:

> **Exhibitor** exhibits **Feature.**
> *(Exhibition sentence)*
> **Feature** characterizes **Exhibitor.**
> *(Characterization sentence)*

---

[9]   Meta-OPD, discussed in Section 13.4, is an OPD that describes how OPM works. When we build a metamodel of OPM there, we model **Thing, Object** and **Process** as objects, because at the metamodeling level we think of them as concepts that exist.

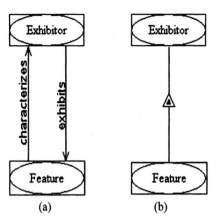

Figure 7.10. The Exhibition-Characterization structural relation (a) as two anti-symmetric unidirectional relations, (b) with the equivalent dedicated symbol

Figure 7.10(b) is the equivalent dedicated symbol for the relation – a small black triangle inside a larger white one.[10] Note that unlike the rest of the fundamental structural relations, Exhibition-Characterization allows to relate objects to processes and processes to objects. The other three fundamental structural relations allow only objects to be linked to objects and processes to processes: objects can be parts, specializations or instances only of objects, while processes can be parts, specializations or instances only of processes.

7.3.3 Attribute and Operation Are Features

What makes a thing a feature is the fact that it characterizes another thing.

> *A feature is a thing that characterizes another thing.*

As two simple examples, **Address** is an object in its own right, but it is also a feature of **Person**, as it is one of the things that characterize it. **Printing** is a process, hence a thing, but it is also a feature of **Printer**, as it is one of the things that characterize it.

A feature is a specialization of thing. In addition to its being a thing, it also characterizes another thing. Recall that we have defined object as a thing, whose perseverance value is *static*, while a process has been defined as a thing, the perseverance value of which is *dynamic*. Hence, being a thing, feature may be either an object or a process, as the two examples above show. This gives rise to two types of features: static features and dynamic features.

---

[10] This symbol can be thought of as a combination of Aggregation-Participation (black triangle) and Generalization-Specialization (white triangle).

> An **attribute** is a static feature.
> An **operation** is a dynamic feature.

More verbosely, an attribute is a feature whose perseverance value is *static*, while an operation is a feature whose perseverance value is *dynamic*. Attributes and operations exist in OO methods. In the OO terminology, an attribute is also referred to as a *data member*, while an operation is also referred to as a *method* or a *service*. In the traditional procedural third generation programming languages, operation is called function, procedure, or routine. All these words are meant to express *"something that the thing can do"* or *"a way in which the thing behaves."*

Table 7 summarizes the definitions of attribute and operation as specialization of feature. Feature and its specializations are recorded lower than their thing counterparts to denote the fact that feature is a specialization of thing, and hence attribute and operation are the respective specializations of object and process.

Table 7. The specializations of thing and feature by perseverance

Perseverance	Thing	Feature	OO Synonym(s)
static	Object	Attribute	Data member
dynamic	Process	Operation	Method, Service

The OPL sentence that relates an **Exhibitor** to three attributes called **Attribute 1**, **Attribute 2**, and **Attribute 3** is:

> **Exhibitor** exhibits **Attribute 1, Attribute 2**, and **Attribute 3**.

The OPL sentence that relates an **Exhibitor** to three operations called **Operation 1, Operation 2**, and **Operation 3** is:

> **Exhibitor** exhibits **Operation 1, Operation 2**, and **Operation 3**.

The OPL sentence that relates an **Exhibitor** to three attributes and to three operations is:

> **Exhibitor** exhibits **Attribute 1, Attribute 2**, and **Attribute 3**, as well as **Operation 1, Operation 2**, and **Operation 3**.

7.3.4 Exhibition Complexity

We have defined aggregation complexity as an attribute of a thing that specifies whether it consists of parts that are meaningful to the system or are needed to fully understand it. We noted that a thing may be atomic (aggregationally simple) or compound (aggregationally complex). A thing is compound if it consists of parts, otherwise it is atomic. Exhibition complexity is defined below in a manner analogous to aggregation complexity.

> **Exhibition complexity** is an attribute of a thing that specifies whether it exhibits features that are needed to fully specify the system.

The exhibition complexity attribute can have one of two values: *dull* and *colorful*.

> A **colorful** thing exhibits one or more features.
> A **dull** thing does not exhibit any feature.

Table 8. The state-space of aggregation and exhibition complexities

Aggregation Complexity Exhibition Complexity	atomic	compound
Dull	atomic-dull	compound-dull
Colorful	atomic-colorful	compound-colorful

Exhibition complexity and aggregation complexity are orthogonal: a thing can assume any one of the two values of the former regardless of the value of the latter.[11] Table 8 enumerates the four points in the state-space of aggregation and exhibition complexities. All possible combinations are possible.

Examining Figure 7.12(b), we can find examples for three of the four points in the state space. **Person** is atomic-colorful, **Name** and **Address** are compound-dull, and **Zip Code** is atomic-dull. Continuing the analogy between aggregation complexity and exhibition complexity, we note that just as the former is relative, so is the latter. A thing that is dull in one system may be colorful in another system. It is up to the system developer to decide whether a thing needs to exhibit features. The decision depends on the role that the thing plays in the system.

7.4 Features in OO vs. OPM

OPM is unique in its treatment of features as things that have their own right of existence, regardless of the fact they characterize a higher-level thing. Aggregation-Participation and Generalization-Specialization are bona fide OO relations. However, OO methods do not recognize attributes as objects, neither do they recognize Exhibition-Characterization as an explicit relation.

---

[11] It may seem paradoxical that even a dull thing, which by definition does not exhibit any feature, exhibits the attribute exhibition complexity. However, we should bear in mind that **Exhibition Complexity** is an attribute of thing at the meta-level, where **Dull** and **Colorful** are its values.

Paradoxically, although OPM does not attempt to be "purely" object-oriented, in its treatment of characterization, OPM is more object-oriented than the object paradigm. It is not clear what attributes are in OO. On one hand, OO methods claim that the world should be modeled in terms of objects, so everything ultimately has to be some kind of an object. In OO, attributes and methods are encapsulated, or embedded, within objects. Are attributes not objects, but rather "different animals" that reside within the object? If an attribute is not an object, then what is it? Does the world consist not only of objects but also of attributes (and methods)? OPM does not encounter this dilemma, since it defines feature generically as a thing that describes a thing.

To demonstrate the problem caused by not treating attributes as objects, consider a "classical" example of **Name** and **Address** as attributes of the object class **Person**, and **Moving** as an operation of **Person**.[12] As Figure 7.11(a) shows, UML (and many of its predecessors, such as OMT (Rumbaugh et al., 1991) lists the attributes and operations in the second and third class box compartments, respectively. Figure 7.11(b) shows the corresponding OPM notation: **Name** and **Address** are objects, and **Moving** is a process. Since they are linked to **Person** with the Exhibition-Characterization symbol, **Name** and **Address** are attributes of **Person**, and **Moving** is an operation of **Person**. A side benefit of this notation is that we can connect **Moving** to **Address** to denote the fact that **Moving** has an effect on **Address** of **Person**.

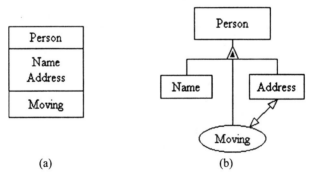

(a)   (b)

Figure 7.11. Expressing attributes and operations: (a) in UML. (b) in OPM

The OPL paragraph of the OPD in Figure 7.11(b) is:

Person exhibits **Name** and **Address,** as well as **Moving.**
Moving affects **Address.**

---

[12] We assume here that **Person** is capable of **Moving** without the need for external objects, such that **Moving** can be considered an operation of **Person**.

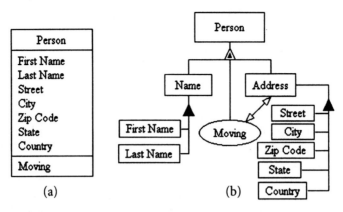

Figure 7.12. Expressing parts of attributes (a) in UML (b) in OPM

In OPM, **Name** and **Address** are attributes of **Person**, and **Moving** is an operation of **Person**. Note that objects and processes in a list are separated by the
reserved phrase as well as. If the exhibiting thing is an object (as in this example),
the list of objects precedes the list of processes, and vice versa. Outside the context
of **Person**, both are objects in their own right. Moreover, each one of them consist
of parts (**Name** consists of **First Name** and **Last Name**; **Address** consists of
Street, City, Zip Code, State and **Country**).

Since OO attributes are not objects, they cannot be decomposed into their parts,
neither can they have attributes of their own. Hence, declaring **Date of Birth** and
Address as non-object attributes of **Person** is a dead end in terms of having the
option to further decompose or characterize them. If we need to refer to **Last Name**
separately from **First Name** and to **Street** separately from **Zip Code**, then each of
them must become a direct attribute of **Person**. As Figure 7.12(a) shows, the result
is that **Last Name** and **Zip-Code** are immediate attributes of **Person** and are at the
same level, directly under **Person**. Continuing along this line, we end up with a
flat, one-level hierarchy, which, in real-life applications, can have hundreds of
attributes (or "fields" as they are called in relational databases). OPM is contrasted
with this approach, as shown in Figure 7.12(b). The OPL paragraph of the OPD in
Figure 7.12(b) follows.

> **Person** exhibits **Name** and **Address,** as well as **Moving.**
> **Moving** affects **Address.**
> **Name** consists of **First Name** and **Last Name,** in this order.
> **Address** consists of **Street, City, Zip Code, State** and **Country,** in this
>   order.

The parts of the objects **Name** and **Address**, which are attributes of **Person**, are
arranged in a hierarchy that puts aggregates higher in the whole-part hierarchy, providing for quick understanding and better manageability.

7.5 The Four Thing-Feature Combinations

The definition of Exhibition-Characterization as the relation between an exhibitor and its features gives rise to four object-process combinations of colorful things and their features. In other words, since both thing and its feature can be either an object or a process, the 2×2 Cartesian product yields a state-space of four different combinations of a thing and the feature that characterizes it. These combinations are an attribute of an object, an attribute of a process, an operation of an object, and an operation of a process.

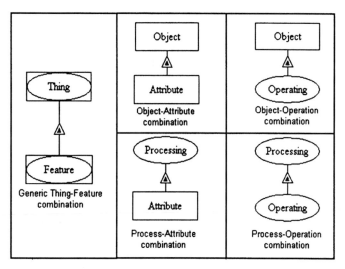

Figure 7.13. The generic (left) and the four concrete thing-feature combinations

To visualize the various combinations, Figure 7.13 presents five OPDs. The left-most OPD, titled *Generic Thing-Feature combination*, is the most general one, showing the symbol of thing, ⊂⊃. The thing is connected through the Exhibition-Characterization relation symbol to another thing, which is its feature. To the right of this generic OPD are the four concrete OPDs, with the combination listed below each OPD. In the following subsections we elaborate on each one of the four Thing-Feature combinations.

The attribute that specifies whether a structural link connects things with the same **Perseverance** (**static** or **dynamic**) is called **Homogeneity**. The values of **Homogeneity** are **Homogeneous** and **Non-homogeneous**, with **Homogeneous** being the default. Using this metamodeling terminology, **Exhibition-Characterization** is **Non-homogeneous**, while the other three fundamental structural relations are **Homogeneous**.

7.5.1 The Object-Attribute Combination

The first thing-feature combination – object and its attribute – is the sequel custom-ary attribute of classical object-oriented approaches. Here we refer to an object B_2 – the attribute – that *characterizes* (describes) a higher level object B_1. Conversely, we say that B_1 *exhibits* B_2. A few examples for such pairs of objects and their attributes are **Material – Specific Weight, Person – Age, Chemical Element – Atomic Weight, Laptop – Manufacturer, Book – Author, Officer – Rank,** and **Dog – Breed**. The first four of these examples are depicted in the OPDs in Figure 7.14. The OPL sentence pairs that correspond to the OPDs in Figure 7.14 appear below.

> **Material** exhibits **Specific Weight.**
>   **Specific Weight** characterizes **Material.**
> **Person** exhibits **Age.**
>   **Age** characterizes **Person.**
> **Chemical Element** exhibits **Atomic Weight.**
>   **Atomic Weight** characterizes **Chemical Element.**
> **Laptop** exhibits **Manufacturer.**
>   **Manufacturer** characterizes **Laptop.**

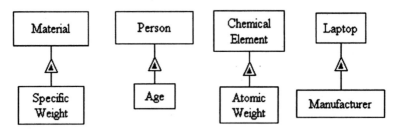

Figure 7.14. Examples of attributes of objects

As is the case with structural relations in general, each sentence in the pair is equivalent to its counterpart. The first sentence in each pair is the forward relation, which uses the reserved word exhibits, while the second is the backward relation, which uses the reserved word characterizes. The default form that is generated automatically is the forward relation sentence. This parallels the Aggregation-Participation relation. The default direction for both is the forward (or downward, as seen in an OPD) direction.

7.5.2 The Object-Operation Combination

The second thing-feature combination is object and its operation. As noted, in object-oriented approaches an operation is also called *method* or *service*. Here we refer to a process P_1 – the operation – that characterizes a higher level object B_1. Conversely, we say that B_1 exhibits the operation P_1.

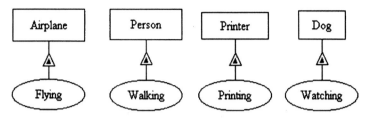

Figure 7.15. Examples of operations of objects

An operation of an object is a process that affects only objects that are parts, features, specializations or instances of that object. In other words, an operation of an object B_1 has no side effect on, nor does it require, any object that is outside of B_1. Under this condition, the operation can be identified as being "owned" by B_1. Object-oriented methods view *all* processes as operations that are encapsulated within and owned by objects. This encapsulation is a major source of confusion and impediment of faithful system modeling. In OPM, encapsulation is valid only when the process is internal to the object. In cases like this, the process is defined as an operation of the encapsulating object.

A few examples of pairs of an object and its operation are **Airplane – Flight, Person – Walking, Printer – Printing, Officer – Commanding,** and **Dog – Watching.** Figure 7.15 presents four OPDs that correspond to these pairs. The OPL sentence pairs that correspond to the four OPDs in Figure 7.15 follow.

Airplane exhibits **Flying.**
  Flying characterizes **Airplane.**
Person exhibits **Walking.**
  Walking characterizes **Person.**
Printer exhibits **Printing.**
  Printing characterizes **Printer.**
Dog exhibits **Watching.**
  Watching characterizes **Dog.**

An operation is a specialization of a process. As such, its name is a gerund, i.e., a verb form ending with the **"ing"** suffix. As we will discuss in Section 10.3, when referring to the function of systems, many objects, in particular physical and artificial ones, exhibit a major operation that expresses the main *function*, or *service*, which the object is designed to perform. Thus, for example, the function or service that the object **Printer** supplies is **Printing**, the service of **Airplane** is **Flying**, the service of **Crane** is **Lifting**, the service of **Dryer** is **Drying**, etc. This is in line with our definition of an artificial system as an object that carried out a function. While processes or operations often attain functions, one should not confuse the distinction between a process and a function.

7.5.3 The Process-Attribute Combination

Like objects, processes require adequate representation in the model of any system. Among other things, processes, just like objects, require attributes – objects that describe them. The idea of attributes for processes is a natural extension to attributes for objects and it poses no special conceptual difficulty.

So far, we have seen that the first and second thing-feature combinations – object describing an object and process describing an object – are the corresponding object-oriented attribute and operation (or service, or method). However, the third thing-feature combination – object describing a process – is not explicitly defined in object-oriented approaches. Here we refer to an object B_1 _ the attribute – that characterizes a higher level process P_1. Conversely, we say that the process P_1 *exhibits* the attribute B_1. Few examples of pairs of a process and its attribute are **Diving – Depth, Commanding – Language, Printing – Quality, Striking – Duration, Manufacturing – Quantity, Watching – Effectiveness, Singing – Volume, Skiing – Location,** and **Flying – Speed.** Figure 7.16 presents OPDs that correspond to the first four process-attribute pairs.

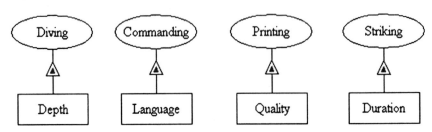

Figure 7.16. Examples of attributes of processes

The OPL sentence pairs for the OPDs in Figure 7.16 are:

> **Diving** exhibits **Depth.**
>   **Depth** characterizes **Diving.**
> **Commanding** exhibits **Language.**
>   **Language** characterizes **Commanding.**
> **Printing** exhibits **Quality.**
>   **Quality** characterizes **Printing.**
> **Striking** exhibits **Duration.**
>   **Duration** characterizes **Striking.**

Each of these process-attribute pairs in Figure 7.16 can be embedded in a natural language sentence. Here are possible examples, where the processes are bold and their attributes are italicized:

(1) **Diving** at a *depth* of 30 meters or more requires decompression stops.
(2) The *language* the office was using for **commanding** was foreign and strange.
(3) The **printing** of this device is of poor *quality*.
(4) The employees have been engaged in **striking** for *duration* of over two weeks.

Note that while all the processes in these examples are nouns having the gerund form, they can be easily converted into sentences where the processes are verbs, with the exact semantics as before:

(1) A diver **diving** at a *depth* of 30 meters or more is required to make decompression stops.
(2) The officer **commands** in a foreign *language*.
(3) This device **prints** with poor *quality*.
(4) The employees have been **striking** for *duration* of over two weeks.

As these examples show, this extension to the OO attribute and operation concepts is a direct consequence of recognizing processes as bona fide things beside objects.

7.5.4 Process-Operation Combination

The fourth and last thing-feature combination – process and its operation – is the second one that is not explicitly defined in object-oriented approaches. It is the least prevalent combination and may be somewhat difficult to grasp. Here we refer to a process P_2 _ the operation – that characterizes a higher level process P_1. Conversely, we say that the process P_1 exhibits the operation P_2. Only a process can induce a change in a thing. In other words, the process is the thing, which is "responsible" for this change. That process can be an operation. An operation of an object changes the object that exhibits ("owns" in OO terms) that operation. Likewise, an operation that a process exhibits changes that process.

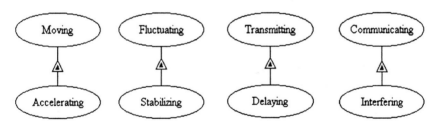

Figure 7.17. Examples of operations of processes

In daily life, we do not think so much about operations of processes. The best way to understand the meaning of an operation of a process is to look at time. A change of an object along the timeline means that the state (or value) of an object inspected at time t is different from its state at a later time $t + \Delta t$. If we sample the thing at two different points in time, we may notice a change in that thing. The change can be manifested as a difference in the value of at least one of the attributes of the thing. **Accelerating** is a process that changes the **Moving**, which is a process. **Stabilizing** is changing of the **Fluctuating** process. **Delaying** is a process that changes **Transmitting**, and **Interfering** changes **Communicating**. Figure 7.17 shows four such examples.

The corresponding OPL sentence pairs are:

> **Moving** exhibits **Accelerating**.
>   **Accelerating** characterizes **Moving**.
> **Fluctuating** exhibits **Stabilizing**.
>   **Stabilizing** characterizes **Fluctuating**.
> **Transmitting** exhibits **Delaying**.
>   **Delaying** characterizes **Transmitting**.
> **Communicating** exhibits **Interfering**.
>   **Interfering** characterizes **Communicating**.

Two additional examples of pairs of a process and its operation are **Decaying**, which changes **Radiating**, and **Damping**, which changes **Oscillating**. In mathematical terms, a change of an object along the timeline is a *first order derivative* of some quantity (which is an attribute value of that object) *with respect to time*. In an analogous manner, since a process is a change agent, an operation of a process is a *change of a change*, or, in mathematical terms, *second order derivative* (derivative of the derivative) of some quantity with respect to time. Indeed, the examples of pairs of a process and its operation shown in Figure 7.17 have the notion of changing a process and can be quantified mathematically using second order derivatives.

7.6 The Feature Hierarchy

Feature is a relative term. A thing is a feature if it describes another thing. This feature itself can be further described by another, lower level feature. Thus, every feature is a thing, but not every thing is a feature. Since the Exhibition-Characterization is a fork relation, we can get a hierarchy of things that describe each other.

Consider the object **City**, whose feature hierarchy is depicted in Figure 7.18. The attribute **Name** of **City** can be used as the *identifier* of **City** within its **Region**. While not specified in this OPD, a **Region** can be **State, Province, Territory**, etc. Three important attributes of **City**, in addition to its **Name**, are **Location, Population**, and **Climate**. Besides being attributes of **City**, **Location, Population** and **Climate** are objects in their own right. Note that while **Name** is an informatical object, the rest are physical ones. As an object, each **City** attribute may have its own set of features. **Location** can have the attributes of the **Continent, Country, Region**, and **Coordinates**, which consist of **Longitude** and **Latitude**. **Population** exhibits the attributes **Size** and **Demographics** and the operations **Growing, Aging, Earning**, and **Studying**. As an object **Demographics**, in turn, consists of **Average Age, Average Education**, and **Average Income**. While **Growing** affects **Size, Aging, Earning**, and **Studying** affect the three parts of **Demographics**, respectively. **Growing** is the root of the aggregation hierarchy depicted in Figure 7.8(a). Figure 7.18 and Figure 7.8(a) are two OPDs that belong to the same OPD set; **Growing** from the OPD in Figure 7.18 unfolds in Figure 7.8(a). Another set of features of

City is related to **Climate**. To keep **Climate** simple, we just consider **Temperature** and **Precipitation** as attributes of **Climate**. **Precipitating** is an operation of **Climate** that affects **Precipitation**, which, in turn, consists of **Rain** and **Snow**.

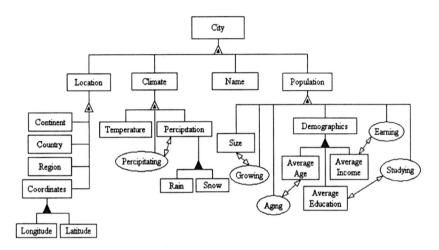

Figure 7.18. The attribute hierarchy of **City**

7.7 Feature-Related Natural Language Issues

Thinking of attributes in terms of natural languages on one hand and in OPM terms on the other hand raises a series of interesting issues. These issues expose dilemmas and shortcuts that natural languages present.

7.7.1 Attribute Naming Dilemmas

Natural languages often provide us with a definite noun for naming the attribute. For example, the attribute whose two extremes are the adjectives **"short"** and **"long"** is called **Length**. The attribute whose two extreme adjectives are **"narrow"** and **"wide"** is called **Width**, and the attribute whose two extremes are **"heavy"** and **"light"** is called **Weight**. Sometimes, the attribute name (the noun) is from the same radical as the value (the adjective). Examples for such attribute-value (noun-adjective) pairs are **Length – long, Width – wide, Readiness – ready**, and **Beauty – Beautiful**. Of these pairs, the original, root word (the radical) may be either the noun (e.g., **Beauty**) or the adjective (e.g., **ready**).

Some attributes are neutral nouns, while others are taken from and are biased towards one of the extreme values of the attribute. The attribute **Shape**, for example, is neutral. Its values may be **round, square, elliptic**, etc. There is no bias in **Shape** toward any of its values. Conversely, **Length** is biased towards the **long** extreme of the **short – long** value spectrum. Hence, a sentence such as "The shape of the house is square." makes perfect sense, whereas "The length of the stick is long," while syntactically correct, is semantically awkward. Skipping the name of the attribute, we would rather say "The stick is long." In this case, the attribute **Length** is implicit in the sentence. We could also skip the attribute name of the attribute **Shape** in the sentence as "The shape of the house is square." and say "The house is square." We call such a sentence *implicit attribute sentence*. Implicit attribute sentences need to be used when the attribute values are of quantitative nature (such as **short – long**) and the name of the attribute is drawn from one of the extreme quantities. The ability to skip the name of the attribute and make direct reference to the object exhibiting that attribute value is most prevalent. In fact, the use of implicit attribute sentences in natural language is the rule rather than the exception.

Implicit attributes are so widespread, that in many cases the natural language does not have a dedicated noun for the attribute itself. The adjectives, which are the values or states of that attribute, do have widely recognized and used names. Consider the implicit attribute sentence "This book is interesting." The adjective interesting refers to an attribute of this book, whose possible values may be "**interesting**" and "**boring**." There is no single noun for an attribute whose values are **interesting** and **boring**. Plausible names of this attribute may be either **Interest Level** or **Boredom Level**. However, each is biased toward one of the extremes of the spectrum or the other. Ideally, we would like a word that is neutral and not biased toward any one of the possible attribute values.

We have already encountered this problem while trying to name the attribute of **Thing** whose values are **natural** and **artificial**. Recall that we have called this attribute **Origin**. We have also called **Essence** the attribute of **Thing** whose values are **physical** and **informatical**. **Perseverance** has been chosen as the name for the attribute whose values are **static** (in which case the thing is an object) and **dynamic** (in which case the thing is a process). The choice of these attribute names points to some difficulty in finding the right word to name an attribute whose values are prevalent. For example, **static** and **dynamic** are widely used, while **Perseverance** is not recognized in conjunction with these adjectives. The American Heritage Dictionary (1996) defines perseverance as "steady persistence in adhering to a course of action, a belief, or a purpose; steadfastness," as well as a concept in theology. **Origin**, **Essence**, and **Perseverance** are neutral.

To further demonstrate the dilemma, consider the implicit attribute sentence "An attribute can be qualitative or quantitative". In OPM concepts, this statement means that the object **Attribute** has another, lower-level attribute, the two possible values of which are **qualitative** and **quantitative**. What might be an adequate name of that attribute? It may be **Qualitativeness** or **Quantitativeness**. However, both are biased.[13] Nevertheless, both may be understandable. A formal, but too cumbersome a name to be practical, would be **Qualitativeness – Quantitativeness Spectrum**.

---

[13] Interestingly, **Quantitativeness** appears in Webster's Dictionary, Ninth Edition, but **Qualitativeness** does not!

7.7.2 Reserved Objects and the Measurement Unit Reserved Object

OPD symbols are translated into OPL reserved phrases. For example, the triangular Exhibition-Characterization relation symbol studied in this chapter is translated in OPL to the reserved word exhibits. An OPD *reserved object* is an object with a special name, which OPM recognizes as a reserved object name and treats in a manner analogous to an OPD symbol. When a reserved object appears in an OPD, it gives rise to an OPL sentence that is suitable to the situation being modeled.

> *A reserved object is an object that is treated as an OPD symbol, which gives rise to a special OPL sentence.*

The reserved object concept is introduced here because we need to use it in order to present the Measurement Unit reserved object and demonstrate its use. Note that like OPL reserved phrases, a reserved object is denoted by non-bold Arial font. Since many attributes are expressed by values, which themselves exhibit measurement units, Measurement Unit is a reserved object.

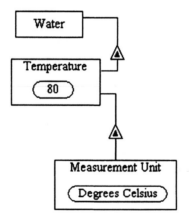

Figure 7.19. Measurement Unit as a reserved OPL phrase with the value of **Temperature** specified.

To explain the motivation for defining Measurement Unit as a reserved object, let us first construct the OPL paragraph of Figure 7.19 without considering **Measurement Unit** as a reserved object.

> **Water** exhibits **Temperature**.
> Value of **Temperature** is **80**.
> **Temperature** exhibits **Measurement Unit**.
> Value of **Measurement Unit** is **Degrees Celsius**.

The last two OPL sentences are cumbersome, unintuitive expressions. Considering Measurement Unit as a reserved object, the last two sentences in the OPL paragraph above are combined into the following sentence:

> **Temperature** is measured in **Degrees Celsius.**
> *(Measurement unit specification sentence)*

As noted, a reserved object invokes a special OPL reserved sentence that may involve a special reserved phrase. In our case, the reserved object Measurement Unit in Figure 7.19 invokes the sentence above. This sentence contains the reserved phrase "is measured in," which is inserted after the attribute whose measurement unit are specified and before the value of the measurement unit. The updated OPL paragraph includes the following two sentences:

> **Temperature** is measured in **Degrees Celsius.**
> *(Measurement unit specification sentence)*
> Value of **Temperature** is **80 Degrees Celsius.**
> *(Unit specified value enumeration sentence)*

Since the quantity in the value rountangle is usually not one, the name of the Measurement Unit is provided in plural, as in "**Degrees Celsius**" or "**US Dollars.**" As a phrase, the two bottom objects in Figure 7.19 are specified in OPL as "**Temperature** with value **80 degrees Celsius.**"

7.7.3 Continuous Values and Multi-Valued Attributes

Attribute values, such as water temperature, are often continuous rather than discrete. Moreover, many discrete values, such as an employee's salary, are too numerous to be explicitly enumerated. In practice, however, continuous values are made discrete by rounding them to whatever level of accuracy they can measured or needs to be specified. In other cases, such as an employee's salary, the value can be specified with accuracy that is dependent on the smallest unit available (e.g., cent) or desirable (e.g., dollar). Hence, for all practical purposes, we can refer to any continuous attribute value as discrete. We may describe that attribute by a second attribute that specifies the level of accuracy of the value.

Even with discrete values, though, we can end up with an infinite or a very large number of possible values. Such attributes are *multi-valued attributes*. For obvious reasons, we cannot and do not want to specify the values of a multi-valued attribute one by one in an OPD or an OPL sentence. Instead, OPD denotes all the values of such a multi-valued attribute by a single rountangle, which is the value/state symbol, inside the attribute box. This rountangle can remain empty, implying that any value is possible, or it can contain the lower and upper bounds of the values that the attribute can take.

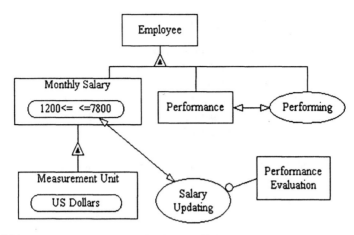

Figure 7.20. Salary as a bounded multi-valued attribute

Figure 7.20 is an OPD that demonstrates how multi-valued attributes are symbolized. **Salary** is a multi-valued attribute of **Employee**. The fact that it can assume numerous values is denoted by the single routangle inside the **Salary** attribute object box. This slot can remain empty, implying that any salary value is possible, or, as Figure 7.20 shows, it can contain the lower and upper bounds of the values **Salary** can take. The semantics of the effect link between the **Salary Updating** process and the **Salary** routangle is that the process changes the value of **Salary** from an unspecified "old value" to an unspecified "new value." The corresponding OPL sentence is the second of the following two sentences.

> **Employee** exhibits **Salary** and **Performance,** as well as **Performing.**
> *(Exhibition sentence)*
> **Salary** is measured in **US Dollars.**
> Value of **Salary** is equal to or greater than **1200 US Dollars** and equal to or less than **7800 US Dollars.**
> *(Value enumeration sentence)*
> **Salary Updating** affects the value of **Salary.**
> *(Value effect sentence)*

The value routangle inside the **Salary** box denotes the fact that **Salary** is a multi-valued object. If the effect link from **Salary Updating** would be to the **Salary** box rather than the routangle of the **Salary** value, the value effect sentence above would become the familiar effect sentence "**Salary Updating** affects **Salary.**" If rather than containing the text "**1200<= <=7800**" the **Salary** routangle in Figure 7.20 were blank, the second sentence would simply not be written.

7.7.4 Mathematical Inequalities in OPM

As we have seen in Figure 7.20 and its corresponding OPL paragraph, the various kinds of inequalities are expressed as mathematical symbols in OPDs and are expanded from their OPD mathematical form into full text in the corresponding OPL sentences. Thus, for example, the OPD **Salary** value "**1200 < < 7800**" in Figure 7.20 is equivalent to the mathematical expression "**1200 < Salary < 7800**", and in OPL it is expressed as: "… **Salary** with value equal to or greater than **1200** and less than **7800**."

Table 9 presents the various combinations of mathematical inequalities and their corresponding OPD text and OPL phrase.

Table 9. Mathematical inequalities and their OPD text and OPL phrase

Mathematical inequality	Text of OPD value	OPL Phrase
V < Vmax	**< Vmax**	**V** with value less than **Vmax**
V ≤ Vmax	**≤ Vmax**	**V** with value equal to or less than **Vmax**
V > Vmin	**> Vmin**	**V** with value greater than **Vmin**
V ≥ Vmin	**≥ Vmin**	**V** with value equal to or greater than **Vmin**
Vmin < V < Vmax	**Vmin < < Vmax**	**V** with value greater than **Vmin** and less than **Vmax**
Vmin ≤ V < Vmax	**Vmin ≤ < Vmax**	**V** with value equal to or greater than **Vmin** and less than **Vmax**
Vmin < V ≤ Vmax	**Vmin < ≤ Vmax**	**V** with value greater than **Vmin** and equal to or less than **Vmax**
Vmin ≤ V ≤ Vmax	**Vmin ≤ ≤ Vmax**	**V** with value equal to or greater than **Vmin** and equal to or less than **Vmax**

7.8 Reflective Metamodeling of an Attribute

Reflective metamodeling is modeling of OPM using OPM concepts. In this section we apply reflective metamodeling to model several aspects of an OPM **Attribute**. Sometimes we refer to **Feature**, which is a generalization of **Attribute** and **Operation**, in which case the metamodel applies to both.

7.8.1 The Size of an Attribute

Some attributes are **unary** while others are **binary** or **multi-valued**. A **unary** attribute has exactly one value, a **binary** attribute has exactly two, and a **multi-valued** attribute has more than two, usually many, and possibly an infinite number of values. Notable examples of binary attributes are those whose values are "0" and "1", "black" and "white", "pass" and "fail", "true" and "false", "yes" and "no", and "stop" and "go".[14] The name of the attribute of **Attribute** that specifies whether it is **binary** or **multi-valued** is **Size**.

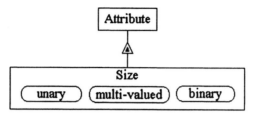

Figure 7.21. The **Size** of **Attribute** and its **unary, binary** and **multi-valued** values

> **Attribute** exhibits **Size**.
> **Size** can be **unary, binary** or **multi-valued**.
> *(State enumeration sentence)*

Frame 13. The OPL paragraph of the OPD in Figure 7.21.

Using the reserved word which,[15] the above OPL sentences can be joined into the single state enumerated exhibition sentence:

> **Attribute** exhibits **Size,** which can be **unary, binary** or **multi-valued**.
> *(State enumeration sentence)*

7.8.2 The Mode of an Attribute

Some attributes are **qualitative** while others are **quantitative**. We have seen the example of the attribute **Shape** of **House**, where possible values can be **round, square,** and **rectangular**. These values cannot be quantified by a numeric value. They are just qualitatively different from each other. We say therefore that **Shape** is a qualitative attribute. Other examples of qualitative attributes include **Mood**, with states **happy, sad, angry,** etc., **Health**, with states **healthy** and **sick**, and **Marital Status**, with states **single, married, divorced,** etc.

> *A **qualitative attribute** is an attribute whose values or states are non-numerical.*

---

[14] Following the Beatles, who sing "You say yes, I say no, you say stop and I say go, go, go..." OPM has adopted "yes" and "no" as the names of the two states of the Boolean object, which is a special type of a binary object.

[15] As discussed in Section 13.5, the reserved word "which" can be used to join any two OPL sentences in which the first sentence ends with the name of the thing with which the second sentence starts.

Most of the attributes are quantitative. **Size, Height,** and **Weight** are simple examples. Quantitative attributes assume numeric values.

> A *quantitative attribute* is an attribute whose values or states numerical.

Since an attribute can be qualitative or quantitative, **qualitative** and **quantitative** are values of an attribute of **Attribute.** This raises again the problem of finding an appropriate name for this attribute, whose values are **qualitative** and **quantitative.** This attribute is called **Mode.**

> *Mode* is an attribute of an attribute whose values are *qualitative* and *quantitative.*

The OPD in Figure 7.22 and the OPL paragraph in Frame 14 express these definitions.

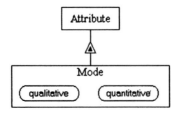

Figure 7.22. The **Mode** of **Attribute**

> **Attribute** exhibits **Mode.**
> **Mode** can be **quantitative** or **qualitative.**

Frame 14. The OPL paragraph of the OPD in Figure 7.22.

Combining the two sentences in Frame 14 with which we get:

> **Attribute** exhibits **Mode,** which can be **quantitative** or **qualitative.**

7.8.3 The Touch of an Attribute

Depending on whether an attribute is can be computed from others or not, a attribute can be hard or soft.

> An attribute is *hard* if its value cannot be deduced or computed from other attributes.
> An attribute is *soft* if its value can be deduced or computed from other attributes.

For example, **Date of Birth** of a **Person** is a hard attribute, while **Age** of **Person** is a soft attribute. By knowing the **Date of Birth** of a **Person** and the current value of **Date**, **Age** of **Person** can be computed. As anther example, the **Weight** of each part of **Airplane** is a hard attribute, while the total **Weight** of **Airplane** is a soft attribute since it can be computed by summing the weights of the individual parts. The name of the attribute of **Attribute** whose values are **hard** and **soft** is **Touch**. The OPL sentence that expresses this is:

> **Attribute** exhibits **Touch,** which can be **hard** or **soft.**

Deciding whether a soft attribute should be pre-computed has practical implications during the detailed design stage of an information system. Pre-computed values can be stored for quick response time at the cost of storage space. Alternatively, soft attributes can be computed on demand, saving space but also delaying the response time of the information system. Collecting the three state enumerated exhibition sentences we have discussed, whose exhibitor was Attribute, we get:

> **Attribute** exhibits **Size,** which can be **unary, binary** or **multi-valued.**
> **Attribute** exhibits **Mode,** which can be **quantitative** or **qualitative.**
> **Attribute** exhibits **Touch,** which can be **hard** or **soft.**
> *(State enumerated exhibition sentences)*

These three OPL sentences can be combined into a single sentence:

> **Attribute** exhibits **Size,** which can be **unary, binary** or **multi-valued,**
> **Mode,** which can be **quantitative** or **qualitative,** and **Touch, which** can be **hard** or **soft.**
> *(Compound state enumerated exhibition sentence)*

7.8.4 The Source of a Feature

Depending on whether a feature is exhibited only by the object as a whole or by one or more of its parts, a **Feature** (an **Attribute** or an **Operation**) can be **inherent** or **emergent**.

> *A feature of an object is **emergent** if no one of the object's parts exhibits it.*

To understand the difference between emergent and inherent features, consider **Airplane**'s attribute **Weight** and its operation **Flying**. **Weight** of **Airplane** is the sum of the individual **Weight** values of each one of the parts that make up the **Airplane**. **Flying**, on the other hand, was not an operation that any part of **Airplane** could exhibit on its own. Rather, this feature emerges from the unique ensemble of the parts of **Airplane** that endows **Airplane** with the ability to carry out the **Flying** operation. As Bar-Yam (1997) noted, the collective behavior is contained in the behavior of the parts if they are studied in the context in which they are found.

Hence, we call **Flying** an *emergent* feature of **Airplane**, while **Weight** is an *inherent* feature of **Airplane**. In systems, operations are frequently emergent, because systems are built with the intent of achieving some function that is not localized in any part of the system. The attribute, for which **inherent** and **emergent** are values, is called **Source**. The OPL sentence that expresses this metamodel fact is:

> **Feature** exhibits **Source,** which can be **inherent** or **emergent.**

Bar-Yam (1997) distinguishes between simple and complex systems and claims that complexity can emerge from a collection of simple parts that comprise a system. The converse can be true as well: a system composed of complex parts may exhibit simple behavior at a larger scale. For example, planet Earth is a highly complex system, but when viewed from the perspective of its movement around the sun, it is relatively simple, pointing to the relativity of the term complexity.

7.8.5 The Operation a Feature Carries

Each system exhibits function. In our metamodel of **Feature**, the function of **Feature** is to describe a **Thing**. The **Feature** of a **Thing** describes the **Thing**. Hence, as the meta-OPD in Figure 7.23 specifies, **Describing** is the operation of **Feature**. **Feature** is designed to describe a **Thing**, for which it is an **Attribute**. In OPL,

> **Feature** exhibits **Describing.**

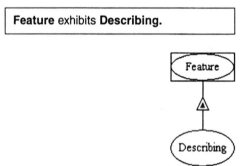

Figure 7.23. Feature exhibits **Describing.**

Summary

- We have introduced two fundamental structural relations: Aggregation-Participation and Exhibition-Characterization. Both are transitive and use a fork, giving rise to a hierarchy.
- Aggregation-Participation is the relation between a whole and its parts.
  - The shorthand name of this relation is aggregation and its symbol is ▲.

- The OPL aggregation sentence is:

> **Whole** consists of **Part 1, Part 2,** and **Part 3.**

- The parts comprising the aggregate can be listed orderly, either implicitly, by ordering them vertically, or explicitly, by writing ordered next to the Aggregation symbol.
- Aggregation can be used to relate objects to each other as well as to relate processes to each other, but a process cannot be a part on an object or vice versa.
- Every thing (object or process) is either atomic (i.e., consists of no parts) or complex (consists of parts).
- Exhibition-Characterization is a relation between a thing and the features that characterize it.
  - The shorthand name of this relation is characterization and its symbol is ⚠.
  - The OPL characterization sentence is:

> **Exhibitor** exhibits **Attribute 1, Attribute 2,** and **Attribute 3,** as well as **Operation 1, Operation 2,** and **Operation 3.**

  - Characterization is the only fundamental structural relation for which all four combinations of an object and a process, as an exhibitor and a feature, are possible.
  - A feature, which is an object, is called an attribute, while a feature, which is a process, is an operation.
  - An attribute value can be a range.
  - The complexity of every thing is either dull (i.e., the thing exhibits no features) or colorful (if it exhibits one or more features).
  - Every feature exhibits source, which can be inherent (i.e., exhibited by part or parts of the exhibitor) or emergent (it is only exhibited by the whole exhibitor).
  - An attribute exhibits the following attributes:
    - Size, which can be unary, binary, or multi-valued;
    - Mode, which can be qualitative or quantitative; and
    - Touch, which can be soft (i.e., computable from other attribute values) or hard (cannot be computed from other attribute values).
  - An OPM reserved object is an object that gives rise to a reserved phrase in an OPM sentence. **Measurement Unit** is an example of a reserved object.

Problems

1. Draw OPDs and write OPL sentences describing two objects for which the parts are ordered and two for which they are not.
2. Write the OPL paragraph of the OPD in Figure 7.6.

3. Create an OPD of the aggregation hierarchy of your bedroom to at least five levels.

4. Add attributes to at least a third of the objects in the hierarchy you created in problem 3.

5. List three adjectives and three adverbs that you've heard recently. What things were these modifiers describing? What attributes were implicitly referred to?

6. Draw two OPDs for an object consisting of at least eight different parts at the first participation level, with non-comprehensive aggregation. A subset of the parts should appear in one OPD and another subset in the other OPD such that the union of the subsets is comprehensive.

7. Give two examples of aggregation over-specification and two for aggregation under-specification.

8. Use parameterized participation constraints to create the aggregation hierarchy of a high rise building. The building has a certain number of floors, each having two types of apartments, standard and luxury. In each floor from floor 4 and above there are three standard and two luxury apartments. In the first three floors, there is one small and two large offices. Decide how many floors there are and how many faucets are required for each unit, and create the appropriate OPD with participation constraints. Using your diagram, compute the number of faucets the contractor needs to order.

9. Show two examples of an object, which is atomic in one system and compound in another. Draw the corresponding OPDs.

10. Use an OPD to describe the organization of the last school you graduated from or the school you attend now, whether K-12 or college, or the company you have been working for. Make School or Company the root (top) of the aggregation hierarchy and Student or Employee the bottom.
    a. How many levels are there in the aggregation hierarchy?
    b. Is there a justification to this number of levels? Is it adequate, too high, or too low?
    c. If inadequate, design a new organizational structure using a new OPD.

11. Compare your design to the current structure in terms of speed and ease of communication as well as manageability. Provide two examples for each one of the four points in the state space of aggregation and exhibition complexities (Table 8.). Draw OPDs and write OPL sentences.

12. Write the OPL script of the OPD set of **City**, which consists of Figure 7.8 and Figure 7.18(a).

13. For each one of the examples in question 1, find an example of a system where the attribute values are different for the same thing. For example, if an object was atomic-dull in the first system, it will be compound-colorful in the other.

14. Based on the specification of **Structural Relation** in Section 6.1, construct a metamodel of **Structural Relation**, which would include the **Reciprocity** attribute and its values.

15. Create an OPD of the parts and features of the chariot according to the description on page 133.

16. Draw the meta-OPD of a feature, according to Section 7.8 and include all the features and their states.

17. The meta-attribute that specifies whether a structural link connects things with the same **Perseverance** is called **Homogeneity**. Create a metamodel of **Link** with its **Homogeneity** attribute, which can take on the values **Homogeneous, Heterogeneous,** with **Homogeneous** being the default. Express the fact that **Exhibition-Characterization** is **Heterogeneous,** while the other three fundamental structural relations are **Heterogeneous.**

Chapter 8

Generalization and Instantiation

> *As this term is most commonly used, a generalization is an "all" statement, to the effect that all objects of a certain general kind possess a certain property.*
>
> E.J. Lowe (1983)

While discussing Aggregation and Exhibition, we talked about entire groups of objects or processes – any scientific paper, any employee, any running. However, what if we wanted to consider the example of a specific paper, written by a certain John Doe? Or if we wanted to consider a group of employees, namely managers, who receive a certain salary out of the range of salaries available for the company? Perhaps we would like to discuss running in a marathon, as opposed to just any kind of running? We need to be able to pay particular attention to a specialized group, which belongs to a more general group, or even a specific instance out of a class of objects. As its name clearly points out, Generalization-Specialization is the relation between a general and a specialized case of a thing. Classification-Instantiation links between a class of things and a unique instance from the class. Since these two of the four fundamental relations are important to systems modeling, we consider them in more detail now; and since they are intimately related, they are explained together in this chapter.

8.1 Generalization-Specialization: Introduction

Let us first consider several simple examples to set the stage for discussing Generalization-Specialization, or "gen-spec"[1]. Recall that the first time we encountered the gen-spec relationship was in the wedding system introduced in Chapter 1, where we saw that

Man and **Women** are **Persons.**

The **Person** is the general case, while **Man** and **Woman** are its special cases. Other examples are "**Dog** and **Cat** are **Pets.**", "**Pascal, Java,** and **C++** are **Programming Languages.**", "**Airplane** and **Car** are **Vehicles.**", and "**Ketchup** and **Mustard** are **Condiments.**"

---

[1]   The shorthand term "gen-spec" is borrowed from Coad and Yourdon (1991).

8.1.1 Specialization Symbol and Sentence

The symbol of the Generalization-Specialization relation is the equilateral white triangle. As Figure 8.1 shows, it denotes the fact that the two objects **Digital Camera** and **Analog Camera** are specializations of **Camera**.

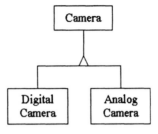

Figure 8.1. Camera and its **Analog Camera** and **Digital Camera** specializations

The OPL paragraph of Figure 8.1 could be:

Digital Camera is a **Camera**.
Analog Camera is a **Camera**.
  (Specialization sentences)

Better yet, these two OPL sentences can be joined into one:

Digital Camera and **Analog Camera** are **Cameras**.
  (Plural specialization sentence)

A more "professional" way of expressing the last sentence might have been "**Digital Camera** and **Analog Camera** specialize **Cameras**." However, sticking to the principle of keeping the OPL language as natural and as simple as possible, OPL uses the clearer and more intuitive reserved phrases "is a" or "are" rather than "specializes" or "specialize". As for the reserved word generalizes, OPL would rather use a more common word, too. Unfortunately however, no simpler adequate English word can replace generalizes in the top-to-bottom direction of the Generalization-Specialization relation. We could use "can be", but this phrase is reserved for states in a state enumeration sentence. Another option is "stands for," but it does not really capture the semantics of generalizes.

Any number of specializations is possible. The following is an example of three specializing objects:

Cucumber is a **Vegetable**.
Tomato is a **Vegetable**.
Carrot is a **Vegetable**.

The reserved phrase "is a" (or "are" for plural) denotes the gen-spec relation from the reverse, or bottom-up direction, from the specialized thing to the generalizing thing. Therefore these sentences are called specialization sentences. When the OPL sentences above are expressed from the forward, top-down direction, they sound much less natural:

> **Vegetable** generalizes **Cucumber.**
> **Vegetable** generalizes **Tomato.**
> **Vegetable** generalizes **Carrot.**
> *(Generalization sentences)*

Therefore the default gen-spec direction is the bottom-up direction with its reserved phrase "is a". We can combine the three specialization sentences above into one:

> **Cucumber, Tomato,** and **Carrot** are **Vegetables.**
> *(Plural specialization sentence)*

Note that in this multiple specialization sentence the phrase "is a" has been replaced by "are." In summary, the bottom-up version of the gen-spec, which uses the "is a" or "are" form, is the most natural. It is therefore adopted as the default way of expressing the gen-spec relation. Hence, in the above **Camera** example, rather than "**Camera** generalizes **Digital Camera** and **Analog Camera.**" the correct OPL sentence is: "**Digital Camera** and **Analog Camera** are **Cameras.**"

Generalization-Specialization is a transitive relation, meaning that if **A** is a **B**, and **B** is a **C**, then **A** is a **C**. More concretely, consider the following two specialization sentences:

> **Tomato** is a **Vegetable.**
> **Vegetable** is a **Plant.**

Since Generalization-Specialization is transitive, we can deduce that:

> **Tomato** is a **Plant.**

8.1.2 Process Specialization

Not only objects are subject to Generalization-Specialization. The same relation applies to processes as well. An example is:

> **Boiling** is **Cooking.**
> **Frying** is **Cooking.**
> **Grilling** is **Cooking.**
> *(Process specialization sentences)*

In order to follow the English grammar, the process specialization sentence is slightly different than the (object) specialization sentence in that instead of the reserved phrase "is a," the reserved word "is" is used. A slight difference also exists between the multiple (object) gen-spec sentence and the following multiple process gen-spec sentence. The difference is that in the multiple (object) specialization sentence the generalizing object is plural, as in **Vegetables**; in multiple process specialization sentence however, it is singular, as in **Cooking**, since nouns ending with "**ing**" cannot be plural.

Boiling, Frying, and **Grilling** are **Cooking.**
(Plural process specialization sentence)

Specializations of objects and processes can be combined to specify procedural links between the object and process specializations. Figure 8.2(a) shows **Cooking Tool** as an instrument of **Cooking**. Figure 8.2(b) shows three specializations of both **Cooking Tool** and **Cooking**. Each **Cooking Tool** specialization is an instrument of a specialization of **Cooking**.

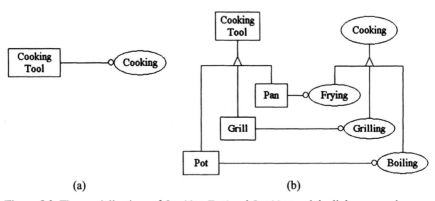

(a)   (b)

Figure 8.2. The specializations of **Cooking Tool** and **Cooking** and the links among these specializations

Figure 8.3 shows a case of link under-specification and a case of link over-specification. The single instrument link from **Cooking Tool** to **Cooking** Figure 8.3(a) result in an under-specified system, since any **Cooking Tool** could be considered as instrument of any **Cooking** process. On the other hand, leaving the generalizing instrument link along with the three specialized links, as in Figure 8.3(b), results in an over-specified system, since the top-level link is now redundant. The right amount of specification is provided in Figure 8.2(b). It is neither under-specified nor over-specified. As we shall see, under-specification and over-specification occur also with structural links.

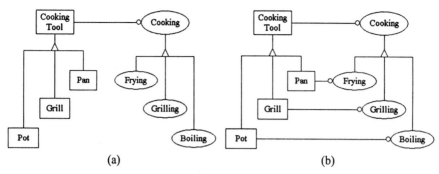

Figure 8.3. Link under- and over-specification. (a) The instrument link from **Cooking Tool** to **Cooking** is under-specified since the links among the specializations are not specified. (b) The instrument link from **Cooking Tool** to **Cooking** is over-specified since the links among the specializations are specified but the link from **Cooking Tool** to **Cooking** was not removed.

8.2 Inheritance

The most prominent immediate benefit gained from using the gen-spec relation is the inheritance it induces. In OO programming languages and design methods, the meaning of inheritance is that attributes, and to some extent also operations, of the generalizing object are inherited to the specialized objects. In OPM, the effect of inheritance is stronger, as it includes inheriting structural and procedural relations, as well as states. The example below shows some of these traits of OPM inheritance.

8.2.1 Feature Inheritance

Figure 8.4 is an OPD of a **Camera**, with the following corresponding OPL paragraph.

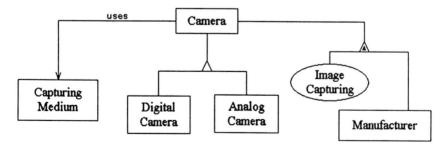

Figure 8.4. Camera and its **Analog Camera** and **Digital Camera** specializations

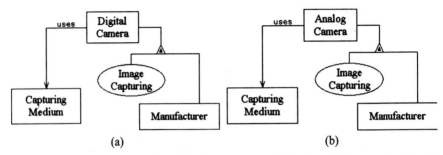

Camera uses Capturing Medium.
Camera exhibits Manufacturer and Image Capturing.
Digital Camera and Analog Camera are Cameras.

(a)   (b)

Figure 8.5. Digital Camera and **Analog Camera** are specializations of **Camera**, therefore it replaces **Camera** of Figure 8.4 in (a) and (b), respectively.

Since, as Figure 8.4 shows, **Digital Camera** and **Analog Camera** are specializations of **Camera**, we replaced **Camera** with its **Digital Camera** and its **Analog Camera** specializations in Figure 8.5(a) and Figure 8.5(b), respectively. This is the basic semantics of inheritance: the specializing thing, the inheritor, inherits features from the generalizing thing, the ancestor. In OPM not only features are inherited; links and states are inherited as well. The inheritor can therefore replace the ancestor. The substitution of the inheritor with the ancestor is expressed in the two OPL paragraphs below. The OPL paragraph of Figure 8.5(a) is:

Digital Camera uses Capturing Medium.
Digital Camera exhibits Manufacturer as well as Image Capturing.

The OPL paragraph of Figure 8.5(b) is:

Analog Camera uses Capturing Medium.
Analog Camera exhibits Manufacturer as well as Image Capturing.

What we see from this OPL script, is that **Digital Camera** and **Analog Camera** inherit not only the features of **Camera**, which are the attribute **Manufacturer** and the operation **Image Capturing**. They also inherit the tagged structural relation **uses** from **Camera** to **Capturing Medium**. Moreover, not only structural relations are inherited; procedural relations are inherited as well. The inheritor may have more features, links, or states. Some of the features, links, and states of the inheritor differ from those of the ancestor. As we shall see, in such cases overriding can be applied.

8.2.2 Structural Relations Inheritance

Consider the OPD in Figure 8.6, in which we specify the parts of **Camera** and the specializations of **Capturing Medium**. The corresponding additional sentences are:

> **Camera** consists of **Lens, Body,** and **Image Capturing Mechanism.**
> **Diskette** and **Film** are **Capturing Media.**

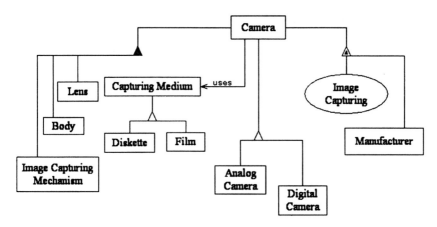

Figure 8.6. The parts, specializations and features of **Camera** are specified along with the specializations of **Capturing Medium**.

This implies that:

> **Digital Camera** consists of **Lens, Body,** and **Image Capturing Mechanism.**
> **Analog Camera** consists of **Lens, Body,** and **Image Capturing Mechanism.**

In other words, the parts **Camera** consists of are inherited to the two **Camera** specializations. Not only the aggregation structural relation is inherited. Any tagged structural relation, such as **uses** is inherited. Since the tagged relation **uses** links **Camera** to **Capturing Medium**, when we specify the specializations of **Camera** and **Capturing Medium** without taking care of the structural relation **uses**, we introduce under-specification. This under-specification, encountered earlier, stems from the fact that the structural relation **uses** from **Camera** to **Capturing Medium** does not specify which **Camera** specialization (**Analog Camera** or **Digital Camera**) uses which **Capturing Medium** specialization (**Diskette** or **Film**). If we take this OPD "on face value," it looks like both **Analog Camera** and **Digital Camera** use both **Film** and **Diskette**. However, we know that this is not the case. To set this straight, we need to specify which **Camera** specialization **uses** which **Capturing Medium** specialization. The OPD of Figure 8.7(a) takes care of this:

> **Camera uses Capturing Medium.**
> **Digital Camera** and **Analog Camera** are **Cameras.**
> **Diskette** and **Film** are **Capturing Media.**
> **Digital Camera** uses **Diskette.**
> **Analog Camera** uses **Film.**

Now, however, we are facing redundancy: the **uses** relation from **Camera** to **Capturing Medium** is an unneeded over-specification. The fact that **Camera uses Capturing Medium** is reflected by the two specialized sentences "**Digital Camera uses Diskette.**" and "**Analog Camera uses Film.**" In the OPD of Figure 8.7(b), we therefore eliminate this link. Note what happens when the specializations of **Camera** or of **Capturing Medium** (or of both, of course), are folded, i.e., eliminated from the OPD along with their gen-spec symbol, as explained in Chapter 8. The two specialized **uses** tagged structural links are replaced by the single **uses** link from **Camera** to **Capturing Medium**.

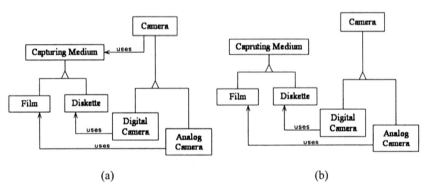

(a)   (b)

Figure 8.7. Inheriting the structural relation **uses**. (a) Matching **Camera** specializations to **Capturing Medium** specializations yields an over-specified system. (b) Eliminating the redundant **uses** relation from **Camera** to **Capturing Medium** removes the over-specification.

8.2.3 Procedural Link Inheritance

Inheritance due to Generalization-Specialization applies to procedural links as well as to structural relations and states. Figure 8.8 demonstrates procedural link inheritance. Each OPD at the top is semantically equivalent to that below it. In Figure 8.8(a), each specialization of the object **Cooking Tool** inherits the instrument link from **Cooking Tool**, while in Figure 8.8(b), each specialization of the process **Cooking** inherits the instrument link to **Cooking**. The OR logical operator next to **Cooking** denotes that any combination of **Cooking Tools** can be an instrument for **Cooking**:

> **Cooking** requires **Pan, Grill,** or **Pot.**

Likewise, the OR logical operator next to **Cooking Tool** denotes that any combination of **Cooking** types can use **Cooking Tool**:

> **Frying, Grilling, or Boiling** require **Cooking Tool.**

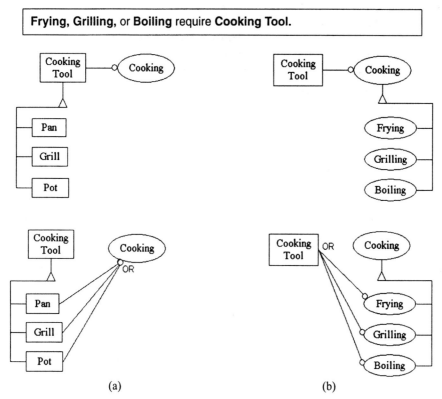

(a)   (b)

Figure 8.8. Procedural link inheritance: Each OPD at the top is semantically equivalent to that below it. (a) Each **Cooking Tool** specialization inherits the instrument link from **Cooking Tool**. (b) Each **Cooking** specialization inherits the instrument link to **Cooking**.

Only when both the specializations of **Cooking** and **Cooking Tool** are specified, as in Figure 8.2(b), it is possible to show which **Cooking Tool** specialization is an instrument to which type of **Cooking**. The OPL paragraph is:

> **Frying, Grilling,** and **Boiling** are **Cooking.**
> **Pan, Grill,** and **Pot** are **Cooking Tools.**
> **Frying** requires **Pan.**
> **Grilling** requires **Grill.**
> **Boiling** requires **Pot.**

The last three sentences can be joined into one:

> **Frying, Grilling,** and **Boiling** require **Pan, Grill,** and **Pot,** respectively.
> *(Respective instrument sentence)*

A similar pattern is applicable to other procedural links. An example that combines a respective agent sentence and a respective instrument sentence follows.

Pilot, Sailor, and **Driver** are **Occupations.**
Airplane, Boat, and **Truck** are **Transportation Systems.**
Flying, Sailing, and **Driving** are **Transporting.**
Pilot, Sailor, and **Driver** handle **Flying, Sailing,** and **Driving,** respectively.
  (Respective agent sentence)
Flying, Sailing, and **Driving** require **Airplane, Boat,** and **Truck,** respectively.
  (Respective instrument sentence)

8.2.4 State Inheritance

In OPM, states are inherited too. Prior to the **Image Capturing** process in the **Camera** example, the **Capturing Medium,** which the **Camera uses,** is **blank.** After the process **Image Capturing** occurs, **Capturing Medium** is **recorded.** Hence, **blank** and **recorded** are two states of **Capturing Medium.** Consider the OPD in Figure 8.9 and the following two OPL sentences:

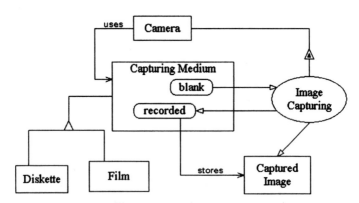

Figure 8.9. State inheritance: **Diskette** and **Film** inherit the states and the input and output links to and from **Image Capturing.**

Capturing Medium can be **blank** or **recorded.**
Film and **Diskette** are **Capturing Mediums.**
Image Capturing changes **Capturing Medium** from **blank** to **recorded.**

Combining the first two OPL sentences above, we can deduce that "**Diskette** can be **blank** or **recorded.**" and that "**Film** can be **blank** or **recorded.**" In other words, the states **blank** and **recorded** of **Capturing Medium** were inherited to both **Diskette** and **Film.** In general, the states of a generalizing object are inherited to its specializations. Inheritance from **Capturing Medium** to **Film** and **Diskette** applies not only to the states and structural relations of **Capturing Medium,** but also to the procedural links to and from the process **Image Capturing.**

Since **Film** and **Diskette** are **Capturing Mediums,** we could substitute **Capturing Medium** by each one of its two specializations, as was done in Figure 8.10, and deduce the following two OPL sentences:

> **Image Capturing** changes **Film** from **blank** to **recorded.**
> **Image Capturing** changes **Diskette** from **blank** to **recorded.**

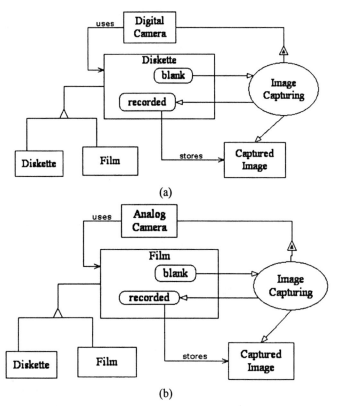

(a)

(b)

Figure 8.10. State inheritance: (a) For **Digital Camera, Capturing Medium** of Figure 8.9 is substituted by **Diskette.** (b) For **Analog Camera, Capturing Medium** is substituted by **Film.**

8.2.5 State Specialization

States of specialized objects may be specializations of the ancestor states. To show this on our **Camera** example, we note that camera film is used for general photography. The silver-halide compounds in the emulsion layer of the film are *exposed* to light and react with it to produce a negative through chemical reactions (Oregon

State University Archives, 2001). Digital cameras, on the other hand, *record* the image through an array of Charged Couple Device (CCD) elements, called pixels (Archeology World, 2001). Through analog-to-digital signal conversion, the signal from the CCD quantizes the individual pixel radiance into a positive value (Lillesand and Kiefer, 1994), which is directly recorded on a magnetic storage medium.

In our example, the image storage medium is the **Diskette**. It is appropriate to specify that both **Film** and **Diskette** can be **blank**, and that **Diskette** can be **recorded**. However, **recorded** is not the most appropriate state for the name of the state of **Film** that has been used to capture an image by exposing it to light. We wish to distinguish between **exposed**, which is a state of **Film**, and **recorded**, which is a state of **Diskette**. To express this, in the OPD of Figure 8.11(b) we have added a Generalization-Specialization symbol, which expresses the fact that **exposed Film** is a (specialization of) **recorded Capturing Medium**. This is *state specialization*: The state **exposed** of **Film** is a specialization of the state of **Capturing Medium**. We say that **exposed** *overrides* **recorded**. First, we need to specify that it is a state of **Film**.

Film can be **exposed**.

The OPL syntax of a state specialization sentence follows.

Exposed Film is a **recorded Capturing Medium**.   *(State specialization sentence)*

Since for **Film** the state **exposed** overrides **recorded**, the OPL sentence for **Film** is:

Image Capturing changes **Film** from **blank** to **exposed**.

Based on state inheritance, Figure 8.10 also shows that:

Recorded Diskette stores Captured Image. **Exposed Film stores Captured Image.**

A human (and a well-designed system) should be able to deduce the OPL sentence "**Film** can be **blank** or **exposed**." rather than "**Film** can be **blank** or **recorded**." Note that since **exposed** is a name of a state, it should be non-capitalized, but because it is the first word in the sentence, it has to be capitalized. An unsophisticated program may therefore consider **Exposed Film** as a new object. The state-specified structure sentence above could have been phrased using less natural but more easily parsable syntax as "**Film** at state **exposed** overrides **Image**

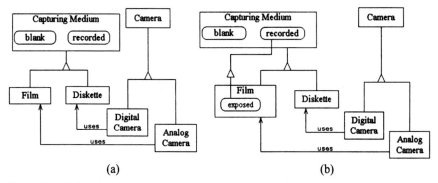

Figure 8.11. Inheriting the states of **Capturing Medium**. (a) Enumerating the specializations of **Capturing Medium**, which are inherited to both **Film** and **Diskette**. (b) **Exposed Film** is a specialization of **recorded Capturing Medium**.

at state **captured**." A sophisticated program, however, can discern that **Exposed** is a name of a state even though it is capitalized as a first word in a sentence.

As Figure 8.13 shows, both **Exposed Film** and **digitized Diskette** are specializations of **Capturing Medium**:

> **Exposed Film** and **digitized Diskette** are **recorded Capturing Media.**
> *(State specialization sentence)*

8.2.6 Process Specialization

Generalization-Specialization in OPM applies not only to objects, but also to processes. So far, our discussion on inheritance focused on objects. This is the traditional inheritance of the OO approach. Since in OPM processes are equivalent to objects, processes are amenable to specialization and inheritance just as objects.

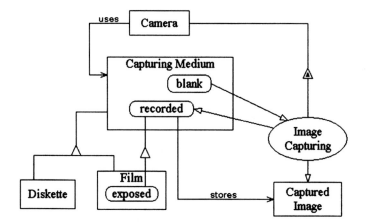

Figure 8.12. Introducing the **Image Capturing** process as an operation of **Camera**

In Figure 8.12, we introduce the **Image Capturing** process into the system. The new OPL sentences are listed in Frame 15.

Camera uses Capturing Medium.
Camera exhibits **Image Capturing.**
Capturing Medium can be **blank** or **recorded.**
Diskette and **Film** are **Capturing Media.**
Film can be **exposed.**
Exposed Film is a recorded Capturing Medium.
Image Capturing changes **Capturing Medium** from **blank** to **recorded.**
Image Capturing yields **Captured Image.**
Recorded Capturing Medium stores Captured Image.

Frame 15. The OPL paragraph of Figure 8.12

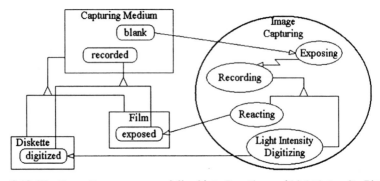

Figure 8.13. The **Recording** process specialized into **Reacting** and **Light Intensity Digitizing**

We model **Image Capturing** as a process that zooms into **Exposing** and **Recording.** **Recording** applied to a **Film** of an **Analog Camera** is different than **Recording** applied to a **Diskette** of a **Digital Camera.** As explained, the former is achieved through a chemical reaction, while the latter involves Analog-to-Digital (A-to-D) Converting. To demonstrate this, consider the OPD in Figure 8.13, in which **Image Capturing** is zoomed into two lower-level processes. The first process, **Exposing**, consumes the input state **blank** of **Capturing Medium**:

Exposing consumes **blank Capturing Medium.**
  (State-specified consumption sentence)

The two other processes, **Reacting** and **Light Intensity Digitizing**, respectively generate the two specialized output states, **exposed** of **Film** and **digitized** of **Diskette**:

Reacting yields **exposed Film.**
Light Intensity Digitizing yields **digitized Diskette.**
  (State-specified result sentences)

The OPL paragraph of Figure 8.13 is listed in Frame 16.

Diskette and **Film** are **Capturing Mediums.**
Capturing Medium can be **blank** or **recorded.**
Film can be **exposed.**
Diskette can be **digitized.**
Exposed Film and **digitized Diskette** are **recorded Capturing Media.**
Image Capturing zooms into **Exposing** and **Recording.**
    (In-diagram in-zooming sentence)
Reacting and **Light Intensity Digitizing** are **Recording.**
Exposing consumes **blank Capturing Medium.**
Exposing invokes **Recording.**
Reacting yields **exposed Film.**
Light Intensity Digitizing yields **digitized Diskette.**

Frame 16. The OPL paragraph of Figure 8.13.

Note that the state specialization sentence in Frame 16 is now "**Exposed Film** and **digitized Diskette** are **recorded Capturing Media.**" The **recorded Capturing Medium** specializes into (is overridden by) both **exposed Film** and **digitized Diskette**.

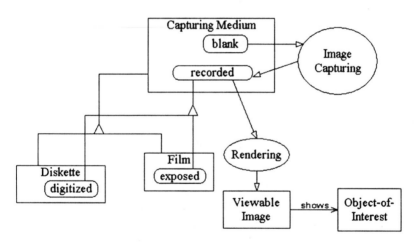

Figure 8.14. Rendering yields **Viewable Image**

In Figure 8.14, we added the process **Rendering**:

Rendering consumes **recorded Capturing Medium.**
    (State-specified consumption sentence)
Rendering yields **Viewable Image.**
Viewable Image shows **Object-of-Interest.**

Figure 8.15 shows that the process **Rendering** can specialize into two processes: **Developing & Printing** and **Displaying**. At the same time, the object **Viewable**

Image specializes into two objects: **Printed Image** and **Displayed Image**, respectively. The following OPL sentences reflect this:

> **Developing & Printing** and **Displaying** are **Rendering**.
> **Printed Image** and **Displayed Image** are **Viewable Images**.
> **Film** can be **exposed** or **developed**.

We added the state **developed** to **Film,** which is specific to **Film** alone. Since it is not shared by **Diskette**, it does not appear as a state of **Capturing Medium**. However, not only do the process **Rendering** and the object **Viewable Image** specialize into their respective objects and processes; the consumption and result links associated with these specializations undergo specializations as well:

> **Developing & Printing** requires **Photo Shop**.
> **Developing & Printing** changes **Film** from **exposed** to **developed**.
> **Developing & Printing** yields **Printed Image**.
> **Displaying** requires **digitized Diskette** and **Computer**.
> **Displaying** yields **Displayed Image**.

Figure 8.15 reflects these changes.

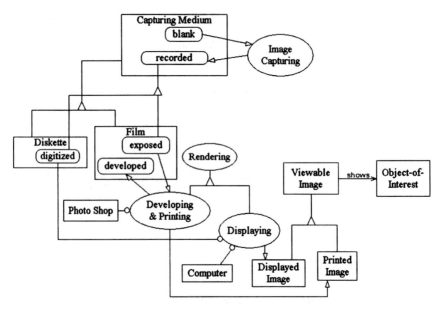

Figure 8.15. The specializations of **Rendering** and **Viewable Image**

Note that we have omitted the consumption links from **recorded Capturing Medium** to **Rendering** and the result link from **Rendering** to **Viewable Image**, because we have a more detailed specification that links the corresponding specializations of these objects and processes. As was the case with the tagged structural link **uses** in Figure 8.7, leaving these links at the generalized level along with the ones at the specialized level would result in over-specification. When these specializations are folded, the omitted links reappear.

8.2.7 Generalization Complexity

Recall that in Chapter 7 we have defined aggregation complexity as an attribute of a thing that specifies whether it consists of parts that are meaningful to the system or are needed to fully understand it. We have also defined exhibition complexity as an attribute of a thing that specifies whether it exhibits features that are needed to fully specify the system. Applying the same train of thought, we define the attribute generalization complexity and its two values, sterile and productive, as follows:

> *Generalization complexity is an attribute of a thing that specifies whether it generalizes one or more things in the system, in which case it is **productive**. Otherwise, the thing is **sterile**, it does not generalize any other thing in the system.*

The terms sterile and productive hint to the inheritance that is induced by the generalization: If a thing generalizes other things, it inherits its features to those things, which means it was productive in giving birth to those things, otherwise it is sterile.

8.3 Qualification

States or values of an attribute that an object exhibits can qualify objects as specializations of the exhibiting object. This is expressed through the qualification link. An example of qualification is shown in the OPD in Figure 8.16 and the OPL paragraph that follows. The qualification link uses the white triangle, but with a different source than gen-spec.[2] In qualification, the tip of the triangle is linked to the value or state and the base of the triangle is linked to the qualified object.

As Figure 8.16 and the OPL paragraph that follows show, each **Size** value of **Mattress** gives rise to a specialized object. The specialization in this case lies in differences in the values of the **Size** attribute of **Mattress**.

For mattress manufacturers and stores, these are four object classes of mattresses. The instances of these classes are the actual mattresses in the store or those sold to customers or ordered from the factory. For the "sleep engineers" in that factory, who are concerned only with the design of the mattresses, and not in each individual physical mattress, the four mattress types can be considered *instances* rather than classes of **Mattress**. This is an example of the relativity of the terms specialization and instance, which will be discussed in more detail later in the chapter.

---

[2]   There are two other types of links that are represented by the white triangle: the role-playing link and the compound state link. They are defined in Chapter 12.

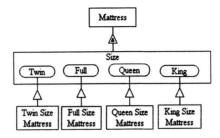

Figure 8.16. The four **Mattress** specializations

Mattress exhibits **Size.**
Values of **Size** are **Twin, Full, Queen** and **King.**
Twin Size Mattress is a **Mattress,** the **Size** of which is **Twin.**
Full Size Mattress is a **Mattress,** the **Size** of which is **Full.**
Queen Size Mattress is a **Mattress,** the **Size** of which is **Queen.**
King Size Mattress is a **Mattress,** the **Size** of which is **King.**
    (Qualification sentences)

Note that unlike the usual gen-spec link, which connects an ancestor generalizing object to a specialized one, the entity to which the tip of the gen-spec triangle symbol is attached is a state rather than an object. Since the link is not between two objects, but between a state ancestor and a specializing object, this is a special type of gen-spec link, called *qualification link*. The white triangle symbol is, therefore, context-sensitive: it depends on the entities it links.

> A ***qualification link*** *is a specialization of the specialization link, in which the generalizing entity is a state.*

A qualification sentence is extracted from an OPD in which an object exhibits an attribute, which has a value that is linked to an object via a qualification link.

8.3.1 Qualification Inheritance

The object that exhibits the attribute is the exhibitor. The qualified objects are connected to the values of an exhibitor's attribute. They are specializations of the exhibitor.

Qualification gives rise to inheritance of attribute values or states. The inheritance that qualification induces is called *qualification inheritance*. To understand this concept, let us examine the OPD in Figure 8.17, which expresses the facts that **Person** is qualified as a **Woman** if the **Gender** of that **Person** is **female,** and as a **Man** if the **Gender** is **male.**

Qualification inheritance is more similar to inheritance in biology than the "regular" inheritance of attributes we have encountered. An organism inherits not only a set of attributes (and behaviors, or operations), but also specific values of the inherited attributes, which are defined by Mendel's laws of classical genetics.

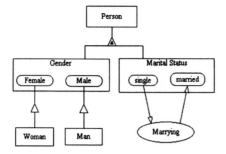

Figure 8.17. Woman is a qualification of **Person** whose **Gender** is **Female**. **Man** is a qualification of **Person** whose **Gender** is male.

Analogously, in our example, **Woman** is a **Person** that inherits the value **Female** of the attribute **Gender**. In the **Mattress** example, the object **King Size Mattress** inherited the value **King Size** of the attribute **Size** of **Mattress**. Qualification inheritance is specific to OPM and does not exist in OO systems. In OO inheritance, only attributes are inherited, but not specific values of these attributes.

Since a qualified object is a specialization of object, it inherits everything that a specializing object would normally inherit. For example, Figure 8.17 shows that since **Man** and **Woman** are specializations of **Person**, each can be **single** or **married**. Moreover, both **Man** and **Woman** inherit the input and output link to and from **Marrying**. **Marrying** affects both **Woman** and **Man** by changing their **Marital Status** from **single** to **married**.

8.3.2 Multiple Qualification Inheritance

A thing can be a specialization of more than one thing at the same time. This gives rise to *multiple inheritance*. To gain insight into multiple inheritance, let us examine Figure 8.18 and its OPL paragraph.

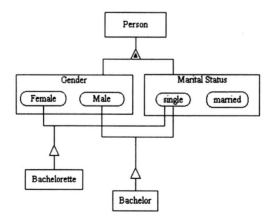

Figure 8.18. Bachelor and **Bachelorette** demonstrate multiple inheritance, inheriting from both the **Gender** values and from **single Marital Status**.

> **Bachelor** is a **Person,** the **Gender** of which is **Male** and the **Marital Status**
>   of which is **single.**
> **Bachelorette** is a **Person,** the **Gender** of which is **Female** and the **Marital**
>   **Status** of which is **single.**

Through qualification inheritance, **Bachelorette** and **Bachelor** acquire the
values **Female** and **Male** from **Gender**, respectively, and both acquire the state **single** from **Marital Status**. In this example, both **Bachelorette** and **Bachelor** inherit
particular values or states from their ancestors.

Multiple inheritance can result from any combination of inheritance and qualification inheritance. Figure 8.19 is an example where regular inheritance and qualification inheritance are mixed: **Wife** and **Husband** are **Spouses**. **Spouse** is a
qualification of **Marital Status** with state **married**. **Wife** and **Husband** are qualifications of **Person** with specified **Gender** values. Both of them inherit from both
Spouse and the corresponding **Gender** value. An incomplete OPL paragraph of
Figure 8.19 is:

> **Person** exhibits **Gender** and **Marital Status.**
> Values of **Gender** are **Female** and **Male.**
> **Marital Status** can be **single** or **married.**
> **Spouse** is a **Person,** the **Marital Status** of which is **married.**
> **Wife** and **Husband** are **Spouses.**

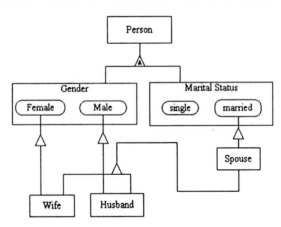

Figure 8.19. Spouse is a qualification of **married Marital Status**. **Wife** and **Husband**,
which are qualifications of **Person** with specified **Gender** values, inherit multiply from both
Spouse and the corresponding **Gender** value.

Through the qualification links between **Wife** and **Female**, between **Husband**
and **Male**, and between **Person** and **Gender**, we get:

> **Wife** is a **Person,** the **Gender** of which is **Female.**
> **Husband** is a **Person,** the **Gender** of which is **Male.**

Generalization-Specialization is a transitive relation. Applying this to our example, since **Wife** is a **Spouse** and **Spouse** is a **Person**, we could deduce that **Wife** is a **Person** even without the qualification through **Gender**. Focusing on **Wife**, for example, we can derive the following OPL sentence:

> **Wife** is a **Person**, the **Gender** of which is **Female**, and a **Spouse**, which is a **Person** whose **Marital Status** is **Married**.

We can eliminate **Marital Status** and **Spouse**, and instead make **single** and **married** implicit states of **Person** directly (as we did with the **on** and **off** states of **Lamp**). We can then qualify **Husband** as a **married Person**, and get a more succinct OPL sentence:

> **Husband** is a **married Person**, the **Gender** of which is **Male**.

8.4 Classification-Instantiation

An instance is an actual object of some class of objects bearing the same name. For example, **Lassie** in Figure 8.20 is an instance of **Dog**. **Dog** is the class of all the dogs, and **Lassie** is an actual exemplar of that class. As depicted in Figure 8.20, the symbol of instantiation is a black inverted triangle inside a larger white triangle.

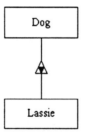

Figure 8.20. The instantiation symbol links a class (**Dog**) to one or more of its instances.

The OPL sentence of Figure 8.20 is:

> **Lassie** is an instance of **Dog**.
>   *(Instantiation sentence)*

The preferred, default direction is the bottom up, which gives rise to an instantiation sentence, such as the one above. An instantiation sentence uses the reserved phrase "is an instance of." In spoken English, the sentence "Lassie is a Dog" is more natural, but the phrase "is a" is reserved for the specialization sentence, so to avoid conflicts and be explicit, the phrase "is an instance of" links an instance with its class in an instantiation sentence. The plural version, used for more than one instance, is "are instances of," as in "**Bach, Beethoven** and **Brahms** are instances of **Composers.**", or in "**Britney Spears** and **Christina Aguilera** are instances of **Pop Singers.**"

As in specialization sentences, the direction of the Classification-Instantiation relation in this sentence is backward, or bottom-up. The Classification-Instantiation sentence in the top-down or forward direction is the classification sentence "**Dog** classifies **Lassie.**" The next example is a more elaborate application of the instantiation concept.

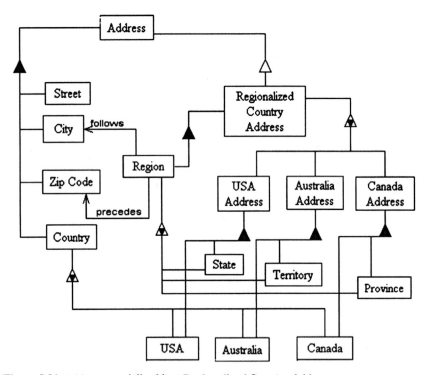

Figure 8.21. Address specialized into **Regionalized Country Address**

Address consists of **Street, City, Zip Code** and **Country,** in this order.
Regionalized Country Address is an **Address.**
Regionalized Country Address consists of **Region.**
Region follows City.
Region precedes Zip Code.
USA, Australia, and **Canada** are instances of **Country.**
State, Territory, and **Province** are instances of **Region.**
USA Address, Australia Address, and **Canada Address** are instances of
      Regionalized Country Address.
USA Address consists of **State** and **USA.**
Australia Address consists of **Territory** and **Australia.**
Canada Address consists of **Province** and **Canada.**

Frame 17. The OPL paragraph of Figure 8.21.

Consider the object **Address** in an automated postal office service, as presented in Figure 8.21. An **Address** consists of **Street, City, Zip Code** and **Country**, in this order. In some "big countries," including **USA, Canada**, and **Australia**, the **Address** includes a **Region**. We define **Regionalized Country Address** as a specialization of **Address**. Unlike a "regular" (non-regionalized) country, **Regionalized Country Address** includes a **Region** following **Zip Code** and preceding **Country** in the **Address** hierarchy. Instances of **Countries** that exhibit **Regionalized Address** include **USA, Australia**, and **Canada**. Instances of **Regions** include **State, Territory**, and **Province**. The instances of **Region** are **State** for **USA, Province** for **Canada**, and **Territory** for **Australia**. **USA Address, Australia Address**, and **Canada Address** are instances of **Regionalized Country Address**. **USA Address** consists of **State** as **Region** and **USA** as **Country**. These ideas are accurately reflected in both the OPD in Figure 8.21 and in its corresponding OPL paragraph in Frame 17.

8.4.1 Classes and Instances

The things we have encountered while discussing Generalization-Specialization are *classes* of things, either object classes or process classes. When we talked about objects, we were actually referring to a typical example of its object class, a pattern of objects from which objects could be generated. We have not discussed instances. The following OPM definitions of class and instance are more general than their OO counterparts, as they refer to things rather than to objects.

> A *class* is a template of a thing.
> An *instance* of a class is an incarnation of a particular identifiable member of that class.

Since a **Thing** is an **Object** or a **Process**, **Class** specializes into an **Object Class** and a **Process Class**. Likewise, **Instance** specializes into an **Object Instance** and a **Process Instance**. An **Object Instance** is an incarnation of the pattern specified by the **Object Class** and a **Process Instance** is an incarnation of the pattern specified by the **Process Class**.

The template that the class defines includes everything that is inherited. As we have seen, in OPM it means that not only features, but also structural relations and procedural link are included. For object classes it also includes values and states. Unlike a specialized class, an instance cannot exhibit any feature that its class does not exhibit, nor can an instance of an object be at a state that is not a state of its class.

An object instance can be uniquely identified in the system, so at any given point in time it is possible to examine it and specify the states or values of its attributes.

8.4.2 The Relation Between Instantiation and Specialization

Generalization-Specialization is a transitive structural relation that gives rise to a hierarchy tree. Each level in the hierarchy contains specializations of the level above it. The "leaves" of that hierarchy are the *instances* of the class. Thus we can say that instantiation is a special case of specialization, which, in the context of the system under study or development, cannot be specialized further. To see an example, consider Figure 8.22 and its following OPL paragraph.

Figure 8.22 shows a specialization hierarchy that starts with **Car** as its top level and presents increasingly specialized object classes until it gets to **Mom's Car**. This is the first object that is physical and unique. It has a **VIN** (vehicle identification number) that uniquely identifies it, and at any given moment the values or states of all its attributes, such as **Color, Location, Mileage** and **Speed**, can be specified.

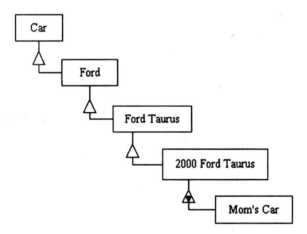

Figure 8.22. The specialization hierarchy of **Car** all the way to the instance **Mom's Car**

Ford is a **Car.**
Ford Taurus is a **Ford.**
2000 Ford Taurus is a **Ford Taurus.**
        (Specialization sentences)
Mom's Car is an instance of **2000 Ford Taurus.**
        (Instantiation sentence)

8.4.3 The Relativity of Instance

Like many other concepts we have encountered, the term instance is relative to the system of discourse. What for a certain system is considered instance of a class, can for another system be just a sub-class of a super-class. To demonstrate this, let us look at a few examples from the world of cars. We have seen that **Ford Taurus 2000** is an object class of all the instances of cars of model **Ford Taurus** manufactured in the year **2000**. Suppose that the system we are now concerned with is a system for comparing and evaluating cars of model year 2000. One of the instances in this system is **Ford Taurus 2000**, and it is an instance of the object class **Model Year 2000 Car**. Physical cars with specific **VIN** do not exist and have no meaning in this system. In the gen-spec hierarchy tree, **Ford Taurus 2000** is one of the leaves: it has no further specializations beneath it.

As another example, consider a national highway system, in which the system architects are interested in the various types of vehicles that use the roads. What matters to them about the vehicles are their size, weight, average speed, and average annual distance each type of vehicle travels. The designers of this system therefore decided to categorize vehicles into three types: cars, trucks and buses. While these three types are specializations of vehicle, for the system under consideration they are also the three instances of the object class vehicle. The architects are not interested in each individual car, bus or truck, so the number of each vehicle type, its average speed, mileage, etc. are attributes of vehicle that are inherited to its three instances. Inheritance of features from a class to its instances is exactly the same as the inheritance of features from a super-class to its sub-class anywhere along the Generalization-Specialization hierarchy. The only difference is that an instance cannot have further specializations, because it is at the bottom of the hierarchy.

Consider now a different system of the **Motor Vehicles Taxation Office** in some country, which, for taxation purpose differentiates between **Taxation Classes** of **Motor Vehicles** as follows: **Commercial Van, Sedan, Collector Car, Sports Utility Vehicle**, and **Luxury Car**. For this system, cars are differentiated into these types based on their **Market Value** and **Application**. Furthermore, the system maintains and constantly updates a list of each **Vehicle Manufacturer** and each **Vehicle Model** by **Year Model**, with an indication of which **Vehicle Model** belongs to which **Taxation Class**. Here, the **Taxation Class** is an attribute of **Vehicle**. **Commercial Van, Sedan**, etc., are values of the **Taxation Class** attribute of **Vehicle**. The instances of the class **Vehicle** in this system are the various **Year Models**, because the system is only concerned with setting tax levels on cars by **Taxation Class** and does not care about individual cars. Finally, consider a car dealership. Here, of course, each individual car has its own record, including its VIN, make, model, year, owner, etc. This is the "classical" case of instance, where each instance is a physical entity with its unique identifier. However, as we have seen, instances can be informatical, such as car models, vehicle types or records in a file.

In summary, we have seen that the term instance is relative to the system under consideration and that an instance in one system may be a class that has instances or that further recursively specializes into more refined classes, which ultimately have instances.

8.4.4 Instance Qualification

Figure 8.23 shows an attribute hierarchy of **Car, Manufacturer, Model**, and **Model Year**. The qualification pattern is applied to instance, giving rise to an instance qualification sentence, demonstrated in Frame 18.

Car exhibits **VIN** and **Manufacturer**.
Manufacturer exhibits **Model**.
Model exhibits **Model Year**.
Mom's Car is an instance of **Car**, the **VIN** of which is **8XCV7246**, the **Manufacturer** of which is **Ford**, the **Model** of which is **Taurus**, and the **Model Year** of which is **2000**.
(Instance qualification sentence)

Frame 18. Instance qualification demonstrated

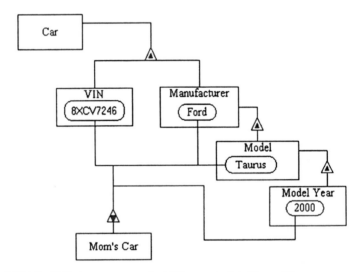

Figure 8.23. The OPD that represents the OPL paragraph in Frame 18

Instance qualification and qualification relations are very similar in both syntax and semantics. The difference in the OPD is that the qualifying state is linked to the qualified instance via the Classification-Instantiation symbol rather than via the Generalization-Specialization one. The corresponding difference in the OPL is that an instance qualification sentence contains the reserved phrase "is an instance of," as in **Mom's Car** is an instance of **Car**, the **VIN** of which is **8XCV7246**.

Figure 8.23 also shows that attribute values can be modeled as instances. For example, **Ford** is an instance of **Manufacturer** can replace Value of **Manufacturer** is **Ford**. Similarly, **Taurus** is an instance of **Model** could replace Value of **Model** is **Taurus**.

8.4.5 Process Instances

As noted earlier in this section, OPM instantiation applies not just to objects but also to processes. The processes we have encountered so far are actually process classes: they are patterns of happenings that involve object classes.

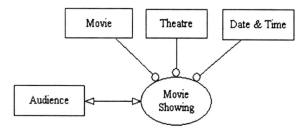

Figure 8.24. Movie Showing as an example of a process class

A process instance is a concrete occurrence of a process class, whose preprocess and postprocess objects are themselves object instances. In particular, a process instance has a time stamp, a specific date and time at which the process started or ended. Figure 8.24 depicts **Movie Showing** as an example of a process class, with **Movie, Theatre,** and **Date & Time** as instruments and **Audience** as the affectee. Figure 8.25 shows **Gone With The Wind Gala Premiere Showing** as a process instance of the **Movie Showing** process class. **Gone With The Wind** is an instance of **Movie, Atlanta Theatre** is an instance of **Theatre,** and **December 15 1939 8:00 PM** (Dirks, 2001) is the value of **Date & Time** at which the process instance took place.

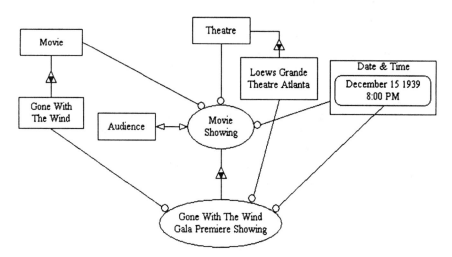

Figure 8.25. Gone With The Wind Gala Premiere Showing as an instance of **Movie Showing** process class

8.4.6 Classification Complexity

So far, we have defined a complexity attribute for each one of the fundamental structural relations we discussed. *Aggregation complexity* is an attribute of a thing that specifies whether it consists of parts. *Exhibition complexity* is an attribute of a thing that specifies whether it exhibits features that are needed to fully specify the system. *Generalization complexity* is an attribute of a thing that specifies whether it generalizes one or more things in the system. Classification-Instantiation, the last of the four fundamental structural relations, is no exception. Recall that when we say that a thing classifies other things, we mean that the thing is a template from which instances of that class are generated. We define the attribute classification complexity and its two values, abstract and concrete, as follows:

> *Classification complexity is an attribute of a thing that specifies whether the thing classifies one or more things in the system, in which case it is concrete. Otherwise, the thing is abstract, it does not classify any other thing in the system.*

The terms abstract and concrete were chosen because are used in a very similar way in the world of object orientation: An abstract (object) class is a class that has no instances, while a concrete class is a class that does have instances. An abstract class is usually productive, i.e., it has specializations. If the specialization is sterile, i.e., it has no specializations, then it should be concrete. A concrete class can be productive or sterile.

8.5 Modifiers and Instances

From a syntactic viewpoint, it may be tempting to think of parts of speech as the syntactic counterparts of OPM entities. In particular, objects and processes can be thought of as nouns and verbs, respectively. As we have seen, however, this is not quite the case. Both objects and processes are nouns, but while processes are related to nouns that are usually derived from verbs, objects are nouns that usually do not have a verb root. In an analogous manner, modifiers, i.e., adjectives and adverbs, are related to OPM features, i.e., attributes and operations. Simplistically, adjectives might be thought of as the parallels of attributes, while adverbs, the parallels of operations. However, the relations are not so straightforward. Now that we know about instantiation we can explain the relations between modifiers, features, and instances.

8.5.1 Natural Language Modifiers and Shortcuts

Tacitly building on the assumption that humans have accumulated a rich knowledge base throughout their lifetime, natural languages provide for a variety of shortcuts. For example, in natural language we simply say "This person is heavy" rather than "The value of the weight attribute of this person is heavy" or "This person is characterized by a heavy weight." As another example, the sentence "This is a complicated task" is shorter and far more natural than the sentences "The complexity attribute value of this task is complicated" or "This task is characterized by a complicated complexity."

Using two types of modifiers, adjectives and adverbs, while referring to instances and specializations, natural languages tend to skip the name of the attribute and indicate only the name of the attribute value or state that applies to the instance being referred. The attribute value is explicit while the attribute itself is left implicit in the sentence. The human processor of the natural language unconsciously infers the missing attribute of which only the value is given. Since OPM is intended to be understood not only by humans, but also amenable to compilation by machines, implicit assumptions of prior knowledge are not acceptable. Just as programming languages need objects, data members and methods to be declared, so does OPM require that knowledge be explicitly specified. As the following two sections compare adjectives with attributes and adverbs with operations, reference is made to particular natural language shortcuts and their OPM comparable constructs.

8.5.2 Adjectives and Attributes

The American Heritage Dictionary (1996) defines "adjective" as follows:

> *Adjective: Any of a class of words used to modify a noun or other substantive by limiting, qualifying, or specifying and distinguished in English morphologically by one of several suffixes, such as -able, -ous, -er, and -est, or syntactically by position directly preceding a noun or nominal phrase, such as white in a white house.*

Unlike OPM, natural languages incorporate many tacit assumptions about prior knowledge that provide for useful shortcuts. For example, we all know that houses have colors. What we do not know is what is the color or height of the house instance being referred to. In OPM, no assumptions regarding prior knowledge can be made, so we need to set the stage before we can specify that a certain house is white or tall. OPL sentences are needed to lay down the prior knowledge required before we can make an assertion on some attribute value of an object instance.

A natural language adjective is the value or state of an attribute of an instance. Consider the natural language sentence "The Empire State Building is tall." "Tall" is an adjective of Empire State Building. This sentence is built on humans' prior knowledge that the Empire State Building is an instance of a building, and that each building has a certain height. OPM cannot make such assumption and must lay down the groundwork to enable specifying that the Empire State Building is tall. Figure 8.26(a) is an OPD that describes this.

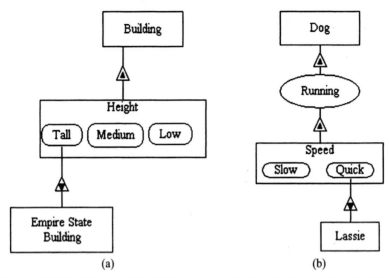

(a)   (b)

Figure 8.26. Modifiers in OPM. (a) The adjective "tall" of the Empire State Building is the value **Tall** of the attribute **Height** of **Building**, of which **Empire State Building** is an instance. (b) The adverb quickly of Lassie's running is the state **quick** of the attribute **Speed** of the operation **Running** of **Dog**, of which **Lassie** is an instance.

The OPL paragraph of Figure 8.26(a) is:

> **Building** Object[3] exhibits **Height.**
> Values of **Height** are **Tall, Medium,** and **Low.**
> **Empire State Building** Object is an instance of **Building** Object, the
> **Height** of which is **Tall.**
> *(Instance qualification sentence)*

Generalizing from this example, adjective is defined below in OPM concepts.

> *An **adjective** is an OPM state or value of an attribute of an object class,*
> *which can be used to describe an instance of that class.*

The natural language *adjective-induced shortcut* avoids mentioning both that a particular object is an instance of its class and that the class exhibits an attribute. Rather, it refers directly to the attribute value of the object instance as an adjective. In the **Building** example, both the fact that **Empire State Building** is an instance of **Building** and that **Height** is an attribute of **Building** are implicit. While processing

---

[3]   Recall that according to OPM's naming conventions, since the object **Building** ends with **ing**, we must append the reserved word **Object** to it to distinguish it from the process **Building.**

the sentence "The Empire State Building is tall," people unconsciously complete these facts in their minds.

8.5.3 Adverbs and Operations

The correspondence between natural language adverbs and OPM operations is somewhat more complicated than that between adjectives and attributes. The American Heritage Dictionary (1996) defines "adverb" as follows:

> *Adverb: (1) A part of speech comprising a class of words that modify a verb, an adjective, or another adverb. (2) A word belonging to this class, such as rapidly in "The dog runs rapidly."*

Consider the natural language sentence "Lassie runs quickly." Just as we all know that buildings have heights, we also know that that Lassie is a name for a dog, dogs run, and that running is characterized by speed, which can be slow or rapid. Therefore, a natural language sentence does not need to state all these facts to set the stage for a statement such as "Lassie runs quickly." Quickly is an adverb in this sentence. We model **quick** as a state of the object **Speed**. **Speed** is an implicit attribute of the process **Running**. **Running**, in turn, is an operation that **Lassie**, being an instance of **Dog**, exhibits. The **Speed** of **Running** of **Lassie** is **quick**. What the natural language sentence tells us is just whether the particular dog being referred to runs slowly or quickly. However, OPM cannot make prior knowledge assumptions, so this sentence needs to be expressed in the OPD in Figure 8.26(b) and the corresponding OPL paragraph in Frame 19.

Dog exhibits **Running**.
Running exhibits **Speed**.
Speed can be **quick** or **slow**.
Lassie is an instance of **Dog**, the **Speed** of **Running** of which is **quick**.

Frame 19. The OPL paragraph of Figure 8.26(b)

Note that we omitted the adverb suffix "ly" from quickly because we refer to a value of the attribute **Speed** of the operation **Running** of **Lassie**. Generalizing from this example, following is a definition of adverb in terms of OPM concepts.

> *An **adverb**, stripped of its "ly" suffix, is an OPM state or value of an attribute of an operation of an object class. Through qualification, it can be used to describe an instance of that class.*

Adverbs give rise to the *adverb-induced shortcut*, which is even more significant than the adjective-induced shortcut discussed earlier. In the sentence "Lassie runs quickly" the facts that **Dog** exhibits **Running**, that **Running** exhibits **Speed**, that **Speed** can be **rapid** or **slow**, and that **Lassie** is an instance of **Dog**, are all made implicit due to this shortcut.

8.6 Specializations of the Involved Object Set Members

The various procedural link types we have encountered in Chapter 5, as well as the various members of the involved object set, are related to each other through Generalization-Specialization. We now apply our knowledge of Generalization-Specialization to model these relations among the members of the involved object set. To conform to the gerund form of process naming, the generic process in the OPD of Figure 8.27 is called **Processing**. As the OPD in Figure 8.27(a) shows, the **Involved Object Set** consists of two types of objects, **Enablers** and **Affectees**. The OPL paragraph that is equivalent to the OPD of Figure 8.27(a) is:

> **SD paragraph:**
> **Involved Object Set** consists of at least one **Transformee** and optional **Enablers**.
> **Enabler** handles **Processing**.
> **Processing** affects **Transformee**.

In the OPD of Figure 8.27(b), **Enabler** and **Transformee** are generalization unfolded. Each of the two specializations of **Enabler** and the three specializations of **Transformee** are connected to **Processing** with a distinctive transforming link. The corresponding OPL paragraph is:

> **Enabler** from **SD** unfolds in **SD1** to specialize **Agent** and **Instrument**.
> **Transformee** from **SD** unfolds in **SD1** to specialize **Affectee, Resultee,** and **Consumee**.
> **SD1** paragraph:
> **Involved Object Set** consists of at least one **Transformee** and optional **Enablers**.
> **Agent** handles **Processing**.
> **Processing** requires **Instrument**.
> **Processing** affects **Affectee**.
> **Processing** yields **Resultee**.
> **Processing** consumes **Consumee**.

The OPL reserved word optional in the first sentence of **SD1** paragraph above means that the **Involved Object Set** may contain no **Enabler**. The reserved phrase at least one means that the **Involved Object Set** must contain at least one **Transformee** (an object that the process transforms). This conforms to the definition of process, according to which there has to be at least one **Transformee**, but the presence of **Enabler** is not mandatory. As explained in Section 6.2, the reserved phrase optional is denoted by "0..m" in the OPD and could be denoted by * for short. The reserved phrase at least one is denoted by "1..m" and could be denoted by + for short.

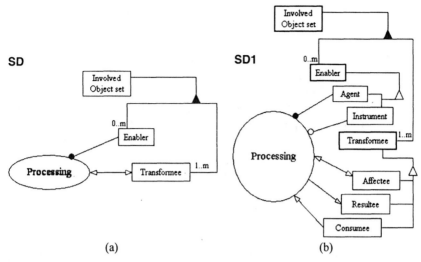

Figure 8.27. The specializations of the **Involved Object Set** members. (a) **SD** showing **Enablers** and **Transformees** as members of the **Involved Object Set**. (b) **SD1** showing the specializations of **Enablers** and **Transformees**

8.7 Non-Comprehensiveness

In Chapter 7 we have encountered the concept of non-comprehensiveness for the Aggregation-Participation. Non-comprehensiveness is not restricted to this relation, but it applies to any fork link, and in particular to the fundamental structural relations, as discussed in this section.

8.7.1 Non-Comprehensiveness of Fundamental Structural Relations

For Exhibition-Characterization, the corresponding reserved phrase is "and additional features." The following OPL sentence is an example:

> **Person** exhibits **Name, Date Of Birth, Address, Marital Status,** and **additional features.**
> *(Exhibition sentence)*

For Generalization-Specialization, the reserved phrase is "and others." The following OPL sentence is an example:

> **Lion, Sheep, Wolf, Elephant,** and others are **Animals.**
> *(Specialization sentence)*

For Classification-Instantiation, the non-comprehensiveness symbol is not needed, because systems are typically designed to accommodate an undetermined large number of instances. A software environment that supports OPM can keep track of all the locations where a specification of a particular fork relation for some object exists. Recall that the teeth set of a fork relation is the set of objects that are linked to the handle of the fork. Wherever the teeth set is not the union of the teeth sets of all the appearances of the object in the various OPDs, the software should add the non-comprehensiveness symbol and update the OPL sentences accordingly. This ensures that the human reader is not misled to think that whatever is presented in a particular OPD is the entire teeth set of the object. The reader should be able to present a query to see all the things that are related to some thing.

8.7.2 Non-Comprehensiveness of States and Values

For states and values, a small blank state or value symbol (possibly with dash or ellipsis in it) is added inside the object rectangle containing a subset of the set of states the object can be at or the values it can assume. Figure 8.28 and Figure 8.29, followed by their respective OPL sentences, demonstrate this.

Figure 8.28. A blank state symbolizes that the states depicted inside the object are not all the states the object can be at.

Product can be **designed, manufactured, tested,** or at other states.

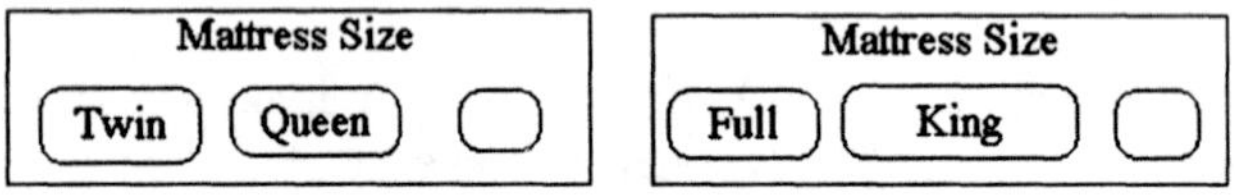

Figure 8.29. A blank value symbolizes that the values depicted inside the object are not all the values the object can assume.

The values of **Mattress Size** are **Twin, Queen,** and others.

Summary

- Generalization-Specialization is a fundamental structural relation between a thing and its types.
- Generalization-Specialization gives rise to inheritance from the generalized thing to the specialized one(s).
- Inheritance is of features (attributes and operations), structural relations and procedural links. For objects, states and values are inherited too.
- OPM processes specialize in a manner similar to objects.
- States of specialized objects can override inherited states.
- Qualification is a type of specialization, in which the specialized object is at a specific value or state of an attribute of the generalizing object.
- Multiple inheritance is inheritance of features from more than one generalizing thing.
- Classification-Instantiation is the fourth fundamental structural relation, denoting the relation between a class of things and an instance of that class.
- A class is a template, from which things that instantiate the class can be generated as members of that class.
- Instance is a relative term. A specialization in one system can be an instance in another.
- A process instance is a particular occurrence of a process at a given point in time that involves a set of preprocess and postprocess object instances.
- There are specific relations between the modifiers adjective and adverb and the attribute values of objects or their operations.
- Non-comprehensive symbols for fundamental structural relations and states denote the fact that not all the descendants or states are depicted in the OPD.

Problems

1. Provide two examples of object specializations and two of process specializations. Specify them in OPDs and OPL.
2. Create a specialization hierarchy of sports games, which would include as a minimum volleyball, basketball, soccer, football, tennis, and baseball. Apply OPM to show what features are common and inherited, and what are game-specific. Use multiple inheritance if needed.
3. Repeat problem 3 for a specialization hierarchy of track and field sport types, which would include at least three types of running, three types of swimming and three types of throwing.
4. What do the two effect links in Figure 8.30 (between **Woman** and **Marrying** and between **Man** and **Marrying**) symbolize?
5. Considering the inheritance of procedural links, are the effect links redundant? Why or why not?

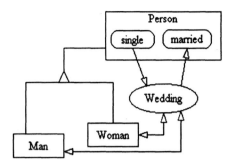

Figure 8.30. Man and **Woman** are affected by **Marrying.**

6. Draw the OPD expressed in the OPL paragraph below.

> **Pilot, Sailor,** and **Driver** are **Occupations.**
> **Airplane, Vessel,** and **Truck** are **Transportation Systems.**
> **Flying, Sailing,** and **Driving** are **Transporting.**
> **Pilot, Sailor,** and **Driver** handle **Flying, Sailing,** and **Driving,** respectively.
> *(Respective agent sentence)*
> **Flying, Sailing,** and **Driving** require **Airplane, Vessel,** and **Truck,** respectively.
> *(Respective instrument sentence)*

7. Give examples of two systems where instances in the first system are specializations in the second. Draw the OPD and write the OPL of these systems.
8. Referring to Section 8.8, draw the OPD from which the sentence

> **Husband** is a **married Person,** the **Gender** of which is **Male.**

is derived.
9. The main types of welding are: (1) Gas – Uses gas flame over metals until molten puddle is formed. Most popular fuels used with oxygen include acetylene and hydrogen. (2) Arc – Two metals are joined by generating an electric arc between a covered metal electrode and the base metal. (3) Oxygen and Arc Cutting – Metal cutting in welding is the severing or removal of metal by a flame or arc. Use OPM to describe these welding types.
10. Specify three instances of electrical appliances at your home. For each one describe its object class with at least three levels of aggregation-participation hierarchy and the operations it performs. Use the instantiation symbol to denote your appliance and provide an attribute that uniquely identifies it.

Chapter 9

Managing Systems' Complexity

> *The field of study of complex systems holds that the dynamics of complex systems are founded on universal principles that may be used to describe disparate problems ranging from particle physics to the economics of societies.* Y. Bar-Yam (1997)
>
> *The human mind, after all, can only juggle so many pieces of data at once before being overwhelmed.* C. Downton (1998)

Complexity is inherent in real-life systems. An integral part of a system development methodology must therefore be a set of tools for controlling and managing this complexity. Like most classical engineering problems, complexity management entails a tradeoff that must be balanced between two conflicting requirements: completeness and clarity. On one hand, completeness requires that the system details be stipulated to the fullest extent possible. On the other hand, the need for clarity imposes an upper limit on the level of complexity of each individual diagram and does not allow for a diagram that is too cluttered or loaded. OO development methods, notably the UML standard (Object Management Group, 2000), address the problem of managing systems complexity by dividing the development of a different model for each one of the important aspects of the system – structure, dynamics, state transitions, etc.

The approach OPM takes is orthogonal, advocating the integration of the various system aspects into a single model. Rather than applying a separate model for each system aspect, OPM handles the inherent system complexity by introducing a number of abstracting-refining mechanisms. These enable presenting and viewing the system, and the things that comprise it, at various detail levels. The entire system is completely specified through its OPD set – a mosaic-like set of compatible OPDs that together provide a full picture of the system being investigated and/or developed. Along with it goes the automatically generated OPL system specification. This chapter elaborates on these complexity management issues and specifies the various abstracting-refining mechanisms.

9.1 The Need for Complexity Management

The very need for systems analysis and design strategies stems from complexity. If systems or problems were simple enough for humans to be grasped by merely glancing at them, no methodology would have been required. Due to the need for tackling sizeable, complex problems, a system development methodology must be equipped with a comprehensive approach, backed by set of reliable and useful tools, for controlling and managing this complexity. This challenge entails balancing two forces that pull in opposite directions and need to be traded off: completeness and clarity. *Completeness* means that the system must be specified to the last relevant detail. *Clarity* means that to communicate the analysis and design outcomes, the documentation, be it textual or diagrammatic, must be legible and comprehensible. To tackle complex systems, a methodology must be equipped with adequate tools for complexity management that address and solve this problem of completeness-clarity tradeoff by striking the right balance between these two contradicting demands.

OPM achieves clarity through abstracting and completeness through refining. Abstracting, the inverse of refining, saves space and reduces complexity, but it comes at the price of completeness. Conversely, refining, which contributes to completeness, comes at the price of loss of clarity. There are "no free meals" – as is typically the case with engineering problems, there is a clear tradeoff between completeness of details and clarity of their presentation. The solution OPM proposes is to keep each OPD simple enough, and to distribute the system specification over a set of consistently inter-related OPDs that contain things at various detail levels. Abstracting and refining are the analytical tools that provide for striking the right balance between clarity and completeness.

Analysis and design are first steps in the lifecycle of a new system, product or project. Creating (sometimes-unconscious) resistance on the side of the prospective audience to accept the analysis and design results, because they look too complex and intimidating, may have an adverse effect of jeopardize the success of subsequent phases of the product development. The severity and frequency of this detail explosion problem calls for an adequate solution to meet the needs of the systems analysis community. A major test of any analysis methodology is therefore the extent to which it provides reasonable tools for managing the ever-growing complexity of analysis outcomes in a coherent, clear, and useful manner. Such complexity management tools are extremely important for organizing the knowledge the architect accumulates and generates during the system architecting process. Equally important is the role of complexity management tools in facilitating the communication of the analysis and design results to other humans, including customers, peers, superiors and system developers down the development cycle road – implementers, testers, etc.

9.1.1 Middle-Out as the De-Facto Architecting Practice

Analyzing is the process of gradually increasing the human analyzer's knowledge about the system's structure and behavior. Designing is the process of gradually increasing the amount of details on the system's architecture, i.e., the structure and behavior combination that enables the system to attain its function. For both analysis and design, managing the system's complexity therefore entails being able to present and view the system at various levels of detail that are consistent with each other. Ideally, analysis and design start at the top and make their way gradually to the bottom – from the general to the detailed. In real life, however, analysis typically starts at some arbitrary detail level and is rarely linear. The design is not linear either. Usually, these are iterative processes, during which knowledge, followed by understanding, is gradually accumulated and refined.

The system architect usually cannot know in advance the precise structure and behavior of the very top of the system – this requires analysis and becomes apparent at some point along the analysis process. Step by step, the analyst builds the system specification by accumulating and recording facts and observations about things in the system and relations among them. Using OPM, the accumulated knowledge is represented through a set of OPDs and their corresponding OPL paragraphs. The sheer amount of details contained in any real world system of reasonable size overwhelms the system architect soon enough during the architecting process. Trying to incorporate the details into one diagram, the amount of drawn symbols gets very large, and their interconnections quickly become an entangled web. This information overload happens even if the method advocates using multiple diagram types for the various system aspects. Because the diagram has become so cluttered, it is increasingly difficult to comprehend it. System architects experience this detail explosion phenomenon on a daily basis, and anyone who has tried to analyze a system will endorse this description. The problem calls for effective and efficient tools to manage this inherent complexity.

Due to the non-linear nature of these processes, linear, unidirectional "bottom-up" or "top-down" approaches are rarely applicable to real-world systems. Rather, it is frequently the case that the system under construction or investigation is so complex and unexplored, that neither its top nor its bottom is known with certainty from the outset. More commonly, analysis and design of real-life systems start in an unknown place along the system's detail level hierarchy. The analysis proceeds "middle-out" by combining top-down and bottom-up techniques to obtain a complete comprehension and specification of the system at all the detail levels. It turns out that even though architects usually strive to work in an orderly top-down fashion, more often than not, the de-facto practice is the middle-out mode of analysis and design. Rather than trying to fight it, we should build software tools that provide facilities to handle this middle-out architecting mode. Such facilities cater also to both top-down and bottom up approaches. Indeed, Systemantica® (Sight Code, 2001) is able to handle these three modes.

During the middle-out analysis and design process, facts and ideas about objects in the system and its environment, and processes that transform them, are being gathered and recorded. As the development proceeds, the system architect tries to *concurrently* specify both the structure and the behavior of the system in order to enable it to fulfill its function. For an investigated (as opposed to an architected) system, the researcher tries to make sense of the long list of gathered observations and to understand their cause and effect relation. In both cases, the system's structure and behavior go hand in hand, and it is very difficult to understand one without the other. Almost as soon as a new object is introduced into the system, the process that transforms it or is enabled by it begs to be modeled as well. By supplying the single object-process model, OPM caters to this structure-behavior concurrency requirement. It enables modeling these two major system aspects at the same time within the same model without the need to constantly switch between different diagram types.

If the OPD that is being augmented becomes too crowded, busy, or unintelligible, a new OPD is created. This descendant OPD repeats one or more of the things in its ancestor OPD. These repeated things establish the link between the ancestor and descendant OPDs. The descendant OPD does not usually replicate all the details of is ancestor, as some of them are abstracted, while others are simply not included. This new OPD is therefore amenable to refinement of new things to be laid out in the space that was saved by not including things from the ancestor OPD. In other words, there is room in it to insert a certain amount of additional details before it gets too cluttered again. When this happens, a new cycle of refinement takes place, and this goes on until the entire system has been completely specified.

9.1.2 Determining the Extent of Refinement

While discussing the four fundamental structural relations, we have seen that each one of them gives rise to a corresponding complexity attribute. Table 10 summarizes the names of the four complexity attributes and the names of their corresponding values. Each complexity attribute has two values. One is the simple and the other is the complex. For example, the Aggregation-Participation fundamental structural relation gives rise to the complexity attribute called **Aggregation Complexity**, whose **Simple** value is **Atomic** and whose **Complex** value is **Compound**. **Atomic** means *a thing with no parts*, while **Compound** means *a thing with parts*.

Table 10. The names of the complexity attributes of the four fundamental structural relations, the names of their values and their meanings

Fundamental Structural		Attribute Values			
Name	Complexity Attribute	Simple		Complex	
		Name	Meaning	Name	Meaning
Aggregation-Participation	Aggregation Complexity	Atomic	*A thing with no parts*	Compound	*A thing with parts*
Exhibition-Characterization	Exhibition Complexity	Dull	*A thing with no features*	Colorful	*A thing with features*
Generalization-Specialization	Generalization Complexity	Sterile	*A thing with no specializations*	Productive	*A thing with specializations*
Classification-Instantiation	Classification Complexity	Abstract	*A thing with no instances*	Concrete	*A thing with instances*

A thing can exhibit a variety of combinations of the complexity attribute values. However, a thing that is both **Dull** and **Productive** makes no sense because if a thing has no feature, its specializations will be meaningless.

Regarding the name **Concrete**, while it may seem that concrete is simple and abstract is complex, following our definitions, abstract specifies no instances, so it contains fewer details, making it simpler than concrete, which does specify the instances.

A frequent problem that a system architect encounters while specifying a system is to what level of detail should each thing be specified from the viewpoint of each one of the four fundamental structural relations. In other words, what should the value of each of a thing's complexity attributes be? What is the right level of refinement for each thing in the system? Each time we refine a thing, we add to it parts, features, specializations, or instances. When each new thing is added, it is simple, because we have not yet had the chance to specify its parts, features, etc. Theoretically, there is always a yet more detailed level of specification than the simple one, at which the thing can be described. This would result in an endless process, which we certainly want to avoid. The specification job must be finite!

The specification level is system-dependent and is determined by the objective of the system under construction. Consider, for example, an **Airplane**. In an air traffic control system, the Participation Complexity of an **Airplane** can be Atomic. In an airplane maintenance system, an **Airplane** is a Compound object, which is decomposed into its parts that may occupy many levels of the participation hierarchy. In a city traffic control system, the instances of **Vehicle** may not be important, so a **Car**'s Classification Complexity would be Abstract, while in a Registry of Motor Vehicles, **Vehicle** instances are essential, so **Vehicle** would be Concrete.

The specification level of a thing in the system should reflect the role that thing plays in that system. Over-specification occurs when the thing is detailed to a higher extent than is required in the system, while under-specification is a result of not providing a level of detail that is sufficient for understanding and/or constructing that thing. The architect should compare the value added to the system specification by adding a detail with the price this adds to the system complexity. Details should be added only as long as their value exceeds their price.

9.1.3 Towards Quantifying Complexity

System complexity has been often referred to as the amount of information required to describe the system – the amount of elements and interconnections (Bar-Yam, 1997) or the Shannon entropy (Lloyd, 1989). The complexity of a system specified through OPM can be quantified using a different approach. Since the entire system is explicitly specified, the amount of elements is known. These can be the entities (objects, processes, and states). The connections are the structural and procedural relations, which can be counted. A number of complexity types can be computed from the OPM model. For example, one can compare the structural complexity to the procedural one by accounting for the number of processes and procedural links to the number of objects and structural links. The reference point can be the theoretical Ultimate OPD or its corresponding OPL script. The number of refinement levels is anther aspect of complexity that can be directly found from the OPM model. Assigning weights for the various complexity factors, a standard measure of complexity can be established.

9.2 Divide and Conquer: By Aspects or by Details?

Rooted in the military art, the decomposition principle, also known as the "divide and conquer" strategy, has been recognized for a very long time and in many domains as an effective means to overcome complexity and enable solving complex problems. The idea is basically to break a complex problem (such as understanding and/or designing a complex system) into smaller, manageable pieces, solve each of them separately and combine the partial solutions to obtain a complete solution.

System development methods have adopted the decomposition principle, either intentionally or not. Most methods apply this strategy by breaking the system into a number of models, each dealing with a different aspect of the system, such as structure, behavior, and function. Each model applies a different set of symbols and concepts, and together they are expected to convey a complete system specification. This *aspect decomposition* is at the heart of standard, state-of-the-art object-oriented development methods like UML (Object Management Group, 2000). As we have seen, OPM's approach is entirely different. A basic principle of OPM is that structure and behavior within a system are so intertwined that effectively separating them is extremely harmful, if not impossible. Therefore, aspect-based decomposition is unacceptable, as it inevitably violates the singularity of the OPM model. The alternative OPM has adopted is *detail decomposition*: rather than decomposing a system according to its various aspects, the decomposition is based on the system's levels of abstraction.

Figure 9.1 shows the two orthogonal divide-and-conquer strategies. In the aspect-based decomposition, two thick, solid, vertical lines separate the structure, behavior and state transition aspects from each other. The thin bidirectional horizontal arrows across these lines symbolize difficult transition among the various models. The detail-based decomposition is represented by the two thin, dashed, horizontal lines that separate the various levels of detail – abstract, detailed and concrete – from each other. The thick bidirectional vertical arrows symbolize easy transition among the detail levels. The diagram is schematic. It by no means implies that horizontally there are only three levels of abstraction in OPM. In fact, this number is not bounded. The diagram should also not be interpreted as if vertically there are only three diagram types in a multi-model method. In fact, the number of diagram types in UML (Object Management Group, 2000) is nine.

Two system decomposition strategies

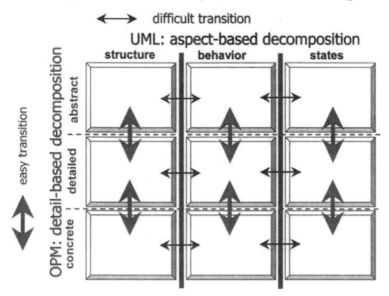

Figure 9.1. The two orthogonal divide-and-conquer strategies

Figure 9.2 illustrates the basic difference between aspect and detail decompositions using a concrete example. The two vertical dotted lines denote the system's aspect-based decomposition: they divide the system into three aspects – structure, behavior and state-transition, which are depicted in a left-to-right order. This decomposition is comparable to that applied in OMT: the structure is OMT's Object model, the behavior is OMT's Functional (DFD-based) model, and the state-transition is OMT's dynamic (statechart-based) model. The horizontal line, on the other hand, denotes the system's detail-level-based decomposition, which OPM's strategy: it divides the system into two detail levels – the abstract level and the detailed level, which are depicted in a top-to-bottom order.

The dotted vertical lines represent the aspect-based decomposition. The left-hand side, Figure 9.2(a), contains the static-structural aspect of the system – the fact that the **Car-Driver Complex** comprises the **Car** and the **Driver**. The dynamic-behavioral aspect, which occupies Figure 9.2(b), the center part of the OPD, represents the fact that the process **Moving** zooms into **Driving** and **Rolling**. Finally, the state transition aspect – the fact that the **Location** attribute of **Car** is changed from **New York** to **Boston,** is on the right hand side, Figure 9.2(c). The horizontal dashed line represents OPM's detail-based decomposition. The top part of the OPD, above the dashed line, shows only three things: the object **Car-Driver Complex, Moving** and **Location.** In the bottom part these things are refined. The parts of **Car-Driver Complex, Car** and **Driver,** are unfolded. The subprocesses of **Mov-**

ing, Rolling and **Driving,** are in-zoomed. Finally, the values of the **Location** attribute, **New York** and **Boston,** are expressed.

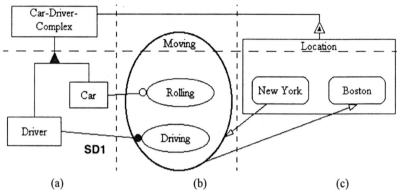

(a)   (b)   (c)

Figure 9.2. The customary aspect-based (vertical dotted line) vs. OPM's detail-based (horizontal dashed line) decomposition: (a) Structure. (b) Behavior. (c) State transition. (top) Abstract. (bottom) Refined.

Figure 9.3 is an OPD resulting from abstracting the three things at the top-level of **SD1,** the OPD in Figure 9.2: **Car-Driver Complex** has been unfolded, **Moving** has been out-zoomed, and the values of **Location** have been suppressed. The outcome of these three abstraction operations is **SD,**[1] the top-level OPD. While containing only three things, **SD** retains the structural and behavioral aspects of the system and the structural and procedural links among them.

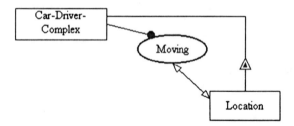

Figure 9.3. A scaled-down OPD resulting from abstracting things in the OPD of Figure 9.2

The OPL paragraph that is equivalent to the OPD in Figure 9.3 is:

> **SD Paragraph:**
> **Car-Driver Complex** exhibits **Location.**
> **Car-Driver Complex** handles **Moving.**
> **Moving** affects **Location.**

---

[1]   Recall from Chapter 2 that **SD** stands for **System Diagram,** the top-level OPD.

The OPL paragraph that is equivalent to the OPD in Figure 9.2 is:

Moving from **SD** zooms in **SD1** into **Driving** and **Rolling**.
Car-Driver Complex from **SD** unfolds in **SD1** into **Car** and **Driver**.

SD1 Paragraph:
Car-Driver Complex exhibits **Location**.
Values of **Location** are **New York** and **Boston**.
Driver handles **Driving**.
Rolling requires **Car**.
Moving changes **Location** from **New York** to **Boston**.

Only the sentence **Car-Driver Complex** exhibits **Location** is identical in the two OPL paragraphs. Other sentences in **SD1** Paragraph are refinements of sentences in **SD** Paragraph. For example, **Moving** affects **Location** from **SD** Paragraph is refined in **SD1** Paragraph to **Moving** changes **Location** from **New York** to **Boston**. The two first sentences provide connections between the two System Diagrams.

9.2.1 Why is Detail Decomposition Good?

OPM's detail-based decomposition is seamless, recursive and selective. Detail decomposition is seamless in the sense that we can relate a detailed thing (object or process) in an OPD to its non-detailed version in another OPD simply by its name. Two things with the same name in different OPDs are the same.[2] Copies of the same thing, possibly at various detail levels, have the same name at all the OPDs in which the thing appears. This enables such operations as refining or splitting an OPD that has become too populated into two OPDs while duplicating just one or a few things that serve as references or hooks for a seamless continuum. To avoid running long links, the same thing can appear more than once in the same OPD, in which case the forward slash (/) symbol immediately precedes all the appearances of the thing's name. However, the use of the slash is not recommended unless its use is absolutely necessary. This slash is not part of the name and is not present in the corresponding OPL sentences.

Recursion is the ability of an operation (function, procedure) to call itself any number of times until some halting condition is met. OPM's refining and abstracting, which are inverses, are recursive. One can recursively apply refining on a thing until the halting condition is satisfied. The halting condition for refining is that the system architect considers the level of detail of the thing to be sufficient for it to be understood and unambiguously implemented. Likewise, one can apply a successive series of abstracting operations on sets of things that resulted from previous refining operations. The effect of abstracting is to decrease the level of detail and

---

[2] We qualify this statement by saying that we need to take care of the scope of the names. Scope is discussed in Chapter 13.

obtain a more compact and succinct description of the system. The halting condition for abstracting is that a single, top-level OPD has been obtained, in which the major things of the system and its environment are clearly represented along with the most important structural relations and procedural links among them.

Recursivability is a property of a thing that is amenable to recursion with respect to some operation. Things in OPM exhibit this recursivability with respect to refining and abstracting. Working middle-out, OPM provides for starting with thing that are at arbitrary abstraction levels. By recursive applications of refining and abstracting, the most detailed level and the most abstract one are obtained, along with the entire spectrum of intermediate levels between these two extremes.

OPM's selectivity implies that neither refining nor abstracting need to be applied to all the things in any given OPD at once. It is possible, and sometimes even recommended or required, that while a subset of one or more of the things in a source OPD is detailed in the destination OPD, another subset of one or more of the things in the source OPD is abstracted in the destination OPD. By so doing, we keep the size of each OPD manageable and avoid ending up with individual OPDs that are too complex to read and understand. This selectivity attribute implies concurrent bidirectional abstracting-refining and makes it possible to generate OPD subsets of the same OPD set that emphasize or focus on a certain portion of a complex system. Selectivity enables detail level variability by providing for a subset of things to be at a detailed, fine granularity level, while leaving the rest at a rather abstract, coarse granularity level that is just enough to enable one to see how the details fit in "the big picture." The result is like embedding islands of detailed things within the vast "system ocean."

The selectivity property comes in handy when a system architect wishes to assign the design of parts of a complex system to several teams of designers. Each team has to know its part to the most detailed level, but need not be bothered with the myriad of details that are only relevant to other teams. It does, however, need to be informed about the general structure and behavior of the entire system. Providing each team with an OPD subset that focuses on the details of the task at hand, while leaving the rest of the system at the "big picture" level, is an adequate solution. The various OPD subsets are somewhat analogous to the different views of a database scheme that are provided to each user according to what she is supposed to see and manage.

The complete system specification is usually not included in a single OPD or even in several OPDs. Recall that the OPD set, introduced in Chapter 2, is the collection of all the OPDs in the system. There is a theoretical, single, grand OPD that contains the complete system specification. However, since the ultimate OPD is too unwieldy to be practically drawn for any system of reasonable size, the system specification is distributed among the OPDs in the OPD set of the system. Details in one OPD need not necessarily be duplicated in all or any of the remaining OPDs in the OPD set. Any two OPDs in the OPD set are consistent with one another as long as there is no contradiction between them. Consistency is maintained even though they may differ in the level of detail or amount of details each OPD contains.

9.2.2 When Should a New OPD Be Created?

An OPD set has to be readable and easy to follow and comprehend. The following rules of thumb are helpful in deciding when a new OPD should be created and in keeping OPDs as easy to read and grasp as possible.

- The OPD should not stretch over more than one page or one average-size monitor screen.
- The OPD should not contain more than 20–25 entities (objects, processes or states).
- Things (objects or processes) must not occlude each other. They are either completely contained within higher-level things, in case of zooming, or have no overlapping area.
- The diagram should not contain too many links.
- A link should not cross the area occupied by a thing.
- The number of links crossing each other should be minimized.

9.3 The Attributes of Scaling

Scaling is the name of the process which can be either abstracting or refining. The scaling process exhibits the following attributes: **Purpose, Mode, Scaled Relation, Primary Operand**, and **Secondary Operand**. The following subsections elaborate on each attribute.

9.3.1 The Purpose of Scaling

Purpose describes the objective, or direction, of carrying out the **Scaling** process. The two opposing values of the **Purpose** attribute are **Elaboration** and **Simplification**. **Refining** is the process, the direction of which is **Elaboration**, while **Abstracting** is the process, the direction of which is **Simplification**. This is expressed in the OPD of Figure 9.4 and the following corresponding OPL paragraph, where qualification sentences best express the relationship between **Scaling** on one hand and **Abstracting** and **Refining** on the other hand. **Refining** and **Abstracting** are inverses of each other.

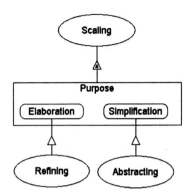

Figure 9.4. Refining and **Abstracting** as qualifications of values of **Purpose** of **Scaling**

Scaling exhibits **Purpose.**
Purpose can be **Elaboration** or **Simplification.**
Refining is **Scaling,** the **Purpose** of which is **Elaboration.**
Abstracting is **Scaling,** the **Purpose** of which is **Simplification.**

The purpose of **Refining** is to provide more details about one or more things in the system. Refining one or more things that appear in an OPD can be done either in the same OPD (in-diagram refining) or in a new OPD (new diagram refining). The same OPD can be used for refining if the refining does not make the OPD too crowded, cluttered or obscured. If this happens, a new OPD is spawned, in which one or more things are refined while other things may be abstracted to preserve the clarity of the new OPD.

The purpose of **Abstracting** is to obtain simpler, clearer, more readable OPD. It is required when the architect notices that the OPD he is working on has become overloaded with details to a point that makes it difficult to read and comprehend. Abstracting means ignoring what is particular or incidental and emphasizing what is general and essential (Pahl and Beitz, 1996). Abstracting yields a new, descendant OPD that presents one or more of the things in the ancestor OPD in a more abstract and compact manner. This abstraction enables the OPD reader to get a more general view of the major things and relations among them before delving into their details.

9.3.2 The Mode of Scaling

Mode is an attribute of **Scaling** that pertains to the graphical mechanism through which **Scaling** is manifested. **Mode** has three values, i.e., there are three scaling modes: **Visibility, Hierarchy,** and **Manifestation.** These are specified in the OPD of Figure 9.5 and the OPL paragraph below.

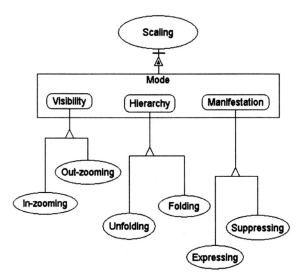

Figure 9.5. The **Mode** attribute of **Scaling** and its qualified processes

---

Scaling exhibits **Mode,** and additional features.[3]
Mode can be **Visibility, Hierarchy,** or **Manifestation.**
Refining is **Scaling,** the **Purpose** of which is **Elaboration.**
Abstracting is **Scaling,** the **Purpose** of which is **Simplification.**
In-zooming and **Out-zooming** are **Scaling,** the **Mode** of which is **Visibility.**
Unfolding and **Folding** are **Scaling, the Mode** of which is **Hierarchy.**
Expressing and **Suppressing** are **Scaling, the Mode** of which is **Manifestation.**

A close relation exists between the two sets of qualified processes. Observe the three pairs of processes, differing by **Mode.** Each pair consists of one **Refining** and one **Abstracting** process. For example, **Folding** is **Abstracting** while **Unfolding** is **Refining.** Both are **Scaling,** with **Hierarchy Mode.** All six **Scaling** processes (under **Mode**) are the actions that the system architect can apply while analyzing or designing the system. The combination of Figure 9.4 and Figure 9.5 is shown in Figure 9.6 and in Frame 20. In the next sections, we will discuss the six different scaling processes.

---

[3]   Notice that the characterization relation is non-comprehensive. This is shown twice: once in Figure 9.5, with the non-comprehensiveness symbol above the triangle, and once in the first sentence of the OPL paragraph. This in itself is a way to manage complexity.

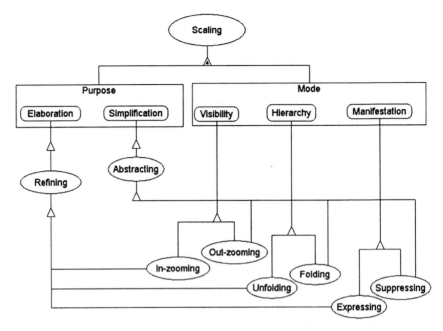

Figure 9.6. The **Purpose** and **Mode** attributes of **Scaling** and their qualified processes

Scaling exhibits **Purpose** and **Mode**.
Purpose can be **Elaboration** or **Simplification**.
Mode can be **Visibility, Hierarchy,** or **Manifestation**.
Refining is **Scaling**, the **Purpose** of which is **Elaboration**.
Abstracting is **Scaling**, the **Purpose** of which is **Simplification**.
In-zooming, Unfolding, and **Expressing** are **Refining**.
Out-zooming, Folding, and **Suppressing** are **Abstracting**.
In-zooming and **Out-zooming** are **Scaling**, the **Mode** of which is
        Visibility.
Unfolding and **Folding** are **Scaling**, the **Mode** of which is **Hierarchy**.
Expressing and **Suppressing** are **Scaling**, the **Mode** of which is
        Manifestation.

Frame 20. The OPL paragraph of Figure 9.6.

The primary operand of a scaling mode is the entity (object, process or state) used by default with that particular scaling mode, while the secondary operand is a thing (object or process) which is rarely used with that mode. Table 11 enumerates the names of the various value combinations of the scaling and purpose attributes.

Table 11. The scaling mode and purpose combinations and their primary and secondary operands

Mode	Purpose		Primary Operand	Secondary Operand
	Refining	Abstracting		
Visibility scaling	In-zooming	Out-zooming	process	object
Hierarchy scaling	Unfolding	Folding	object	process
Manifestation scaling	Expressing	Suppressing	state	–

9.3.3 Controlling Visibility by In- and Out-Zooming

The **Visibility Mode** of **Scaling** operates on a thing by **In-zooming** – zooming into it, and **Out-zooming** – zooming out of it.[4] Zooming into an abstract thing decreases the distance of viewing it such that lower-level things enclosed within the thing become visible. Conversely, zooming out of a refined thing increases the distance of viewing it, such that a set of lower-level things that are enclosed within it become invisible. It is often useful to concurrently apply scaling of two types on two things for the same purpose (refining or abstracting). For example, the process **P2** in Figure 9.7(a) is in-zoomed in Figure 9.7(b), such that its parts, **P2.1** and **P2.2**, are revealed. At the same time, the object **B1** in Figure 9.7(a) is unfolded in Figure 9.7(b), such that its parts, **Part 1** and **Part 2**, are revealed. Conversely, the process **P2** in Figure 9.7(b) is out-zoomed in Figure 9.7(a), such that **P2.1** and **P2.2** are concealed. At the same time, the object **B1** in Figure 9.7(b) is folded in Figure 9.7(a), such that its **Part 1** and **Part 2** are concealed.

Out-zooming results from enclosing a set of related processes with optional intermediate objects and relations among them. Processes can be automatically out-zoomed if they result from in-zooming.

Figure 9.8(a) shows the abstracted **Processing** that is in-zoomed in Figure 9.8(b). Several observations are in order. First, the subprocesses **A Sub-processing**, **B Sub-processing**, and **C Sub-processing**, are *parts* of **Processing**. Hence, in- and out-zooming implicitly assume that Aggregation-Participation is the relation being refined or abstracted. The *scaled relation* of in-zooming and out-zooming is Aggregation-Participation. Second, the objects **Interim Object A** and **Interim Object B** are "encapsulated" within **Processing**. They disappear when **Processing** is out-zoomed. These objects cannot be considered parts of **Processing**,

---

[4] While the names "zooming in" and "zooming out" might be more natural, **In-zooming** and **Out-zooming** are designed so that they end with the gerund form, as proper OPM process names. The other option would be to call them "Zooming-in process" and "Zooming-out process", which is more cumbersome.

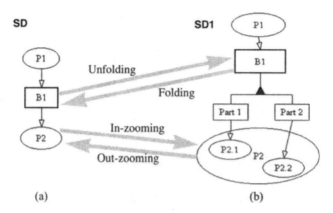

Figure 9.7. Concurrent zoom and hierarchy scaling. (a) **B1** is unfolded and **P2** is in-zoomed in **SD1**. (b) **B1** is folded and **P2** is out-zoomed in **SD**.

because they are of the opposite perseverance. Recall that objects cannot be parts of a process and vice versa. **Interim Object A** and **Interim Object B** are *attributes* of **Processing**. Third, links are refined and abstracted along with the operations on **Processing**. For example, the single agent link in (a) is expanded into two agent links connecting **Agent** with both **A Sub-processing** and **B Sub-processing**.

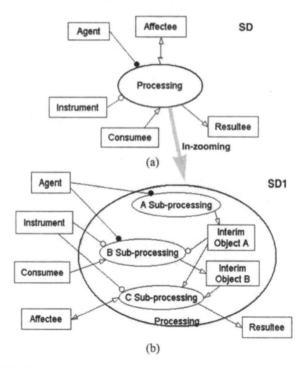

Figure 9.8. The In-zooming process. (a) The abstract **Processing**. (b) The detailed **Processing**.

In-zooming and out-zooming are primarily applied to processes, so, as Table 11 shows, process is the primary operand of these scaling operations. Experience has shown that it is graphically most expressive to depict process refining through in-zooming. By default, in-zooming is applied to processes, but it can also be applied to objects. When a process is zoomed into, lower level subprocesses, which are parts of the zoomed-in process, become visible. While zooming into a process, objects may optionally become visible as well. These objects are attributes of the zoomed in process. Conversely, when an object is zoomed into, its constituent lower level objects are exposed as parts of the zoomed-in object, and processes are optionally exposed as its operations.

9.3.4 The Distributivity of Procedural Links

A procedural link that is attached to the in-zoomed ellipse of a process is considered to be attached to each one of the subprocesses (parts) inside the in-zoomed process. This is demonstrated in Figure 9.9(b), in which **Processing**, zoomed in from Figure 9.9(a), serves as the "graphic parenthesis" for the object Instrument.

The distribution of procedural links implies that the instrument link, attached to the outer **Processing** ellipse, pertains to all three subprocesses of **Processing**: **Inserting, Machining** and **Removing**. This is equivalent to drawing three instrument links, one to each of these three subprocesses. Obviously, there is a saving in graphic symbols and links. Without this convention, one would need to specify many more links, which would cross each other and clutter the diagram. The OPL sentence set equivalent to the OPD set of Figure 9.9 follows.

Both **SD** Paragraph and **SD1** Paragraph are self-contained. Each contains all the information needed to reconstruct the corresponding OPD from which it was derived. Naturally, since **SD1** is more refined, its equivalent **SD1** Paragraph is longer and more specific. However, just as **SD1** does not contradict **SD**, neither do the corresponding paragraphs contradict each other. Procedural link distribution occurs also in conjunction with folding, as Figure 9.10 shows.

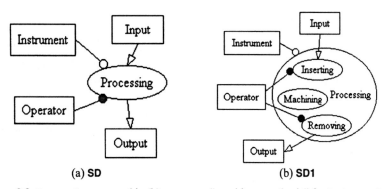

(a) **SD**   (b) **SD1**

Figure 9.9. Processing, zoomed in (b), serves as "graphic parenthesis" for **Instrument**.

> **SD** Paragraph:
> **Operator** handles **Processing.**
> **Processing** requires **Instrument.**
> **Processing** consumes **Input.**
> **Processing** yields **Output.**
>
> **Processing** from **SD** zooms in **SD1** into **Inserting, Machining,** and **Removing.**
>
> **SD1** Paragraph:
> **Operator** handles **Inserting** and **Removing.**
> **Processing** requires **Instrument.**
> **Inserting** consumes **Input.**
> **Removing** yields **Output.**

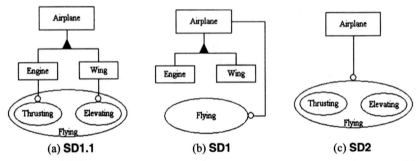

(a) **SD1.1**   (b) **SD1**   (c) **SD2**

Figure 9.10. Distribution of procedural links. (a) Two instrument links connect parts of **Airplane** to subprocesses of **Flying.** (b) When **Flying** is out-zoomed, the two instrument links are joined into one, as there is no point in keeping two links of parts of **Airplane** to **Flying.** (c) When **Airplane** is folded, the two instrument links are joined into one, as there is no point in keeping two links from **Airplane** to subprocesses of **Flying.**

9.3.5 Unfolding and Folding

The **Hierarchy Mode** of **Scaling** operates on a thing by **Folding** and **Unfolding** it. **Unfolding** reveals a set of lower-level things that are hierarchically below a relatively higher-level thing. The hierarchy is with respect to one of the first three fundamental structural relations: Aggregation-Participation, Exhibition-Characterization, or Generalization-Specialization.[5] The result of unfolding is a tree, the root of which is the thing being unfolded. Linked to the root are the things that were

---

[5]   Classification-Instantiation does not generate hierarchy, because, as explained in Chapter 8, instances are at the bottom of the gen-spec hierarchy, constituting the leaves of the gen-spec tree.

exposed as a result of the unfolding. Conversely, folding is applied to the tree, from which the set of unfolded things is removed, leaving just the root.

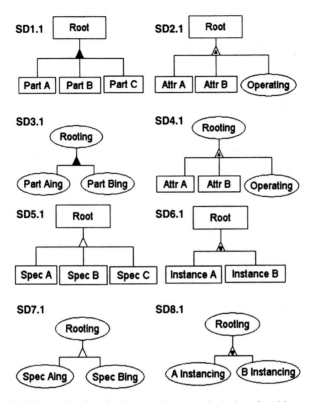

Figure 9.11. Unfolding of the four fundamental structural relations for objects and processes

Figure 9.11 shows unfolding of the four fundamental structural relations for objects and processes. Aggregation unfolding appears in **SD1.1** for objects and **SD3.1** for processes; exhibition unfolding in **SD2.1** for objects and **SD4.1** for processes; generalization unfolding in **SD5.1** for objects and **SD7.1** for processes; and finally, classification unfolding in **SD6.1** for objects and **SD8.1** for processes. Every unfolding has a corresponding folding counterpart. Hence, aggregation folding, generalization folding, exhibition folding, and classification folding eliminate the things at the bottom of the two-level hierarchy, leaving only the root. Frame 21 contains the OPL object unfolding sentences of the four fundamental structural relations that describe the OPDs in the top line of Figure 9.11.

> **Root** from **SD1** unfolds in **SD1.1** to consist of **Part A, Part B,** and **Part C.**
> *(Aggregation unfolding sentence)*
> **Root** from **SD2** unfolds in **SD2.1** to exhibit **Attr A** and **Attr B,** as well as **Operating.**
> *(Exhibition unfolding sentence)*
> **Root** from **SD5** unfolds in **SD5.1** to specialize **Spec A, Spec B,** and **Spec C.**
> *(Generalization unfolding sentence)*
> **Root** from **SD6** unfolds in **SD6.1** to instantiate **Instance A** and **Instance B.**
> *(Classification unfolding sentence)*

Frame 21. The OPL unfolding sentences of the four fundamental structural relations that are applied to the **Root** objects in the OPDs in the top line of Figure 9.11

The OPL process unfolding sentences of the four fundamental structural relations that describe the OPDs in the bottom line of Figure 9.11 are the same, except for the exhibition process unfolding sentence, in which the list of process(es) precedes the list of object(s):

> **Rooting** from **SD4** unfolds in **SD4.1** to exhibit **Operating,** as well as **Attr A** and **Attr B.**
> *(Exhibition process unfolding sentence)*

This is in line with the in-zooming sentence style: when a process is in-zoomed, the list of processes precedes the list of objects, and vice versa. Several observations are in order.

First, unfolding and folding are primarily applied to objects, so, as Table 11 shows, object is the *primary operand* of these scaling operations. Experience has shown that it is graphically most expressive to depict object refining through unfolding. Finally, any fork structural relation, not just one of the fundamental ones, can be used as a basis for unfolding.

Second, three out of the four relations (all except characterization) are *homogeneous* structural relations, which means that for those relations, unfolding of an object reveals only objects and unfolding of a process reveals only processes. Characterization is an exception. It is a *heterogeneous* structural relation, meaning that when a thing undergoes characterization unfolding, the unfolded set, i.e., the set of unfolded things, can be a mixture of features: attributes (which are objects) and operations (which are processes). **Homogeneous** and **Heterogeneous** are two values of the **Homogeneity** attribute of a fundamental structural relation. The **Homogeneity** value of **Characterization** is **Heterogeneous**, while the other three are **Homogeneous**.

The final unfolding type is the tagged structural link. The tag unfolding for any tagged fork will have the same sentence. The tag name is not specified by name in the unfolding sentence, but only later, in the body of the OPL paragraph. This mean that the OPL sentence for both Figure 9.12(a) and (b) is the same:

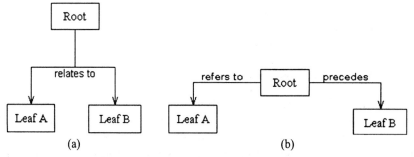

Figure 9.12. Unfolding for a tagged structural link: (a) As a fork. (b) As two different links.

Root from **SD1** unfolds in **SD1.1** to relate to **Leaf A** and **Leaf B**.
(Tag unfolding sentence)

Unfolding of various fundamental structural relation types can be mixed and matched, giving rise to a hierarchy that combines the various relations. For example, if we combine aggregation unfolding with exhibition unfolding, joining **SD1.1** and **SD1.2**, we get the following OPL sentence:

Root from **SD1** unfolds in **SD1.1** to consist of **Part A, Part B**, and **Part C** and to exhibit **Attr A** and **Attr B**, as well as **Operating**.
(Aggregation-exhibition unfolding sentence)

The order of listing the unfolded things in an unfolding sentence is aggregation, exhibition, generalization, classification and finally the tagged structural link. Thus, we can even combine all five and get:

Root from **SD1** unfolds in **SD1.1** to consist of **Part A, Part B**, and **Part C**, to exhibit **Attr A** and **Attr B**, as well as **Operating**, to specialize **Spec A, Spec B**, and **Spec C**, to instantiate **Instance A** and **Instance B**, and to relate to **Leaf A** and **Leaf B**.
(Aggregation-exhibition-generalization-classification-tag unfolding sentence)

9.3.6 State Expressing and Suppressing

Expressing reveals a set of states inside an object. **Suppressing** conceals a set of states inside an object. From a manifestation (state-visibility) viewpoint, an object can be in one of two states: state-expressed or state-suppressed. A state-expressed object is shown on the right hand side of Figure 9.13: the object's states are shown explicitly as rountangles within the object's box. A state-suppressed object is shown on the left hand side of Figure 9.13, where the states are not shown.

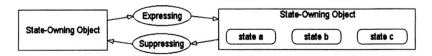

Figure 9.13. Expressing reveals object states while suppressing conceals them.

Expressing is a refining operation that takes a state-suppressed object in the ancestor OPD and presents in the descendant OPD the object rectangle, within which a proper subset (i.e., at least one) of the object's states or values are depicted. Suppressing is an abstracting operation that takes a state-expressed object B in the ancestor OPD and presents in the descendant OPD the object rectangle without any one of its states. The pair of expressing and suppressing operations enables the expression of a state-suppressed object and the suppression of a state-expressed object, respectively.

Figure 9.14 shows how expressing the states of **Water** provides for explicitly showing the state transformations induced by **Melting, Boiling, Freezing**, and **Condensing**.

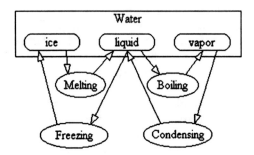

Figure 9.14. Expressing **Water** states provides for explicitly showing the state transformations induced by **Melting, Boiling, Freezing**, and **Condensing**.

The OPL paragraph that is equivalent to the OPD of Figure 9.14 is:

Water can be **ice, liquid** or **vapor.**
Melting changes **Water** from **ice** to **liquid.**
Boiling changes **Water** from **liquid** to **vapor.**
Freezing changes **Water** from **liquid** to **ice.**
Condensing changes **Water** from **vapor** to **liquid.**

What happens to the procedural links that were attached to the states before they were suppressed? Clearly we would like to maintain as much procedural information as possible, even though the suppressed states are no longer visible. We continue our **Water** example to show how procedural links are preserved. Figure 9.15(a) is an intermediate step that explains what happens to the procedural links that existed in Figure 9.14: since **Water**'s states (**ice, liquid** and **vapor**) are about to vanish, we disconnect the effect links from these states and connect them to the "parent" object itself, i.e., to **Water**. Note that each pair of effect links incoming to and outgoing from the same process gets diffused into a transformation link – a single, bidirectional arrow.

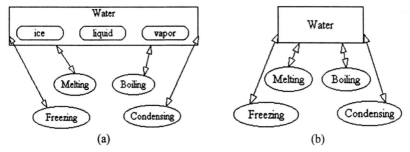

(a)   (b)

Figure 9.15. From input and output links to effect links. (a) In preparation for state suppression, as an intermediate step, the input and output links from and to the states of **Water** in Figure 9.14 migrate to the **Water** rectangle. (b) The states of **Water** are suppressed.

Having done this, we can now safely suppress the states of **Water** while preserving as much as possible from the procedural information. The OPL sentence set of the OPD in Figure 9.15(b) is reduced to:

> **Melting** affects **Water**.
> **Boiling** affects **Water**.
> **Freezing** affects **Water**.
> **Condensing** affects **Water**.

which can be further reduced to

> **Melting, Boiling, Freezing,** and **Condensing** affect **Water**.

All we can tell now, is that each one of the above four processes affects Water somehow, but we cannot tell how, as we could by looking at the OPD of Figure 9.14. This is an example of the price of abstraction. Figure 9.16 shows a system map with two nodes. In Figure 9.16(a) the states of **Water** suppressed, while in Figure 9.16(b) they are expressed. The arrow between the two OPDs is labeled Expressing, since this is the scaling operation that got us from the OPD on the left to the one on the right.

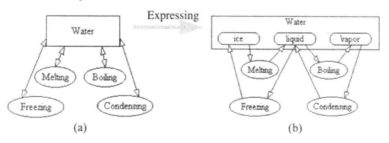

(a)   (b)

Figure 9.16. A system map with two nodes. (a) States of **Water** suppressed. (b) States of **Water** expressed

9.3.7 Primary and Secondary Operands

Operand is the entity (object, process, or state) to which the scaling is applied. **Visibility Scaling** (in-zooming and out-zooming) and **Hierarchy Scaling** (unfolding and folding) can be applied to both objects and processes. As we noted, experience with OPM graphics has shown that it usually makes sense to apply **Hierarchy Scaling** to objects and **Visibility Scaling** to processes.

Primary Operand and **Secondary Operand** are attributes of **Scaling**. **Primary Operand** specifies the type of entity (process, object, or state), to which scaling is applied by default. As Table 11 shows, the **Primary Operand** of **Visibility Scaling** (in-zooming and out-zooming) is **Process**, the **Primary Operand** of **Hierarchy Scaling** (unfolding and folding) is **Object**, and the **Primary Operand** of **Manifestation Scaling** is **State**.

Secondary Operand is an attribute of **Scaling**, which specifies the type of entity to which scaling can be applied, but is not applied by default. As Table 11 shows, the **Secondary Operand** of **Visibility Scaling** (in-zooming and out-zooming) is **Object**, the **Secondary Operand** of **Hierarchy Scaling** is **Process**, and there is no **Secondary Operand** of **Manifestation Scaling**, since **Manifestation Scaling** applies only to states.

9.4 Abstracting

So far we have focused mainly on refining, but as we noted, due to the middle-out nature of the analysis and design processes, analysis proceeds in a way that makes abstractions apparent only after lower level details are explicitly laid out. Abstraction can be done either through out-zooming or through folding. In this section we elaborate on abstracting, the inverse of refining.

9.4.1 Consolidating

Sometimes, we notice that a set of things are closely related, such that it makes sense to define a new, higher level thing that would abstract these things. *Consolidating* is the process of grouping the things that are about to be abstracted.

> *Consolidator is a thing that abstracts a subset of things in an OPD.*
> *Consolidating is determining the subset of things in an OPD, which the consolidator abstracts.*
> *A consolidated set is the subset of things in an OPD, which the consolidator abstracts.*

Consolidating can be a step that precedes abstracting in a new OPD. However, it can also be an operation, the purpose of which is to introduce the consolidator into the existing OPD to emphasize the commonality among the things in the consolidated set in the existing OPD. The consolidator can be an object or a process, and the consolidated set may consist of objects as well as of processes. When the consolidator is an object, the consolidated set consists of least two objects and optional processes. When the consolidator is a process, the consolidated set consists of least two processes and optional objects.

Since abstracting of things can be done either by out-zooming or by folding, we distinguish between two modes of consolidating: zoom consolidating and fold consolidating.

> *In **zoom consolidating**, the consolidator encloses the consolidated set.*
> *In **fold consolidating**, the consolidator is the root of a tree in which each thing in the consolidated set is a leaf.*

9.4.2 Zoom consolidating

In zoom consolidating, a set of things, the consolidated set, are enclosed within a process or an object, the consolidator. The processes **Melting** and **Boiling** are associated with raising the temperature, or **Heating**, while **Freezing** and **Condensing**, with lowering it, or **Cooling**. Starting with the OPD in Figure 9.16(b) and applying generalization, we model this observation by fold consolidating these two pairs of related processes into **Heating** and **Cooling**, as shown in Figure 9.17(a) and the OPL paragraph of SD in Frame 22. **Heating** in Figure 9 17(a) consolidates **Melting** and **Boiling**, while **Cooling** consolidates **Freezing** and **Condensing**. In Figure 9.17(b), folding of both **Heating** and **Cooling** is done concurrently, so **Melting** and **Boiling**, as well as **Freezing** and **Condensing**, disappear. The input/output links migrate to the contours of **Heating** and **Cooling**. Figure 9.17(b) also shows that **Thermal Energy Exchanging** fold consolidates **Heating** and **Cooling**. In Figure 9.17(c), the states of **Water** are suppressed while **Thermal Energy Exchanging** is folded, leaving only one effect link between **Thermal Energy Exchanging** and **Water**.

9.4.3 Paths and Path Labels

When **Heating** and **Cooling** are out-zoomed, the circumferences of their ellipses become the source of two input links and two output links. The state **liquid** is in a similar situation. Its rountangle becomes the source of two input links and two output links. This makes **SD1** in Figure 9.17(b) under-specified and ambiguous, because there is no way of telling which output link should be followed from **Heating**, **Cooling** or **liquid**. To avoid this under-specification, *path labels* are used.

> *A **path label** is a label on a procedural link that removes the ambiguity arising from multiple outgoing procedural links.*

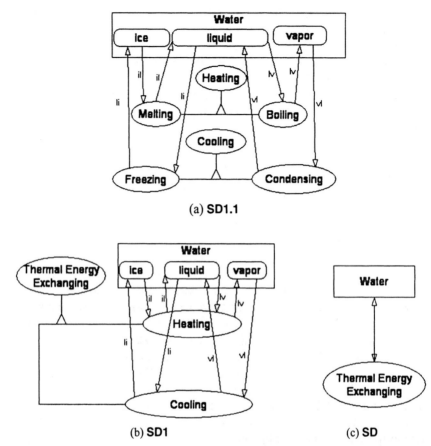

(a) **SD1.1**

(b) **SD1**   (c) **SD**

Figure 9.17. Fold consolidating demonstrated. (a) Starting with the OPD in Figure 9.16(b) and using the generalization fold consolidation, **Heating** consolidates **Melting** and **Boiling**, while **Cooling** consolidates **Freezing** and **Condensing**. (b) **Thermal Energy Exchanging** consolidates **Heating** and **Cooling**, which are folded. To avoid under-specification, **Heating** and **Cooling** now require the path labels. (c) States of **Water** suppressed while folding **Thermal Energy Exchanging** leaves only one effect link.

When procedural links that originate from an entity are labeled, the one that must be followed is the one whose label is identical with the label of the procedural link that arrives at the entity. Four path labels are used in Figure 9.17(b): **il**, which labels the **ice** to **liquid** transformation, **li**, which labels the **liquid** to **ice** transformation, **lv**, which labels the **liquid** to **vapor** transformation, and **vl**, which labels the **vapor** to **liquid** transformation. Eight procedural links appear in **SD1**. Each path label is common to one input link and one output link. Suppose the system arrived at the process **Cooling** from **vapor**. This means that the input link it passed was labeled **vl**. Therefore, of the two output links that emerge from **Cooling**, the one labeled **vl** must be selected, which gets us to **liquid**. From **liquid** only one output link, labeled **lv**, goes to **Heating** and only one output link, labeled **li**, goes to **Cooling**. If we take the **li** path to **Cooling**, we must exit **Cooling** via the **li** path to **ice**.

From **ice** we can only go via **il** to **Heating**. From **Heating** we take **lv** to **vapor**, thereby completing a full traversal of the eight procedural links. On first glance, Figure 9.17(a) may seem to require path labels as well; two input and two output links are connected to **liquid**. However, close inspection shows that the OPD is not under-specified. Each of the four processes is connected to exactly one input and one output state. It is possible to add path labels in this case, but is not at all required. The OPL paragraphs in Frame 22 introduces the path-labeled change sentences syntax, which is a change sentence preceded by the reserved phrase Following path, followed by the link label and a comma.

SD Paragraph:
Thermal Energy Exchanging affects **Water**.

Thermal Energy Exchanging from **SD** unfolds in **SD1** to specialize **Heating** and **Cooling**.

SD1 Paragraph:
Water can be **ice, liquid,** or **vapor**.
Following path **il, Heating** changes **Water** from **ice** to **liquid**.
Following path **lv, Heating** changes **Water** from **liquid** to **vapor**.
Following path **vl, Cooling** changes **Water** from **vapor** to **liquid**.
Following path **li, Cooling** changes **Water** from **liquid** to **ice**.
    (Path-labeled change sentences)

Cooling from **SD1** unfolds in **SD1.1** to specialize **Freezing** and **Condensing**.
Heating from **SD1** unfolds in **SD1.1** to specialize **Melting** and **Boiling**.

SD1.1 Paragraph:
Water can be **ice, liquid,** or **vapor**.
Melting changes **Water** from **ice** to **liquid**.
Boiling changes **Water** from **liquid** to **vapor**.
Freezing changes **Water** from **ice** to **liquid**.
Condensing changes **Water** from **vapor** to **liquid**.

Frame 22. The three OPL paragraphs of the OPDs in Figure 9.17

9.4.4 Zoom Consolidating Pitfalls

While performing zoom consolidating, care must be exercised to draw the contour of the process or object in a way that would enable out-zooming. Figure 9.18(a) shows an example of an illegal zoom consolidation. To see why this consolidation is illegal, we try to zoom out of **P4**. Consequently, the instrument link from **B2** to **P1** now links the two processes **P4** and **P1**. This violates one of the basics of OPM, that a procedural link can only connect an object to a process, but not two processes.[6] Figure 9.18(b) is the correct consolidating: **B2** is depicted out of the boundaries of the consolidating process **P4**. When **P4** is out-zoomed in Figure 9.18(c), **B6** disappears but **B2** remains, and all the procedural links correctly connect an object with a process.

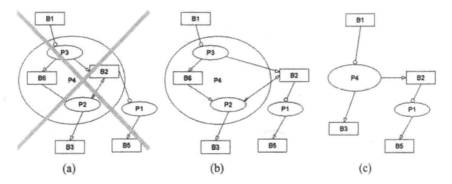

(a)   (b)   (c)

Figure 9.18. Illegal and legal consolidating of **P2** and **P3** into **P4**. (a) The consolidating is illegal because the instrument link from **B2** to **P1** crosses the border of **P4**, so when **P4** is out-zoomed, **B2** disappears and the instrument link illegally connects two processes **P4** and **P1**. (b) The consolidating is legal because all the links that cross the connect process inside **P4** to objects outside it. (c) **P4** of (b) is out-zoomed, yielding a legal OPD.

9.4.5 Zoom Consolidating Conditions

We have seen that not any subset of things is a legal consolidating set. Since zoom consolidating is applied by default to processes, we assume initially that the consolidator is a process. In order for zoom consolidating to be legal, the following conditions must be met:

1. A procedural link that crosses the consolidator ellipse must link a process inside the consolidator to an object outside it.[7]
2. An object that is created by a process inside the consolidator, and is consumed by another process inside the consolidator, must be modeled inside the consolidator.
3. An object that is created by a process inside the consolidator, and is not consumed by another process inside the consolidator, must be modeled outside the consolidator.
4. The contour of the consolidator should not cross any structural link.

Figure 9.18 has demonstrated condition 1. If the consolidator is an object, only two conditions need to be met: condition 4, which remains the same, and the con-

---

[6]   With the exception of the invocation link
[7]   Again, with the exception of Invocation

verse of condition 1 above: Any procedural link that crosses the consolidator rectangle must link an object inside the consolidator to a process outside it.

If an object is consolidated into a higher level process, that object is an attribute of the process. Analogously, if a process is consolidated into a higher level object, that process is an operation of the object.

9.4.6 Fold Consolidating

Fold consolidating takes place when the architect realizes that two or more things that are already in an OPD are parts, features, specializations or instances of a higher-level thing. Fold consolidating takes a set of things and links them to a root. There are four fold consolidating types, one for each fundamental structural relation:

- Aggregation consolidating links a set of parts to a whole.
- Exhibition consolidating links a set of features (attributes and/or operations) to an exhibitor.
- Generalization consolidating links a set of things to a generalized thing.
- Classification consolidating links a set of instances to their class.

The consolidator is respectively the whole, the exhibitor, the general, or the class. It becomes the root of the tree while each thing in the consolidated set is a leaf in that tree.

9.5 What Happens to Procedural Links During Abstracting?

In Chapter 5 we noted that only one procedural link can connect an object and a process. Often, more than one thing in the consolidated set is connected via a procedural link to a thing outside the set. When the consolidated set is abstracted, we need to decide which of the procedural links will "represent" the links that were eliminated along with the things in the set. In Section 9.5 we have seen what happens to procedural links when states are suppressed. To understand the problem, let us return to Figure 9.6 of the **Camera** example in Chapter 8. Suppressing all the states in Figure 9.19, we note that the states of **Film** in Figure 8.15 are suppressed, such that the input and output links between **Developing & Printing** and the states of **Film** become an effect link. While **Developing & Printing** affects **Film**, **Displaying** only requires **Diskette**, but it does not affect it. Thus, **Diskette** is just an instrument: the recording on the diskette remains unchanged by the **Displaying** process. At this refined level of detail we can distinguish between the agent and effect links that connect the two specializations of **Capturing Medium** and the two specializations of **Rendering**. The effect link is between **Developing & Printing** and **developed Film**, while the instrument link connects **Displaying** and **Diskette**.

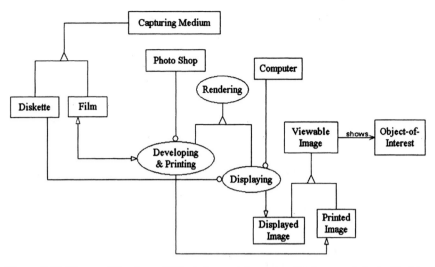

Figure 9.19. The specializations of **Rendering** and **Capturing Medium** are linked with different procedural links.

In Figure 9.20, the specializations of **Rendering** are folded, so the instrument links from **Photo Shop** and **Computer** migrate to **Rendering**. Since **Photo Shop** and **Computer** are required for different specializations of **Rendering**, they are mutually exclusive and therefore terminate at the same point on the rendering ellipse. The corresponding OPL sentence is:

> **Rendering** requires either **Photo Shop** or **Computer**.
> *(XOR instrument sentence)*

For the same reason, the procedural links from **Diskette** and **Film** terminate at the same point, and the OPL sentence is:

> **Rendering** either affects **Film** or requires **Diskette**.
> *(XOR effect-instrument sentence)*

9.5.1 Procedural Link Precedence

When **Capturing Medium** is folded in Figure 9.21, we need to also give up the ability to distinguish between the effect link connecting **Rendering** to **Film** and the instrument link connecting **Rendering** to **Diskette**. The question is which link remains and which is eliminated? The answer is that there is precedence among the procedural links, and effect links take priority over enabling links. Based on this link precedence, the link in Figure 9.20 from **Capturing Medium** to **Rendering** is an effect link, not an instrument link.

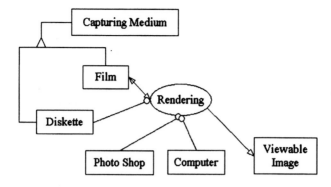

Figure 9.20. The specializations of **Rendering** are folded.

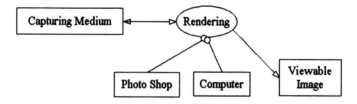

Figure 9.21. The specializations of **Capturing Medium** are folded.

A similar problem occurs when out-zooming a process consolidator, where two or more procedural links between processes inside the consolidator and an object outside it must now become one, because. If the links are of different types, a question arises: What link should remain?

The decision is based on the priorities of the procedural links. Table 12 lists the procedural links and their priorities. Transforming links always prevail over enablers. Consider an object that is an enabler for one subprocess and a transformee of a second. As the enabler, it must not be changed by the first subprocess; as a transformee, it is changed by the second subprocess. When the processes are consolidated, the total result is a change to the object; and so the transforming link remains.

Within the transforming links, consumption and result links are considered the most profound links, because they create or destroy the object in case. These are followed by input and output links, which change the state of an object. If the first subprocess creates an object, and the second subprocess changes it, the creation is more significant than the change. Therefore, while abstracting the two subprocesses, the result link prevails. Effect links are less expressive than their input/output counterparts, since the semantics of the change is not specified, and so their precedence is lower.

Table 12. Procedural links and their various attributes

Type	Name	Priority	Symbol	Source	Destination
Transforming	Consumption	1		Object	Process
	Result	1		Process	Object
	Input	2		state	Process
	Output	2		Process	state
	Effect	3		Object	Process
Enabling	Agent condition	4		state	Process
	Condition	5		state	Process
Condition	Agent	6		Object	Process
	Instrument	7		Object	Process

Among the enablers, condition links have higher precedence. They are state-specified enablers, and therefore carry more information than instruments and agents. Within an agent and an instrument, the agent always prevails, because in artificial systems, the humans are central to the process: they must ensure the system's proper operation. Also, wherever there is human interaction, an interface must be put in place; this information must be carried to higher levels so that the system architect can plan accordingly.

9.5.2 Semi-Folding and Semi-Unfolding

Unfolding is a graphic-intensive operation. To accomplish it we need to allocate diagram space to accommodate the symbols of the structural relation being unfolded of each unfolded thing and the links among them. It is frequently desirable to specify a thing's structural descendents (parts, features, specializations, or instances) without fully adding them. The OPD notation allows us to do so via *semi-folding*, demonstrated in Figure 9.22.

In Figure 9.22(a), the unfolded object **B** has two parts, three features (the two attributes **C1** and **C2** and the Operation **Bing**), one specialization **S1**, and one instance **Inst1**. In Figure 9.22(b), **B** has been semi-folded, and is now shown semi-folded as a single box with two compartments. The top one contains the name **B**, while the bottom one contains the seven things that are structural descendants of **B**. In Figure 9.22(c), **B** is completely folded. The corresponding OPL paragraphs follow.

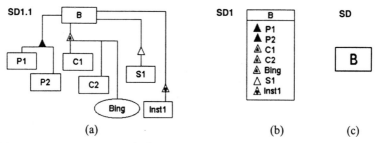

(a)   (b)   (c)

Figure 9.22. Semi-folding and semi-unfolding. Semi-folding is shown from left to right: (a) The object **B** in **SD1.1** is completely unfolded. (b) The object **B** in **SD1** after semi-folding is semi-folded. (c) The semi-folded object **B** in **SD** after another semi-folding is completely folded. Semi-unfolding is shown from right to left: (c) The object **B** in **SD** is completely folded. (b) The object **B** in **SD1** after semi-unfolding is semi-unfolded. (a) The semi-unfolded object **B** in **SD1.1** after another semi-unfolding is completely unfolded.

SD1.1 Paragraph:
B consists of **P1** and **P2**.
B exhibits **C1** and **C2**, as well as **Bing**.
S1 is a **B**.
Inst1 is an instance of **B**.
B from **SD1.1** is semi-folded in **SD1**.
    (New diagram semi-folding sentence)
SD1 Paragraph:
B lists **P1** and **P2** as parts, **C1** and **C2**, as well as **Bing** as features, **S1** as a
    specialization, and **Inst1** as an instance.
    (Listing sentence)
B from **SD1** is semi-folded in **SD**.

Semi-folding of a semi-folded object amounts to its complete folding. Likewise, semi-unfolding of a semi-folded object amounts to its complete unfolding. Thus, if we semi-unfold the semi-folded object **B** in Figure 9.23(b), we get **B** unfolded, as shown in the OPD of Figure 9.23(a). If we semi-fold the semi-folded object **B** in Figure 9.23(b), we get **B** folded, as shown in the OPD of Figure 9.23(c).

Figure 9.23 shows that **Object** can be in one of the following folding states: **folded, semi-folded**, which is the same as **semi-unfolded**, and **unfolded**. **Unfolding** changes the object from **folded** to **unfolded**, while **Folding** does the reverse. **Semi-unfolding** changes the object from **folded** to **semi-folded** and from **semi-folded** to **unfolded**, while **Semi-folding** changes the object from **unfolded** to **semi-folded** and from **semi-folded** to **folded**. Similar to the **Water** example in Figure 9.17, to avoid under-specification, **Semi-folding** and **Semi-unfolding** require path labels, since each of them can accept two input states, and the output state depends on the input state. The path labels enable unambiguous determination of the pairs of input and output states. The OPL sentences in Frame 23 show this.

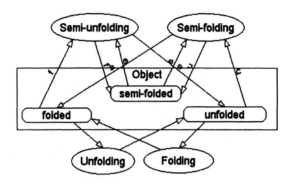

Figure 9.23. The folding states of an object and the processes that change them. Path labels prevent ambiguity.

Object can be **folded, semi-folded,** or **unfolded.**
Folding changes **Object** from **unfolded** to **folded.**
Unfolding changes **Object** from **folded** to **unfolded.**
Following path **u, Semi-folding** changes **Object** from **unfolded** to **semi-folded.**
Following path **s, Semi-folding** changes **Object** from **semi-folded** to **folded.**
Following path **f, Semi-unfolding** changes **Object** from **folded** to **semi-folded.**
Following path **e, Semi-unfolding** changes **Object** from **semi-folded** to **unfolded.**

Frame 23. OPL paragraph of Figure 9.23

9.5.3 Selective Semi-Folding and Semi-Unfolding

As noted in Section 9.3, there are four types of unfolding and folding, one for each fundamental structural relation. The same four types apply to semi-unfolding and semi-folding. Therefore, these operations can be applied selectively to any subset of the four fundamental structural relations.

Figure 9.24 shows selective semi-folding. The object **B** in Figure 9.24(a) is selectively semi-folded Figure 9.24(a): only things that are linked to **B** through its aggregation and characterization relations are semi-folded, leaving the generalization and instantiation unfolded. The corresponding OPL paragraphs follows.

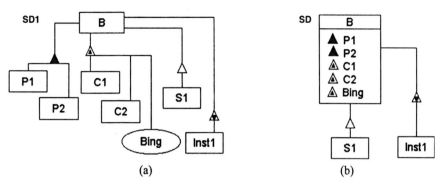

(a)   (b)

Figure 9.24. Selective semi-folding. (a) The object **B** is fully unfolded. (b) The object **B** is shown after aggregation semi-folding and characterization semi-folding.

SD1 Paragraph:
B consists of **P1** and **P2**.
B exhibits **C1** and **C2**, as well as **Bing**.
S1 is a **B**.
Inst1 is an instance of **B**.

B from **SD1** is aggregation and characterization semi-folded in **SD**.
   (New diagram selective semi-folding sentence)

SD Paragraph:
B lists **P1** and **P2** as parts, and **C1** and **C2**, as well as **Bing** as features.
   (Listing sentence)
S1 is a **B**.
Inst1 is an instance of **B**.

9.6 Looking at the Big Picture: The System Map and the OPM Construct Pairs

The system map enables seeing the "big picture" of the entire system at once. It is the collection of all the OPDs we made using OPM's refining and abstracting mechanisms. The system map enables visualization of how the many OPDs work together; how each piece of the puzzle contributes to the whole.

The system map is a directed acyclic graph. Each directed edge in the map is a link from an abstract thing (object or process) to its refined version in the target OPD (another system map node). Each node is a miniature OPD icon showing the general OPD image. Gray and white arrows relate the OPDs to each other. Each semi-transparent gray arrow leads from an abstract process to its refined (in-zoomed) version, while each hollow (white) arrow leads from an abstract object to

its refined (unfolded) version. The OPD labels, such as **SD** and **SD1**, help identify
the level of detail of the thing within the OPD. **SD** is at level zero, and **SD1.1.2.1** is
at level 4.

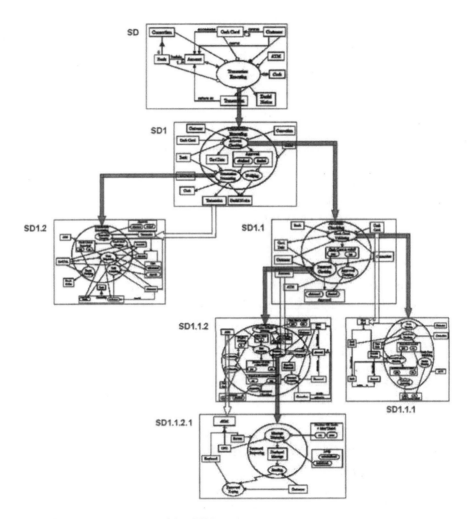

Figure 9.25. The system map of the ATM system

The complete OPM specification is provided by first showing the system map.
Following the system map are the individual OPDs, where the details are readable.
Along with each OPD comes its OPL paragraph. This specification may take many
pages, but each drawing should generally not be bigger than half a page, and its
OPL no longer than a page.

In Figure 2.9 we have shown the system map generated by the first two OPDs in
the OPD set of the ATM system. Figure 9.25 shows the system map of the entire

ATM system, which has seven nodes. Appendix A contains the entire specification, consisting of seven OPDs and seven OPL paragraphs to go with them.

In Chapter 3 we noted that the OPM system specification is a hierarchy of OPM graphic and textual construct pairs. These are defined below.

- An *OPD phrase* is an OPD symbol: entity (object, process or state) or (structural or procedural) link.
- An *OPL phrase* is a collection of one or more words that translate an OPD phrase into OPL as follows:
  - The OPL phrase of an OPD entity is the name of that entity as it appears in the OPD and is therefore a *non-reserved OPL phrase*. **Transaction Executing** is an example of such an an OPL phrase.
  - The OPL phrase of any OPD link, except for the tagged structural link, is the translation of that link into OPL and is therefore a *reserved OPL phrase*. The pair of words "**consists of**," which is a translation of the aggregation symbol, is an example of such an OPL phrase.
  - The OPL phrase of a tagged structural link is the tag of that link and is therefore a *non-reserved OPL phrase*. The pair of words "**refers to**" or the word "**accesses**," which are tags in the ATM system, are examples of such an OPL phrase.
- An *OPM phrase pair* is a pair of an OPD phrase and an OPL phrase that reciprocally describe the semantics of each other.
- An *OPL sentence* is an ordered collection of OPL phrases with specific semantics.
- An *OPD sentence* is a set of one or more OPD phrases arranged in a certain pattern that conveys specific semantics.
- An *OPM sentence pair* is a pair of an OPD sentence and an OPL sentence that reciprocally describe the semantics of each other.
- An *OPD* is the set of all the OPD sentences in a single diagram.
- An *OPL paragraph* is the collection of OPL sentences that together express the same information specified in an OPD.
- An *OPM paragraph pair* is a pair of an OPD and an OPL paragraph that reciprocally describe the semantics of each other.
- An *OPD set* is the collection of all the OPDs that together provide a complete graphic system specification.
- An *OPL script* is the collection of the OPL paragraphs that together provide a complete textual system specification.
- An *OPM system specification* is a pair of an OPD set and an OPL script that reciprocally describe the semantics of each other.
- An *OPD item* is a generalization of an OPD phrase, OPD sentence, OPD, and OPD set.
- An *OPL item* is a generalization of an OPL phrase, OPL sentence, OPL paragraph, and OPL script.

- An *OPM item pair* is a pair of an OPD item and its corresponding OPL item, generalizing an OPM phrase pair, an OPM sentence pair, an OPM paragraph pair, and an OPM system specification.

The following definitions summarize the relations between the ultimate OPD, the OPD set, the OPL script, and the system Map.

- The ultimate OPD is a theoretical OPD obtained from combining all the OPDs in the OPD set.
- The OPL script, which is the union of the OPL sentences that appear in all the OPL paragraphs, is the OPL paragraph of the ultimate OPD.

In Section 13.9 we elaborate on this issue further and provide a meta-model of the OPM construct hierarchy.

Summary

- OPM manages complexity through controlling the detail level of the things described in the various OPDs in the OPD set.
- While architects strive to perform top-down analysis and design, the practice is more often than not a mixture of top-down, bottom-up, and middle-out. OPM provides the tools to do all three.
- Scaling, the process of changing the level of detail, exhibits purpose (elaboration or simplification), mode (visibility, hierarchy, and manifestation), primary operand (the default entity for the scaling mode), and secondary operand (the non-default entity).
- Procedural links exhibit the distributive law. When attached to a zoomed-in process, a procedural link is considered to be attached to each of its subprocesses.
- Elaboration is attained through refining; simplification through abstracting.
- A pair of in-zooming and out-zooming operations controls visibility.
- A pair of unfolding and folding operations controls the hierarchy.
- Manifestation is the expression and suppression of states within objects.
- Consolidating is grouping a subset of things into a more abstract thing.
- Procedural links are preserved during abstracting based on their link priority; the lower the score, the more important the link.
- Semi-folding and semi-unfolding provide for an option of interim detail level that is economical in "real estate" yet provides information about a thing's parts, features, specializations and instances.
- Scaling at the system level is done using the system map, where each OPD is represented as a node and each edge represents a scaling operation that gives rise to a new OPD.
- The system map is a graph in which each node is an OPD in the OPD set and each edge denotes an abstraction-refinement relation between these OPDs.

Problems

1. Explain the importance of complexity management in systems and demonstrate what happens if no mechanism is in place to handle it.
2. Explain in your own words the divide and conquer principle of OPM and how it differs from strategies in other methods. Demonstrate it on a system of your choice.
3. Explain and demonstrate how combinations of values of the **Purpose** and **Mode** attributes of **Scaling** give rise to the various abstracting and refining options.
4. Is consolidating needed only as a preliminary step before abstracting? Why or why not? Exemplify your answer.
5. Create an OPM model of the water cycle in nature. Use at least two levels of detail and the appropriate refining options.
6. Based on Section 9.1.2 and Table 10, list all the possible simple/complex combinations of attribute values. For each combination either provide and example in OPD and OPL or explain why this combination is not feasible or does not make sense.
7. Add to Figure 9.19 **Painting** as a specialization of **Rendering** and add an agent and two instruments that enable it, as well as the resulting **Picture**.
8. Unfold one or two things in the model of Problem 7 and apply link precedence rules to determine which link remains and which disappear.
9. Look at Appendix A and list all the scaling methods you see there.
10. Check the author's work by out-zooming each diagram and verifying that it does not contradict OPDs that are linked to it in the system map.
11. Figure 9.26 is **SD**, the system diagram of a **Marrying** system. Figure 9.27 is **SD1**, a more detailed OPD of the system.
    a. What scaling operation generated **SD1**?
    b. Four links in **SD** touch **Marrying**, but in **SD1** five cross the **Marrying** boundary. How come?
    c. Write the **SD** paragraph and the **SD1** paragraph.
    d. Refine **Ceremony Conducting** in **SD1.1**. Add agents and instruments as needed.
    e. Repeat d. with **Dining** in **SD1.2**.
    f. In **SD1.1.1** differentiate between a **Civil Wedding** and two types of **Religious Weddings**. Create specializations for **Justice** and various aspects of the **Ceremony Conducting**. Use any scaling mechanisms you need. Verify the validity of the four OPDs you have created by scaling them down.
    g. List the different scaling mechanisms you used in modeling this system and explain your choices.

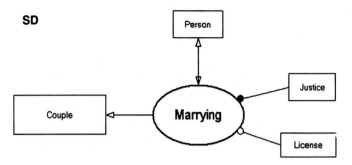

Figure 9.26. System diagram of a **Wedding** system with **Marrying** as the central process

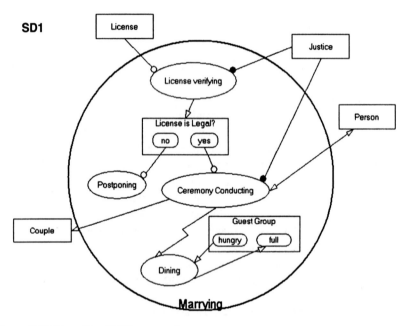

Figure 9.27. Marrying of **SD** in-zoomed

Part III

Building Systems with OPM

Chapter 10

Systems and Modeling

> *A system is a set of variables sufficiently isolated to stay*
> *discussable while we discuss it.*   W. R. Ashby (1956)

Each one of us is a system. We live within systems and are surrounded by them. Systems exist in nature as well as in virtually any conceivable area of human activity. Systems are the focus of research and development in any field of human endeavor. In Chapter 4 we have elaborated on objects and processes as the fundamental building blocks of the universe. In the chapters that followed, we have learned how these building blocks can be combined in a variety of ways to model ever more complex things, which are still objects and processes. Having studied structure and dynamics, we are now ready to discuss systems.

All systems exhibit a common feature: they carry out some *function*. A system consists of a collection of related objects, represented by the system's structure, that interact with each other via processes in a coordinated way, accounting for the system's behavior. The combination of structure and behavior is the system's *architecture*. An artificial system is the outcome of a human *intent*, a goal or objective, translated to function, for which the system has been designed and built. System analysis is a process of increasing human knowledge about the structure and behavior of existing systems. For artificial systems, design typically follows the analysis, yielding an orderly specification of an architecture for attaining the system's desired function. In this chapter we apply our knowledge of OPM to define, study and model systems and related concepts, such as function, product and project.

10.1 Defining Systems

A system can be natural or artificial. It can be physical or informatical, biological, social or symbolic, or comprise more than one of these. Winograd and Flores (1978) noted that "an entity can be explained when its behavior can be described in terms of parts that play functionally defined roles in its operation." This sentence contains many concepts that require crisp definitions. In this section we examine several definitions of the term "system" and then try to formulate our own definition.

10.1.1 Some Existing Definitions

The rich literature on systems contains many definitions of the term "system." Let us examine a few definitions before proposing one that fits our frame of reference.

The definition of a system according to Ashby (1956), provided at the beginning of this chapter, is somewhat cynical and probably too viewer-centered to be useful. In UML (Object Management Group, 2000), a system is defined as a collection of connected units that are organized to accomplish a specific purpose. The US Defense Systems Management College (1999) defines a system is an "integrated composite of people, products, and processes that provide a capability to satisfy a stated need or objective." The purpose, need, or objective is important as a basis for defining the system's function, and function is an essential element in our definition of system.

Chen and Stroup (1993) have defined a system as "an ensemble of interacting parts, the sum of which exhibits behavior not localized in its constituent parts." This definition introduces the Aristotelian principle of synergy, which proclaims that the whole is more than the sum of its parts. It also implies that the behavior of the system is *emergent*, i.e., it cannot be attributed to any one of the system's parts. However, this definition also assumes that a system must consist of more than one part. While this is true for any practical non-trivial system and is therefore an academic issue, it should not be a mandatory condition.

Kerzner (1995) has defined a system as a group of elements, either human or non-human, that is organized and arranged in such a way that the elements can act as a whole toward achieving some common goal, objective or end. While this definition refers to non-human as well as human systems, it is oriented towards artificial systems in the sense that it includes the element of intent, goal, objective or end, which is unique to man-made systems. We would like the definition of system to encompass natural systems as well as artificial ones.

Wand and Weber (1989) define a system as "a set of things [which, in our nomenclature, are objects], for any partitioning of which, interactions exist among things in any two subsets." This definition emphasizes internal interactions among the system's components. Referring to a possible mathematical model of a system, they note that "if we view such a set of things as a graph, where every thing is a node, and every interaction is represented as a link, then a system is a connected graph." This definition captures the fact that any two things in a system must be related, either directly or indirectly. Assuming that a system comprises at least two things, this connectivity requirement is indeed important. If two OPM things are not related through at least one structural or procedural link, it is not possible to say anything about a relationship between these things. Being unrelated, they cannot be considered parts of the same system.

Systems are frequently complex, but since complexity is a relative term, it should not be part of the definition of a system. To see why, we note that simple, mundane objects, like a chair or a knife, are systems. It may be odd to refer to a **Chair** or a **Knife** as systems, because they are so simple, yet it is acceptable to view a more complex object, such as a **Vehicle**, as a system. Contemplating the chair example for a moment, a **Chair** can be more just the classical wooden four-legged object. It can be a sophisticated, orthopedically designed object with many adjustment options and degrees of freedom, in which case we would have no problem calling it a **Sitting System**. A knife could be an elaborate electric cutting machine with varying power, speed, and blade types, a perfect **Cutting System**. Many variations lie in-between the two extremes that span the simple-to-complex spectrum. When does a chair stop being a mere object and starts being qualified as a **Sitting System**, or a knife, a **Cutting System**? Obviously, there is no clear-cut border. These examples demonstrate that a definition of a system that is based on a certain level of complexity is not adequate.

10.1.2 Function

We have defined an object as a thing, which has the potential of existence. A subset of the objects in the universe is capable of doing something, exhibiting some observable phenomenon or providing a service that is useful to some people or organizations. Such objects are said to *function* in a certain way.

> **Function** *is an attribute of object that describes the rationale behind its existence, the intent for which is was built, the purpose for which it exists, the goal it serves, or the set of phenomena or behaviors it exhibits.*

This definition of function applies to natural objects, which exhibit phenomena or behaviors, as well as to artificial ones, which are useful for some purpose, aim or goal. The definition emphasizes the "what" and "why" aspects and is not concerned with the "how." The emphasis on these aspects distinguishes function from dynamics, as dynamics is about *how* the object operates, while function is about *what* the object does and *why* it does it. As we show next, function provides the basis for our definition of a system.

10.1.3 The Various Functions of Stone

To gain insight into the difference between any object and an object that is a system, let us consider a stone lying in the field. This stone would normally be referred to as just an object, but if you take this stone home and use it as a paper holder, it can be considered a (very simple) system. The function of this system is to prevent papers lying on a desk to be swept away by wind. Functioning as an object that holds papers, we can conceive of this object as a *paper holding system*. The same object, a stone, if used in different ways, can play the role of various systems. It can be used for hammering nails – a *nailing system*, for hunting animals – a *hunting system*, or, if it is big enough and shaped conveniently, it could serve as something to sit on – a bench or a *sitting system*. In ancient days, people used stone tablets to store information. Used for this purpose, a stone can be considered *an information storage system*. Figure 10.1 shows the object **Stone** exhibiting an attribute called **Function**. In OPL syntax this is phrased simply as:

```
Stone exhibits Function.
```

The **Function** attribute of **Stone** in our example can be at one of the states **paper holding, hunting, nailing,** and **information storing**. In OPL:

> Function of **Stone** can be **sitting, paper holding, hunting, nailing,** or **information storing**.

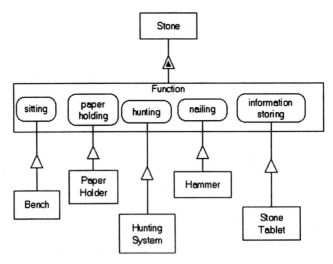

Figure 10.1. Each **Function** of **Stone** qualifies it as a different system.

Bench, Paper Holder, Hunting System, Hammer, and **Stone Tablet** are five different systems, each with its own function. A stone that functions to hold paper is a **Paper Holder**. A stone that functions to store information is a **Stone Tablet**, etc. The following OPL qualification sentences express this:

> **Bench** is a **Stone,** the **Function** of which is **sitting**.
> **Paper Holder** is a **Stone,** the **Function** of which is **paper holding**.
> **Hunting System** is a **Stone,** the **Function** of which is **hunting**.
> **Hammer** is a **Stone,** the **Function** of which is **nailing**.
> **Stone Tablet** is a **Stone,** the **Function** of which is **information storing**.

10.2 System Defined

Consider the following citation of a manager from a HP Desktop Printer supply chain Stanford case study (Kopczak and Lee, 1994): "Localization and customization are manufacturing tasks... We are best at moving products... We don't have the system to support these functions." This excerpt has the right notion of what a system is all about: any system, regardless of whether it is natural or man-made, simple or complex, can be viewed as an object that performs some major function. In other words, *a system is a function-carrying object*. Hence, a compact definition of a system that applies equally well to artificial and natural systems is as follows:

> A *system* is an object that carries out or supports a significant function.

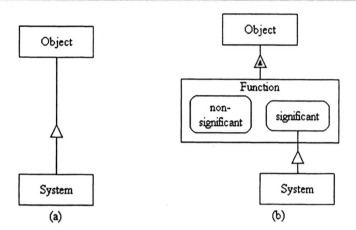

Figure 10.2. System as an **Object**: (a) **System** is an **Object**. (b) A **significant Function** qualifies **Object** to be a **System**.

Figure 10.2 depicts the definition of system in OPM terms. Figure 10.2(a) expresses the fact that **System** is a specialization of **Object**, and conversely, that **Object** is a generalization of **System**. The corresponding OPL specialization sentence is simply:

> **System** is an **Object**.

Function can be either **significant** or **non-significant**. Figure 10.2(b) denotes that **Object** qualifies as a **System** if its **Function** is **significant**.

> **Object** exhibits **Function**.
> **Function** can be **significant** or **non-significant**.
> **System** is an **Object**, the **Function** of which is **significant**.

Note that OPM's definition of system addresses neither the number of components comprising the object in question nor its level of complexity. A system can be simple or complex, as long as it carries out a function. All systems are objects, but not all objects are systems. As we see in the next two subsections, function is relative, as it depends on the context, and it is subjective, as it depends on the viewpoint of the beneficiary.

10.2.1 System as a Relative Term

While object is an absolute term, system is relative to the domain of discourse or task at hand. To see why it is relative, consider the following example. In a certain setting, a magnetic **Disk**, installed inside a **Computer**, performs the processes of **Writing, Storing,** and **Reading** of large quantities of **Data** at a high speed through the elaborate coordination between the **Spinning** of the **Disk** and the **Moving** of its **Read-Write Arm**. Under these circumstances, that **Disk** is definitely a system, within the larger **Computer** supersystem, as it carries out the meaningful and significant **Writing** and **Reading** functions. However, at a coarser level of granularity, in a comprehensive system of networked computers, routers, communication lines, and protocols, we would probably consider the same disk as merely a simple object for data storage and retrieval, without being concerned about its internal structure or operation. Furthermore, since a disk is contained inside a computer, which is the atomic object in the network system, we can even ignore its existence altogether. As another example, the **Car-Driver Complex** in Figure 9.3 is definitely a system when we are interested in its function as a means to move people and goods around. However, in the context of a city traffic system or the interstate highways system, the same **Car-Driver Complex** is a simple object, and can even be considered an atomic one.

These examples demonstrate the *relativity* of the term system. A system can be a super system of any one of its parts and a sub-system of one or more higher-level systems. A system in one setting, context or frame of reference, can be a sub-system, or an object in another. A system in one context can be considered a simple, atomic object in a yet more complex environment, or it can even be altogether ignored because it is visible only at more refined levels. Possible exceptions for this relativity principle are the universal extremes as we know them: sub-atomic elementary particles on the micro side and the known universe on the macro side.

10.2.2 System as a Subjective Term

The term "system" is not only relative but it is also subjective and task-oriented: an object is a system if and only if we, as researchers or developers, are interested in exploring or developing the *function* of that object. An important implication of our definition of system is that *an object is a system in the eyes of the beholder*. The meaning of "significant" in this definition is that the function the object in question exhibits is central, significant, or major, with respect to the domain of discourse or problem at hand. Beer (1999) has noted: "A system is a model that is more or less useful for some purpose.[1] If that *purpose* is not defined, then there is no criterion of utility." The emphasis in the source on the purpose underlines the role of function in systems, as well as the relativity and subjectivity of the term system. Indeed,

---

[1]   The notions of system and model are used here interchangeably, whereas we make a clear distinction between these two terms: an OPM model is a system that specifies a system.

Beer (1999) claims that "any system is a subjective phenomenon. We cannot have an objective system, which means that no system is ever right or wrong." This statement in incomplete agreement with our definition of system. A function is what makes an OPM object a system, and the function is in the eyes of the beneficiary. A human deems an object to be a system if he or she can point to its function, and since function is a subjective attribute, so is system.

10.2.3 The Function of Natural and Artificial Systems

Following our definition of system, the existence of a *significant function* is the criterion that enables distinction between any object and an object that qualifies as a system. While discussing the function of a system, we distinguish between systems of two different origins: natural systems and artificial systems. Both these system types exhibit function, but the function of each is different.

Scientists study the functions of natural systems. We all agree that the solar system is a system, and a natural one. What, then, is its function? The solar system functions to maintain the planets and asteroids in orbit around the sun. Gravity, a natural force, causes this system to function. Planet Earth is a subsystem within the solar system that sustains life. Within Earth, a biological system, an organism, is a system that performs the function of living, or, according to Dawkins (1989), protecting genes and carrying them from one generation to the next. These are examples of natural objects that are clearly systems, since they evidently function. However, without entering a theological debate, which is beyond the scope of this book, it is not possible to associate with a natural system a pre-meditated purpose, intent, or goal, which the function of man-made systems exhibit.

Architects and developers create artificial systems. The function of an artificial system is more straightforward than that of natural ones. An artificial system is supposed to benefit its user. It generates the utility, or benefit, that its user gains, or expects to gain, from using it. The definition of an artificial system is therefore intimately linked with a *beneficiary* – an agent (a human, an organization, or a community of humans) that benefits from the function that the system exhibits, or provides. An object is a system from the beneficiary's viewpoint. If we are unable to identify a (direct or indirect) beneficiary, then we cannot say that the object in question is an artificial system.

10.3 Goal, Concept, and Function

In this section, we discuss three terms: goal, concept, and function, and how they relate to each other. A goal, directed by a human's intent, is what a system is supposed to attain. Function is a translation of the system's goal into practical terms. Concept is the system architect's application of principles underlying the system's architecture, which accomplishes the system's function. A function-oriented definition of system is more robust than one that relies on complexity. As long as an object is constructed to perform some function that is useful to the beneficiary, it should be considered an artificial system. In other words, artificial systems exhibit a specific goal, which they are supposed to attain.

10.3.1 The Intent and Goal of Artificial Systems

As we argued, intent, goal, or purpose characterize artificial systems. Humans are often defined as intelligent, goal-directed, tool building organisms with self-awareness and conscience. These traits are often considered characteristics that distinguish mankind from lower-level organisms.[2] To survive and improve his living conditions, prehistoric man began to build tools and instruments, such as sticks, spears, and arrows, for the purpose of hunting, and ploughs for cultivating the land. These tools, arcane as they may be, are systems, since they are objects that support significant functions.

Man-made systems have since been constantly growing in complexity and sophistication, but most, if not all of them, share in common a goal, or a "super function" of improving human living conditions, either directly or indirectly. The following examples demonstrate this claim. A **Vehicle** is an artificial system for **Moving** people and goods, a **Knife** is used for **Cutting**, an **Accounting System** is used for **Accounting**, a **Power Plant** is built for **Electricity Generating**, and a **Microscope** is used for **Magnifying**. A **Chair** is built with the purpose of **Seating** a person, a **Pen** is for **Writing**; a **Shovel**, **Digging**; a **Spacecraft**, **Space Travelling**. These tools and instruments are artificial systems, built with specific human intentions and aimed at specific goals that are attained by the functions these systems support. Like our earlier **Stone** example, objects can function as more than one system. A **Magnifying Glass** is a system that enables humans to see details that cannot be noticed by the naked eye, but used to focus sunrays, it can also serve as a **Fire Making** system, much like a box full of matches. Exploring the hierarchy of these systems' functions, one can ultimately trace their intent to improve human quality of life, at least from the physical aspect.

10.3.2 Telling System Function and Dynamics Apart

System function and dynamics are often confused as synonyms, but these are two different terms. Normally, when we think about the function of a system, what we have in mind is the system's dynamics. The distinction between function and dynamics (behavior) is at times blurred, as it is often the case that a main process that the system executes obtains the system's function. However, following our train of thoughts, they are not the same. Function is what the system does and why it does it, while dynamics (or the system's behavior) is how the system acts or oper-

---

[2]   Scientific evidence has recently started to accumulate on existence of such characteristics in certain developed animals.

ates to attain its function. Function is a derivative of the system's goal – what service the system is expected to provide. Dynamics, on the other hand, is how the system's structure behaves (changes over time) to perform its function. To clarify this subtle but significant difference, let us consider a couple of examples. The first example is a river-crossing system and the second is a time keeping system.

10.3.2.1 River Crossing System

The function of a **River Crossing System** is to enable humans and goods to cross from one of the riverbanks to the other. Both a **Bridge** and a **Ferry** can implement this system. Both implementations fulfill the function of river crossing. However, the two systems are very different in terms of their architecture, i.e., their structure-behavior combination. **Bridge** is a static, passive object that enables people and goods to cross over it. The only "dynamics" of the bridge is the alternation of stress and shear forces it endures while loads pass over it. The nature of **Ferry**, on the other hand, is dynamic – it performs the **River Crossing** process itself, carrying the people and goods as it goes. Figure 10.3 is an OPD that described the two different system architectures, where **N** and **S** denote the north and south banks of the river, respectively.

The **Bridge** and the **Ferry** are two different **River Crossing Means**. Each is an instrument for a different crossing process: **Bridge Crossing** requires **Bridge** while **Ferry Crossing** requires **Ferry**. The OPD specifies how a **Person** (the beneficiary) at the north bank **N** crosses to the south bank **S**. The subprocesses into which **Ferry Crossing** zooms reveal the dynamic nature of this process. They include **Embarking** of **Person** from **N** onto the **Ferry**, **Cruising** of the **Ferry** from **N** to **S** and **Disembarking** of **Person** from **Ferry** to **S**. Conversely, **Bridge Crossing** requires only that **Person** uses **Bridge** to move from **N** to **S**. The two systems differ in their behavior, yet they attain the same function.

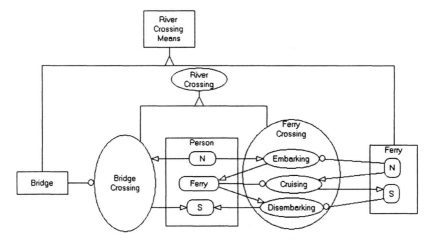

Figure 10.3. Two architectures for a **River Crossing** system. N and S are the north and south banks of the river.

10.3.2.2 Time Displaying System

When asked to refer to an instrument that displays the time, we normally think of a watch or a clock of some kind. A digital watch achieves the function of time keeping by exploiting the accurate frequency obtained from a quartz crystal. The structure of a mechanical clock can be seen as the assembly of cogwheels, springs, axes and display mechanism in a box. Figure 10.4 show **Sundial** and **Clock** as two time keeping systems with the same function but different architectures (structure-behavior combinations). The **Sundial** comprises a **Pole**, whose cast shadow displays the time of day based on the position of the sun in the sky. The function of both **Time Displaying System** is **telling the time of day**. As Figure 10.4(a) shows, **Sundial** consists only of **Display** and **Pole**. It has no moving parts and hence exhibits no dynamic behavior.[3] Unlike the **Mechanical Clock**, it requires no internal **Energy Source**. The **Sundial** achieves the function of keeping and showing time by structure alone, taking advantage of the relative movement of the earth around the sun.

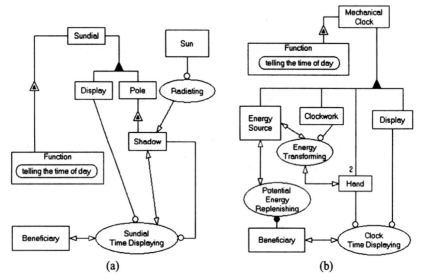

Figure 10.4. Two architectures for a **Time Keeping System**: (a) **Sundial**. (b) **Mechanical Clock**

As presented in Figure 10.4(b), a **Mechanical Clock** consists of an **Energy Source**, **Clockwork**, two **Hands**, and a **Display**. It keeps time by **Energy Transforming**, i.e., transferring potential energy, stored in **Energy Source**, a wound spring or a lifted weight, to movement of the two clock's **Hands** through the elabo-

---

[3] One may wish to consider the **Sun** part of the system, but it is really part of the environment, because the system architect has no control over it.

rate **Clockwork** mechanism. The dynamics of the clock can be studied by examining the way the energy in the spring is released and converted into movement of the interconnected cogwheels. This movement propagates until it ends at the clock's **Hands** that display the time on the dial.

Figuring out function from architecture is a major part of reverse engineering. Imaginary intelligent extra-terrestrial creatures trying to comprehend what a mechanical clock is, will be able to understand the structure and behavior of the clock but will probably have harder time realizing its function. The situation these creatures would be at is not unlike that of the archeologists trying to figure out the function of the giant stone circles of Stonehenge (Witcombe, 2001).

As Figure 10.4 shows, the function of the two versions of **Time Displaying System** is **telling the time of day**. In each system there is a different version (specialization) of **Time Displaying**. One is **Sundial Time Displaying** and the other – **Clock Time Displaying**. In Figure 10.5, the two architectures of **Time Displaying System** are presented in the same OPD. In both systems, **Time Displaying**, which generalizes **Clock Time Displaying** and **Sundial Time Displaying**, affect the **Beneficiary**, who is the person looking at the time displaying system. The effect is changing the state of the **Beneficiary** from not knowing the time of day to knowing it.

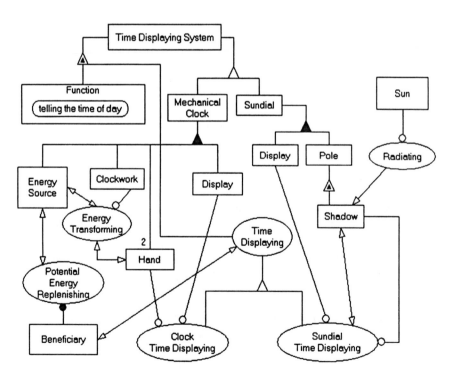

Figure 10.5. The two architectures of **Time Displaying System** presented in the same OPD

10.3.3 Function, Structure, and Behavior

We noted that the function of an artificial system is derived from the intent, goal, or purpose, for which the system was constructed. Function describes *what* the system does or can do. Crawley (2000) defined function as "the behavior of the system, the activities, operations and transformations that cause, create or contribute to performance (i.e. meeting goals), or the actions for which a thing exists or is employed." Indeed, this definition points to the goal, which is at the basis of the function. As we have seen, though, very similar or even identical functions can be achieved by systems that employ different architectures (structure-behavior combinations). The system's function can be architected, understood and explained from two major complementary aspects: structure (or form) and behavior (or dynamics). The system's architecture is the particular combination of structure and behavior that enables materialization of the function.

Structure pertains to the ensemble of the system's components and the long-term relationships among them that are not affected by the flow of time. Crawley (2000) referred to structure as *form* and characterizes it as the "sum of the elements or objects, the structure or arrangement of the physical/logical embodiment" or the "shape or configuration." Behavior, on the other hand, is the dynamic aspect of the system – the way it changes over time. The change is manifested in changes in one or more of the subsystems that comprise the system, which are objects in their own right as well as subsystems – systems at lower levels.

10.4 System Architecture

System architecture is a widely used term that warrants a concise definition. In the context of information systems, the Open Group Architectural Framework (2001) has defined architecture as "a set of elements depicted in an architectural model, and a specification of how these elements are connected to meet the overall requirements of an information system." This definition requires that connections among elements be specified, which is a structural aspect of the architecture. In this section, we define architecture of systems in terms of both their structure and behavior.

10.4.1 Function vs. Dynamics

As the sundial and bridge examples show, function can be the result of a particular structure, with little or no dynamics involved on the system's side. The dynamics that operates the sundial system comes entirely from the sun, which is part of the system's environment. Likewise, the bridge is essentially static. Here, the dynamics is of the people who cross the bridge – the system's users, or beneficiaries.

Dynamics is inseparably coupled with structure, because dynamics is about what happens to the structure and in what sequence. Some combination of structure and behavior enables the system to function properly, such that its goal is attained. This combination is the system's architecture. Often, there is more than one architecture that attains a system's function. For example, the function of a flying system is carrying humans or goods in the air. Such a system can be an airplane, a glider, a Zeppelin, or a hot-air balloon. The set of underlying concepts of each of these systems is different. While airplane and glider are heavier than air, and take advantage of Bernoulli's Law, Zeppelin and hot-air balloon are lighter than air and take advantage of floatation. While airplane and Zeppelin are propelled, glider and hot-air balloon are not. Although each is based on a different concept, all attain the function of carrying humans and goods in the air. Similarly, sending either airmail or electronic mail achieve the same function. This and numerous other examples demonstrate that a system's function can be frequently achieved by a number of different structure-behavior combinations.

The system's function dictates some combination of structure and behavior that enable the system to function – to achieve the goal for which it is designed. As our examples have shown, function can be obtained with little or no system dynamics, just by the system's structure (or form), as with the sundial or bridge, or through a highly dynamic system, as with the clock or ferry. The combination of the system's structure and behavior, as conceived by the system architect is the system's *architecture*.

> *System architecture* is the overall system's structure-behavior combination, which enables it to attain its function while embodying the architect's concept.

Put differently, the system architecture is the embodiment of the concept, which the system's architect conceived. The architecture combines the system's structure (form) that enables its behavior (dynamics, or operation) in such a way that the function is achieved. The goal that the system aims to achieve determines its function. The system's architecture – its structure-behavior combination – is the manifestation of some particular implementation of the system that enables it to carry out its function. Many systems are complex both dynamically and structurally. The term concept in the above definition of architecture is defined next.

10.4.2 The Concept Behind a System

Crawley (2000) refers to *concept* as the product or system vision, idea, notion or mental image which maps form to function and embodies "working principles." He indicates that concept is created by the system architect and it must allow for execution of all functions. It establishes the solution vocabulary and implicitly represents a level of technology.

> *Concept* is the system architect's strategy for a system's architecture.

Indeed, concept is *function-directed*. It is the ensemble of underlying ideas and principles that govern the system's architecture. The architecture is the system's structure/behavior combination that enables it to function as best as possible. Function, in turn, is *goal-directed*. The following account (Henry Ford Museum, 2001) on the development of the first flying machine by the Wright Brothers provides a great example of concept:

> *While nursing Orville, who was sick with typhoid in 1896, Wilbur read about the death of a famous German glider pilot. The news led him to take an interest in flying. On May 30, 1899, he wrote to the Smithsonian Institution for information on aeronautical research. Within a few months after writing to the Smithsonian, Wilbur had read all that was written about flying. He then defined the elements of a flying machine: wings to provide lift, a power source for propulsion, and a system of control. Of all the early aviators, Wilbur alone recognized the need to control a flying machine in its three axes of motion: pitch, roll, and yaw. His solution to the problem of control was "wing warping." He came up with the revolutionary system by twisting an empty bicycle tube box with the ends removed. Twisting the surface of each "wing" changed its position in relation to oncoming wind. Such changes in position would result in changes in the direction of flight. Wilbur tested his theory using a small kite, and it worked.*

The goal of the Wright Brothers was to enable people to fly. At the goal stage there is often no system at all. The function of the flying machine system was to fly a heavier-than-air object in air and be able to navigate it and bring it safely back to the ground. As the above excerpt notes, the concept that the Wright Brothers employed to achieve the function of flight included the elements of wings for lifting, a power source for propulsion, and a controlling mechanism for navigating the flying machine system. A crucial element in this concept, which was not applied in earlier attempts to build a similar system, was the specific profile of the wing, which provided the needed lift to hold the system in the air.

10.4.3 The Origin and Essence of Systems

In Chapter 4 we have seen that things exhibit four attributes: perseverance, origin, essence, and complexity. As a specialization of **Object**, the **Perseverance** value of a **System** is **Static**. The **Complexity** of a non-trivial **System** is **Non-simple**. This leaves **Origin** and **Essence** as the two **System** attributes whose values can vary. **Origin** has two possible values: **Natural** and **Artificial**.

The **Origin** of an artificial system is **Artificial**, while the **Origin** of all other systems is **Natural**.[4] Using OPM qualification sentences, we say that a **Natural Sys-**

---

[4]   Of emerging significant importance are systems that involve both natural and artificial elements. Advents in molecular biology, genetic engineering and cloning technology, which raise major ethical problems, have started to yield such organic systems, in which humans manipulate nature, blurring the distinction between the two **Origin** values.

tem exhibits **Origin**, the value of which is **Natural**, and that that an **Artificial System** exhibits **Origin**, the value of which is **Artificial**. One important difference between artificial and natural systems is that while there is no doubt that artificial systems are built by humans to serve some purpose or attain some goal, the existence of purpose or goal in natural systems is an open question.

With respect to **Origin**, a clear distinction exists between **Natural** and **Artificial** systems. However, with respect to **Essence**, an analogous dichotomy between **Physical** and **Informatical** does not exist. Any system has elements of both matter and informatics, and can therefore be mapped somewhere along the matter-informatics continuum. **Essence** is orthogonal to (independent of) the **Origin** attribute. No informatical object can exist (at least outside the human brain, and most probably also within it) if it is not inscribed on some physical medium. This is true not only for artificial, but also for natural informatics. Therefore, even the most informatics-oriented system must have a physical component to sustain it.

To see an example of two systems with identical function that differ in their location along the physical-informatical spectrum, consider two variants of the system **Car**, one with **Manual Transmission** and the other with **Automatic Transmission**. **Manual** and **Automatic** are two values of the **Gear Shifting** operation of **Car**. While both **Car** systems perform the same function of moving people and goods from one location to another, a **Car** with an **Automatic Transmission** lies closer to the Informatical end of the physical-informatical spectrum than its **Manual Transmission** counterpart. The mechanical operation of **Gear Shifting** in a **Car** with a **Manual Transmission** is done by the human driver, whereas in a **Car** with an **Automatic Transmission**, this operation is delegated to the **Car** system.

10.5 Objects, Systems, and Products

Object, system and product are three closely related terms, between which a specialization hierarchy exists. As we argue in this section, object is the most general term, system is a specialization of object, and product is a specialization of system.

10.5.1 Product Defined

Once an artificial system is put to work, it is expected to start providing value or benefit for its users or beneficiaries. The value can be in terms of economical benefit (monetary profit), defensive or security benefit (e.g., a weapon system), scientific, intellectual, cultural, aesthetic, artistic, spiritual, or entertainment benefit. Focusing on economical benefit, which is the most prevalent and tangible type of value that drives product development, there would be people or organizations – the system users or beneficiaries – who would be willing to pay in some way in return for the perceived value that the system provides. This perceived value, in turn, drives the commercialization of the system. Commercializing a system entails its development, manufacturing and servicing in a way that would generate profit. Profit is generated when the revenues from selling the system exceed, at least in the long run, the expenses of developing, manufacturing, and servicing it.

As soon as a system is commercialized, it becomes a *product*. Product, then, is an artificial system, the user (individual or organization) of which is not identical with the entity (individual or organization) that made that system. This implies that the system was created by some entity with the intent of selling it to a different entity. In yet other words, product is a system that has been commercialized. Commerce implies the exchange of hands or the existence of a supply chain with a minimal length of two elements: the seller and the buyer. Accepting that these entities are individuals or organizations, we can adopt the following simple definition for a product.

> **Product** *is an artificial system that is produced by an entity with the intent of selling it to another entity.*

To support this straightforward definition, let us recall the Wright Brothers. When they created the first flying machine, it was a system but not a product. It fulfilled a function but had almost no economical value. However, as soon as people realized the commercial and military potential of airplanes and figured out how to manufacture them economically, airplanes turned from mere systems to into products – commercialized systems. A similar commercialization process characterizes many systems and materials that were originally developed for such purposes as space exploration or weapons, where the economical consideration was secondary.

As Ulrich and Eppinger (2000) noted, products are originally expressed by customers, in the "customer language," but this leaves too much margin for subjective interpretation. For engineering purposes, these requirements need to be specified in precise, measurable detail of what the product is expected to do.

10.5.2 The Object-System-Product Hierarchy

We have seen earlier that **System** is a specialization of **Object**. Recognizing that **Product** is a specialization of **System**, we can now add **Product** to the hierarchy below **System**. Figure 10.6(a) and its OPL paragraph below express this hierarchy.

> **System** is an **Object**.
> **Product** is a **System**.

Figure 10.6(b) and its OPL paragraph below elaborate on Figure 10.6(a) as it specifies the attributes that qualify **Object** as **System** and **System** as **Product**.

> **Object** exhibits **Function,** which can be **significant** or **non-significant.**
> **System** is an **Object,** the **Function** of which is **significant.**
> **System** exhibits **Commercial Value,** which can be **significant** or **non-significant.**
> **Product** is a **System,** the **Commercial Value** of which is **significant.**

Figure 10.7 elaborates on the **Object-System-Product** hierarchy of Figure 10.6(a). **Commercializing** yields **Product,** and **Benefiting** is the process that requires the **Function** of the **System,** which the **Beneficiary** benefits from. Since **Product** is a **System,** it also exhibits **Function.** The **Beneficiary** of **Product** is a **Customer** – a specialization of **Beneficiary** (not shown in Figure 10.7).

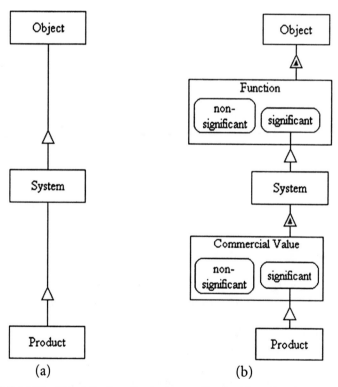

(a)   (b)

Figure 10.6. (a) The **Object-System-Product** generalization hierarchy. (b) The attributes that qualify **Object** as **System** and **System** as **Product**

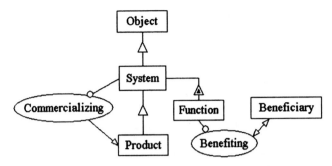

Figure 10.7. The **Object-System-Product** hierarchy of Figure 10.6(a) elaborated with **Commercializing**, which yields **Product**, and the **Function** of the **System**, which the **Beneficiary** benefits from.

Figure 10.8 summarizes the terms discussed so far and put them in context with each other.

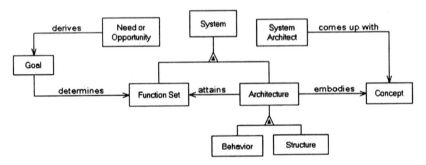

Figure 10.8. The relationships among system attributes and surrounding terms

10.5.3 Goods, Services, and Projects

Products are frequently categorized into *goods* and *services*. They are distinguishable by whether they are objects or processes. Products that are objects are termed "goods," while those that are processes are "services." **Scissors**, for example, are goods, while **Haircutting** is a service. A **Medicine** is a product, while **Health Caring** is a service. The distinction is based on whether the product in question involves an operation, or a process, in which case it is classified as a service. Goods (such as **Scissors** or a **Medicine** or **Food**) are objects, which the customer uses himself, when in need, to provide the required service. A service involves a process (such as **Haircutting, Health caring, Catering**, etc.), that an agent (usually one or more humans) provides. Generalizing these examples, goods and services are two specializations of product that are based on the distinction between objects and processes as two specializations of a thing.

A (commercial) *project* is a specialization of product that is based on entirely different attributes. The distinction between the two is that a product is usually manufactured in mass quantities to make it profitable, benefiting from the economy of scale. Conversely, a **Project** is one-of-a-kind **Product**, usually of significant magnitude in terms of **Cost, Duration** and **Complexity,** that requires careful **Planning** and **Monitoring**. Typical high-end civil engineering project examples of include Boston's "Big Dig" (Central Artery/Tunnel Project, 2001) highway project and China's Three-Gorge Dam (Osburn, 2001). The attributes that enable distinction between a product and a project are **Size** (which should be large for a project) and **Quantity** (one or few for a project).

10.6 Documenting Functions of the System Architecture

By definition, a direct relation exists between the system's function and the architecture that embodies this function. Functions are often not explicitly stated, but they emerge from the system's architecture at various levels. Meaningful phrasing of function names can be used to document the intent of the system architect, which the architecture often does not explicitly express. Lack of such documentation is a problem that designers frequently encounter while reusing an existing design. A certain part or feature, whose function or necessity are not quite understood, is removed just to find out later that it was added to solve a particular problem, which resurfaced once the part was removed. Such intent documentation would prevent these incidents. This section discusses the hierarchy of functions in a system and how to denote it in OPDs. Finally, we delineate the difference between function and functionality.

10.6.1 The Function Hierarchy

The function of the entire system is apparent at the top level of the system, where the system can be seen as a whole. At this level, we can usually point to a main process that functions to attain the system's goal. As we drill down, refine, and increase the level of detail at which this process is described, we deal less with the system's top level function and concern ourselves with lower-level functions that enable the higher-level ones. While increasing the level of detail, we include the system's constituent subprocesses and the objects that are involved in their occurrences. As we do so, we learn more details about the way these processes transform objects. At each level, we should be able to identify the function that the subsystem is supposed to carry out. We can also analyze how this function is embodied in the architecture (structure-behavior combination) of that subsystem.

Analyzing a system, we can see a hierarchical function pattern, in which the top-level description of the system can be decomposed into subsystems, each of which exhibits a distinct function. The function of each subsystem is to attain some sub-goal that serves to ultimately attain the system's goal at the top level. This *function hierarchy*, if properly constructed and understood, can be a powerful tool for designing robust systems. Many system specification and design experts have taken this approach. For example, Hatley and Pirbhai (1988) noted:

> *Requirements and architecture models apply throughout all levels of systems development and both are applied at each level. A set of requirements is generated at a given level, using the requirements model. These requirements are allocated to physical units or modules using the architecture model. The requirements for each module are then expanded to more detail, and the whole process is repeated at the next level.*

In Axiomatic Design (Suh, 1995) this recursive design process is referred to as "zigzagging." Consider, for example, a military system, whose goal is to defend a country. A massive coordinated attack is a process with clear motivation with respect to the military's function. Greasing a tank or filling a cavity in a soldier's tooth, on the other hand, are complex processes within some sub-sub-subsystem of the military that are related to the function of winning a war only remotely and indirectly. The functions that these processes fulfill within the larger system can be inferred by analyzing the adverse effects that refraining from executing them has on the top-level military's function.

10.6.2 Function Boxes and Function Sentences

We demonstrate the function hierarchy concept through a car engine-cooling example, where function boxes and sentences are introduced. A car is a system whose function is to move people and goods. Examining the architecture of a car's water pump, we may be able to understand the structure of the pump and the dynamics of how it circulates water in the car, but this does not explain what role the water pump subsystem plays in moving people and goods. If we analyze (or experience...) the damage caused to the car's engine by failure or malfunction of the water pump, we may be able to conclude that it functions to cool the car's internal combustion engine. The water dissipates the heat generated when the engine functions to produce torque, which, in turn, is transmitted to the car's wheels through a transmission subsystem, enabling the moving of the car.

As Figure 10.9 shows, a function in an OPD is depicted inside a dashed box, called *function box*. A function box encloses at least one object and one process. Since function is at a level that is separate from the structure and behavior, function boxes in OPDs are super-imposed as a layer on top of the object-process layer. The name of a function is phrased as a *command sentence*, or an imperative, written in **Italicized Bold Arial**, which document the function the system architect had in mind for the set of things surrounded by the function box. Function boxes can appear in OPDs at any level of detail. For example, in Figure 10.9 the function at level 0, the top level of the car system, is called "*Enable translation.*" It is recorded within a function box that encloses the object **Car** as an instrument and the process **Moving**, which **Car** enables. The architecture of this top-level function dictates the functions of subsystems at progressively lower levels.

The *Enable translation* function may call for some engine to provide the required propulsion for the translation. Following the above convention for function naming, this level 1 function is called *Generate circular motion*. Applying a different concept, the system architect could have decided that the function would be "*Generate thrust*," which would still achieve the "*Enable translation*" function at the higher level, albeit by different means, or a different *architecture*. A *Generate thrust* function would affect design decisions downstream, possibly yielding a "jet car." This demonstrates the well-known fact that the earlier design decisions affect the product or project more than later ones. The earlier the decision, the more profound is its impact on the final deliverable, because the remaining decisions down the road are derived from it, restricted by the direction it dictates.

Having decided to generate circular motion to achieve translation, the architect now faces a decision of how to achieve it. An engine can generate circular motion. It converts energy, stored in some form, into rotation energy, generating heat as it goes. Taking into account considerations of the energy source for running the engine, an **Internal Combustion Engine** is selected as the propelling subsystem of the **Car** system. This is reflected in Figure 10.9 as the *Generate circular motion* function.

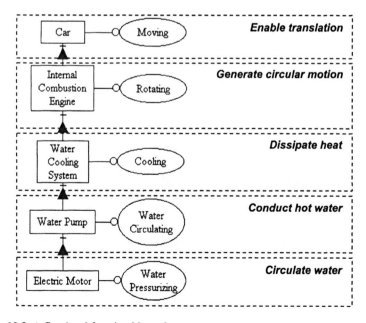

Figure 10.9. A five-level function hierarchy

The heat, which the **Internal Combustion Engine** generates, needs to be dissipated to prevent overheating and enable continuous working. Hence, a function at level 2 is phrased in Figure 10.9 as *Dissipate heat*. This function is attained by **Water Cooling System**, which the **Cooling** process requires. This function calls for a yet lower level function, *Conduct hot water*. The instrument carrying this function is **Water Pump**, which enables **Water Circulating**. To carry out this *Circulate water* function at level 4, the water needs to be driven somehow. The architects elected to do so by pressurizing the water. **Water Pressurizing** is achieved by **Electric Motor**. While not shown in Figure 10.9, the function hierarchy goes on to yet lower levels. For example, to run the electric motor, an electric energy source is required. This function can be phrased as *Run electric motor*. To achieve this, an electric energy source is needed, which needs to be charged, and so on.

Frame 24 contains the OPL paragraph of Figure 10.9. In this simplified example, the architecture that attains the function at each level consists of one process and one object. For example, the top function level is expressed by the instrument sentence **Moving** requires **Car**, which satisfies the *Enable translation* function at that level. At the bottom level, the instrument sentence **Water Pressurizing** requires **Electric Motor** carries out the *Circulate water* function.

Moving and **Car** function to *enable translation*.
   (Function sentence)
Moving requires **Car**.
Car consists of **Internal Combustion Engine** and additional parts.
Rotating and **Internal Combustion Engine** function to *generate circular motion*.
Rotating requires **Internal Combustion Engine**.
Internal Combustion Engine consists of **Water Cooling System** and additional parts.
Cooling and **Water Cooling System** function to *Dissipate heat*.
Cooling requires **Water Cooling System**.
Water Cooling System consists of **Water Pump** and additional parts.
Water Circulating and **Water Pump** function to *conduct hot water*.
Water Circulating requires **Water Pump**.
Water Pump consists of **Electric Motor** and additional parts.
Water Pressurizing and **Electric Motor** function to *circulate water*.
Water Pressurizing requires **Electric Motor**.

Frame 24. The OPL paragraph that corresponds to the OPD in Figure 10.9

A new type of OPL sentence, the *function sentence*, is introduced in this OPL paragraph. To see how a function sentence is constructed, we examine the first function sentence in Frame 24:

Moving and **Car** function to *enable translation*.

The sentence begins with the list of processes followed by objects, to which the reserved phrase "function to" is concatenated, followed by the function name in bold italic Arial font.

10.6.3 Functionality

The term *functionality* is often substituted for a more sophisticated version of function. OPM makes a clear distinction between these two terms. The American Heritage Dictionary (1996) defines functionality as the noun of *functional*. Functional, in turn, is an adjective that means *designed for or adapted to a particular function or use; capable of performing; operative*. In the context of systems development, we will refer to functionality as the extent to which the system performs its function.

> *Functionality is a system attribute that measures the effectiveness of the system's function.*

Consider, for example, a system whose function is to add numbers. The functionality of this number addition system can be measures in terms of speed, accuracy and error rate. An abacus, a laptop computer, and a hand-held calculator are three different architectures of an addition system. The functionality of the hand-held calculator as an addition system is definitely higher than that of the abacus or the laptop computer.

10.7 From Systems to Models

Models are tools that enable us to think and talk about systems by emphasizing essential aspects that are of interest to the system architect. Fox and Gruninger (1998), for example, indicate that "most information systems in use within an enterprise incorporate a model of some aspect of the enterprise's structure, operations, or knowledge." The next sections are devoted to models. Various modeling paradigms are surveyed and the relationships between OPM and these paradigms are discussed.

Regardless of its position along the theory-application spectrum, any scientific research or engineering domain uses some kind of a modeling methodology. The methodology serves as a means to set forth hypotheses, design experiments, explain and communicate research results, and possibly design solutions to a problem that originally triggered the research and development endeavor. Modeling is thus an essential tool of any field of research and development. We build models of complex systems because we cannot comprehend any such system in its entirety. As the complexity of systems increases, so does the importance of good modeling techniques.

10.7.1 Some Model Definitions

Bubenko (1986) and Weber (1987) have noted that if systems analysis and design is to become better understood, better principles and concepts should be developed and a common paradigm must emerge. Bouvier et al. (1997) Claim that a model is a reduction and abstraction of a system. The model is simpler than the system itself. The simplification allows us to handle the problem by focusing only on what is important, leaving out confusing details. The definition of the term model is about as much in argument as the term system. For example, Beer (1999) has indicated that "models are mental constructs of what we rather uncritically call 'reality'." Model building opens exciting possibilities, and in particular liberates creative alternatives in both policy and practice. Indeed, models provide for insightful and potentially inspiring analogies. However, a more basic and not less powerful trait of models is their ability to generate abstractions that help focus our attention on the important aspects of the system being studied and/or developed.

10.7.2 Model Defined

A model is tightly linked with the notion of the system it is aimed to represent. A building, a dam, a highway and an airplane require detailed, scaled, three-dimensional representations of their physical structure, their constituent parts and the way they are combined. Additional models are aimed at designing the strength, capacity, heat transfer, or thrust of these systems. In chemistry and chemical education, molecules are represented as 3-D models that researchers and students can manipulate in space to explore new compounds and study about chemical structure and bonding (Dori, Y.J., 1995). These systems are primarily physical, and their models are therefore mainly spatio-temporal.

When considering software, developing a model for an industrial-strength software system prior to its construction or renovation is as essential as having a blueprint for a large building. However, software models are more abstract since they need to model a system that is primarily informatical rather than physical. We see, then, that different domains often mean different things when referring to the term model. Still, model is a concept that is common to all science and engineering domains. Our definition of model is formulated to capture this:

> A *model* is an abstraction of a system, aimed at understanding, communicating, explaining, or designing aspects of interest of that system.

10.8 Modeling Paradigms

Modeling is the process of generating models. Identifiability, i.e., the ability of a thing to be identified, comes about only from the point of view chosen by the person who models the system – the human modeler. Depending on the modeler's viewpoint, representations of the same system may involve different sets of objects. The point of view of the modeler depends on the aspect the modeler strives to understand, analyze, or design. The identifiable things in the world are called objects. The identifiable causal relationships are "happenings," called processes. Once the modeler has chosen a point of view, the representational scheme for the world is populated with objects and processes that are pertinent to the modeler's perspective.

Approaches to modeling can be classified into three major paradigm categories: natural languages, mathematics, and graphics. In the following sections, we examine each one of these approaches, discuss its virtues and shortcomings, and indicate where OPM is positioned with respect to each modeling approach.

10.8.1 Natural Language as a Modeling Tool

The most common and immediate modeling technique is natural language, free text, or simple prose. Language is the tool of the mind. It enables humans to think freely. Explaining the theory about the role of models, which is attempted in the text you are reading, uses just this tool. In order to think of something, which is a prerequisite of modeling it, that thing needs to have a symbol. The symbol is initially a natural language word, which we can manipulate in our minds and embed in the language we speak while communicating our thoughts to others. As Winograd and Flores (1978) claim, the apparently paradoxical view, that nothing exists except through language, gives us a practical orientation for understanding and designing complex systems, such as computers. This is so because the domain in which people need to understand the operation of computers and computer-based systems goes beyond the physical composition of their parts. The "things" that make up "software," "interfaces," and "user interactions" are clear examples of entities whose existence is generated through language. These concepts exist in the minds of the people engaged in the lingual communication and underline their common understanding by those who think of them and build them.

The topic of whether thought is independent of language, or whether our thinking necessarily requires or involves natural language, is the subject of an ancient debate. Carruthers (1996) has argued for a version of the latter thesis. His view is that much of human conscious thinking is necessarily conducted in the medium of natural language sentences. He claims that the case for the independence of thought from language is by no means as clear-cut as many philosophers and cognitive scientists have assumed.

If thought is independent of such language, then language itself becomes only a medium for the communication of thoughts. According to this communicative conception, the function and purpose of natural language is to facilitate communication and not to facilitate thinking. The communicative conception of language has been widely endorsed in the history of philosophy. However, Carruthers maintains that natural language is involved in our conscious thinking and such thinking is essentially linguistic. We often think in language, and the train of reasoning, which leads to many of our ideas, decisions, and actions, consists of sequences of natural language sentences. This indicates that beside its obvious inter-personal uses, language has an intra-personal cognitive function.

The power of natural language emanates from the culmination of thousands of years of evolution, during which humans, who have been using them, developed the ability to model highly complex systems. Most scientific domains that relate to humans, including psychology, sociology, anthropology, and organizational behavior, use prose as their basic modeling tool, although quantitative assertions about models in these domains are often expressed and tested using mathematics and statistics. Even in the domain of information systems development, some professionals prefer the use of natural language over symbolic or diagramming techniques. This choice is understandable because prose has the expressive power to elicit ideas, assertions and requirements about things in systems. The systems can be of any magnitude and at any level of complexity, and the ideas that can be expressed by free prose are difficult to achieve using any other technique.

10.8.1.1 Natural Language Linearity

Despite their phenomenal expressive power, natural languages do suffer from some basic deficiencies. Two major problems with prose are its *linearity* and *ambiguity*. *Linearity* implies that prose is written and read in a sequential order. However, this is rarely a good thing when it comes to system specification, as links among things in the system under study or development are in general not sequential. Most systems exhibit things that exist or happen concurrently. This non-linearity is difficult to model with prose alone.

Even in systems that have no parallel threads of existence and execution, one must repeat the names of the things in the model for every sentence in which some relation among these things needs to be expressed. Hyperlinks provide for linking concepts and terms that reside apart from each other within a document or across documents. Indexing in traditional paper books, and more so in electronic books, help overcome the problem of text linearity. The advent of hypertext and its use in telecommunication networks, in particular the World Wide Web, alleviates this problem as well.

10.8.1.2 Natural Language Ambiguity

Ambiguity, the second problem with unconstrained prose, is the unavoidable result of the natural evolution and consequent wealth of expression modes these languages exhibit. Many concepts in natural languages have more than one meaning. Concepts may have some extent of overlapping interpretation, which can render precise understanding of the writer's intent, a difficult task. The amusing phrase *time flies like arrow, fruit flies like banana*, whose first and second parts have to be interpreted using different grammatical constructs to make sense, is but an extreme funny example. The following excerpt from Evans (2001) vividly illustrates this point:

> *One of the problems is that we don't have a language (other than English) to describe precisely what the abstractions (associations, dependencies, aggregation, etc.) mean. Another problem is that the abstractions have many possible meanings (each of which has advantages and disadvantages in terms of intuitiveness, simplicity, etc.). We must accept that there is no one right abstraction or semantics. Perhaps if we started with something very basic (e.g., what do we mean by an object!) we would then have something more concrete to build the interesting abstractions on?*

While language variants, connotations and double meanings may be assets for rich and expressive human communication, they can severely hinder unambiguous interpretation, which is essential for precise engineering modeling and scientific inquiry. Using the same wording in prose is considered bad writing style, but many of the synonyms do not span exactly the same range of interpretations. Hence, using synonyms and similes may have an adverse effect on the clarity and conciseness of the text when it serves as a modeling tool. Beer (1999) has phrased this problem eloquently in the context of scientific inquiry:

> *Science is supposed to amass "the facts," and then to form hypotheses based on those facts. But the hypothesis is already covertly implicit in the selection of the facts, and further selection of a less than overt nature occurs during the process of elaborating the hypotheses further. Attention to this kind of absurdity is often dissipated by changes of terminology so that contradictions are accommodated.*

In other words, the use of different terms helps conceal internal inconsistencies, gaps, and possible contradictions among the various statements and requirements in a system requirement specification document. A dictionary that holds accurate definitions of the concepts used in the system's model is an essential ingredient of any model.

Languages usually give up one element in favor of the other: they are either formal and unintelligible to most human beings, or are expressed in free prose, in which case they are susceptible to the pitfalls of potential incoherence and contradictions. OPL, in contrast, is designed such that every sentence has a very concise meaning that stands for a particular constellation of graphic symbols and, by construction, is in line with the rest of the OPL specification. This apparent rigidity is a source of power when it comes to achieving unambiguous human and machine interpretation. Thus, OPM combines the power of a natural language as an instrument of thought with the formality that is required to eliminate ambiguity and enable subsequent automation of the implementation generation process (code, database schema, and human-machine interface).

10.8.2 Mathematical and Symbolic Modeling

Natural science and engineering disciplines enhance their qualitative, prose-based models with rigorous mathematical modeling, which provides for quantitative analysis and prediction ability. As Bouvier et al. (1997) note in the introduction to a paper proposing particle systems as a new paradigm of simulation: "by replacing the real world by equations, one can predict the evolution of the system starting from given initial conditions." Referring to the problem of using natural language in a manner similar to Beer (1999) above, Ashby (1956) has pointed out the value of mathematical form for the study of such complex issues as cerebral mechanisms:

> *I made it my aim to accept nothing that could not be stated in mathematical form, for only with this language can one be sure, during one's progress, that one is not unconsciously changing the meaning of terms, or adding assumptions, or otherwise drifting towards confusion.*

Mathematics encompasses a broad class of modeling approaches under the "quantitative" umbrella. Most mathematical models use symbols (usually alphabetical letters of various languages) put in expressions to specify quantities, relations, and constraints among these symbols. A *mathematical function*[5] transforms entities (which can be considered objects) from a domain to entities in a range. In this sense, it is very close to our definition of process, where the domain entities are the input objects and the range entities are the output ones. Moreover, just as mathematical notation uses symbols, OPM denotes things through symbols. However, the symbols of OPM entities (objects, processes, and states), as well as tags in tagged structural relations, are not arbitrary letters as is usually the case in mathematics. Rather, they are sense making words and phrases that are intelligible to humans and combine to produce sentences that are meaningful to the situation at hand.

Other branches of mathematics use more graphics-oriented notation. A notable example is graph theory, where nodes and edges are respectively used to denote abstract entities and relations among them. OPDs are directed acyclic graphs, whose nodes are classified into two types: objects and processes. Moreover, structural link edges (between pairs of objects) generate tree structures, while procedural link edges (between an object and a process in both ways) generate a bipartite graph. While adjacency matrices provide a complete one-to-one mapping to graphs and are used for computer storage and manipulation, humans prefer the graphic representation for both generation and inspection of graphs. This is because

---

[5]   Note that a mathematical function is different from an OPM function!

humans grasp pictorial information more readily than symbolic one. This is why the main interaction of the majority of humans with OPM is by way of drawing OPDs rather than scripting OPL sentences (which is still much easier than programming in some third- or even fourth-generation programming language).

10.8.2.1 Mathematical Systems

Mathematical functions are OPM systems. A function $f: D \rightarrow R$ is a system that establishes a correspondence between any element (object) in the domain to a unique element (object) in the range. It does so through a specified transformation, which is the **Transforming** process in the mathematical function system.

Generally, some combination of numbers, variables, and parameters are objects in both the domain and the range, and some combination of mathematical operators is the process. Figure 10.10 shows the system **Square Function** $f: x \rightarrow x^2$ as a specialization of a **Mathematical Function** $f: D \rightarrow R$. It specifies that each element in the **Domain** of this function is a **Real Number**, each element in the **Range** is a **Non-negative Number**, and the **Transforming** process is **Squaring**.

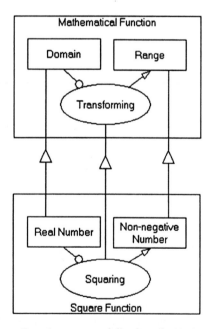

Figure 10.10. The **Square Function** as a specialization of a **Mathematical Function**

Figure 10.11 shows the four basic arithmetic operators as processes, with the conventional mathematical notation below each resulting object. There are operations, such as **Subtracting** and **Dividing**, where the order of objects (variables, or operands) is important. As Figure 10.11 shows, the convention in such cases is that the input link from the first operand terminates at a point in the upper or left half of the process ellipse, and the second operand, at its lower or right half.

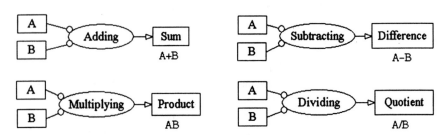

Figure 10.11. The four arithmetic operators as processes

Figure 10.12 show the application for computing the average of three numbers, **A**, **B**, and **C**, in a **Set**, where the **Set** is consumed and not kept.[6] **SD** in Figure 10.12(a) shows **Averaging** as a top-level process, while in **SD1**, **Averaging** is in-zoomed to expose three inner processes: **Adding** and **Counting**. Both processes require all three parts (elements) of **Set** and can be executed in parallel. **Dividing Sum** by **Quantity** yields the **Average**. Note that following our convention, **Sum** is the dividend and **Quantity** is the divider, because the consumption link from **Sum** to **Dividing** is to the left of that from **Quantity** to **Dividing**.

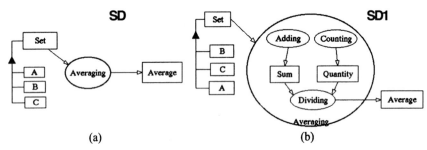

Figure 10.12. Averaging of three numbers in a **Set**. (a) Top-level OPD. (b) Zooming into Averaging shows **Adding** and **Counting** as two processes, which can be done in parallel, and **Dividing Sum** by **Quantity** to get the **Average**.

Figure 10.13 shows **Integrating** and **Differentiating** as inverse processes, or operations. **Integrand** and **Integral** are both **Mathematical Functions**. **Integrating** requires **Integrand** and yields **Integral**. Conversely, **Differentiating** requires **Integral** and yields **Integrand**. We use instrument links to denote the fact that the functions are informatical objects that are not lost in the process.

---

[6] The examples so far assumed that the input objects (variables or numbers) are not lost, therefore an instrument link connected them to the the operation.

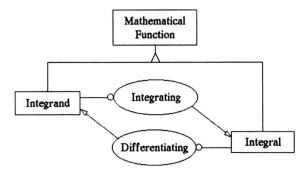

Figure 10.13. Integrating and **Differentiating** as inverse operations

10.8.2.2 Chemical Reactions as Processes

Chemistry is a science that is heavily based on symbolic representations of atoms, molecules and reactions. Chemical reactions are obviously processes, with the reagents being the objects. Consider, for example, the reduction-oxidation (redox) reaction $Ni(OH)_2 + H_2SO_4 \rightarrow Ni\,SO_4 + H_2O$. Figure 10.14 is an OPD specification of this reaction. If a reaction involves a catalyst, it would play the role of instrument with respect to the Reacting process.

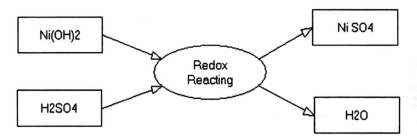

Figure 10.14. An instance of a chemical **Redox** (reduction-oxidation) **Reacting** system

The OPD in Figure 10.15 has exactly the same meta-structure as the one in Figure 10.14. The difference is that each thing in Figure 10.15 is a generalization of its counterpart in Figure 10.14. This correspondence is explicitly depicted in Figure 10.16. Finally, we abstract the systems in Figure 10.14 as **Redox Reacting System** and that in Figure 10.15 as **Nickel-Sulfate Redox Reacting System**. Figure 10.17(a) specifies the latter as a specialization of the former. In Figure 10.17(b), each system is in-zoomed to expose its things and enable the specification of the gen-spec relationships within pairs of things, one belonging to the generic systems and the other – to the specialized one.

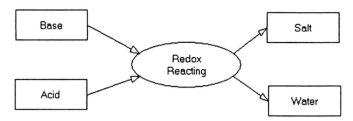

Figure 10.15. A generic **Redox Reacting** system

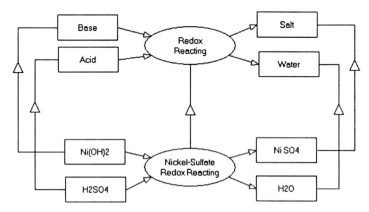

Figure 10.16. The specialization correspondence between things in the generic and specialized **Redox Reacting** systems

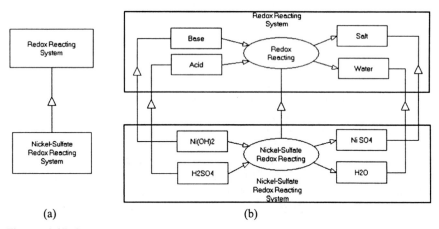

(a)   (b)

Figure 10.17. Out-zoomed (a) and in-zoomed (b) views of the generic **Redox Reacting System** and its specialization

Similar systems, such as elementary particle reactions in physics, can be effectively specified with OPM.

10.8.3 Graphic Modeling and Knowledge Representation

While discussing the question "*Does the knowledge require graphic representation?*" in their book *Knowledge Elicitation*, Firlej and Hellens (1991) noted:

> *The term "graphic representation" ... means anything that can only be adequately represented by using drawings or sketches. Some domains do have a large component of graphic content. ... Sketching and drawing may ... provide a useful way of expressing idea on a paper that is difficult to express in words. ... These drawings may be indispensable for the user and a practical method for the representation of these drawings and plans may have to be found to achieve acceptable functioning."*

Indeed, diagrams are often invaluable for describing models of abstract things, especially complex systems. The fact that most people use some kind of diagramming technique to express their knowledge or ideas about systems is a testimony to the viability of the graphic representation. It is an endorsement of the clear advantage graphics has over the textual representation.

Having acknowledged the merit of the graphic representation, such representation of our knowledge about a system is valuable if and only if it is backed by a comprehensive and consistent modeling methodology. Such methodology is essential if we are to be able to understand complex systems in any domain and communicate our understanding to others. The obtained analysis results can serve as a basis for designing solutions to existing problems, as well as modify and enhance current systems. An accepted diagramming method has the potential of becoming a powerful modeling tool if it constitutes an unambiguous language. Such visual formalism is valuable if each symbol in the diagram has a defined semantics and the links among the symbols unambiguously convey some meaningful information that is clearly understood by those who are familiar with the formalism. OPDs provide this level of formality.

Scientific papers that describe systems of all kinds and domains sorely lack a common, acceptable formalism for describing systems. Examining diagrams that people draw with the intention of expressing some system aspect, we often find implicit, undocumented, arbitrary collection of symbols and links. Graphic representations often use ad hoc symbols that the authors make up spontaneously to be used in some diagram without caring to specify their meaning. The same graphic symbol may be used in contradictory, perplexing ways.

This claim may sound all too familiar and even trivial. Nonetheless, as striking as it may be, there is no widely accepted *single* model for representing systems. The claim of Bubenko (1986) that no generally accepted, workable theory of information systems and their development has evolved, is true for systems in general. If no accepted theory exists, one cannot expect that a graphic notation of systems specification be widely accepted. Many articles contain inexplicable inconsistencies among symbols in different diagrams, or, worse yet, within the same diagram.

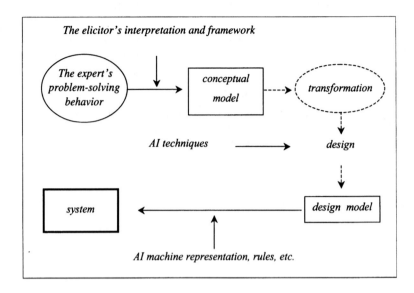

Figure 10.18. The KBS modeling development process

To support this claim, consider, for example, the graphic representation in Figure 10.18 of "The knowledge-based system (KBS) modeling development process", as it appears in *Knowledge Elicitation* by Firlej and Hellens (1991). Examining this diagram, we can try to infer the intentions of the symbols in the diagram. However, regardless of the diagram content, contemplating upon it for a while, many questions come to mind. For example, it is not possible to understand what is the difference (if any) between text surrounded by an ellipse, text surrounded by a rectangle, and non-surrounded text. Is it indeed the case that ellipses represent processes? Why, then, doesn't an ellipse surround design as well? Is it because there is a profound difference between transformation and design? Furthermore, why is one ellipse solid while the other is dashed? Why are some boxes drawn with a thin line while another with a thick one? Why are some arrows solid, while others are dashed? What is the meaning of an arrow anyway? What is the difference between an arrow connecting surrounded text and one that connects non-surrounded text with another arrow? Did the diagram designer have any intention in making distinctions through different graphic constructs to convey particular meanings? If so, how are we supposed to interpret them? Since we are supplied with neither legend nor an explanation within the text, we are left puzzled with these questions. These questions are not meant to criticize the authors; unfortunately, such pictographs abound in the best of textbooks and scientific journals as representations of systems or parts thereof. The scandalous state of affairs in graphic systems modeling is that this is the rule rather than the exception, and few care to change this.

It is not just the case that the richness of graphic symbolism is not taken advantage of – the arbitrary use of graphic symbols potentially adds more confusion than it provides help. Not only are such diagrams not always helpful, they may even obstruct the understanding of a system's architecture. Probably, the most advanced formalism in graphics notation has been achieved and agreed upon in the software engineering community, especially with the acceptance of UML (Object Management Group, 2000) as a standard. UML, however, focuses on software systems rather than on systems in general, and its notation is complex. A comprehensive report on the evolution of software engineering methods, with emphasis on recently developed object-oriented methods, is provided in Chapter 15. The full potential of graphic representation is yet to be realized. A major contribution of OPM to systems science and engineering is the precise semantics and syntax it ascribes to graphic symbols and the unambiguous association of graphic symbols with natural language constructs.

10.9 Reflective Metamodeling

The premise of OPM is that every thing in the universe is ultimately either an object or a process. One may wonder, then, what are the links between objects and processes in an OPD? Are they objects, processes, or a third type of thing we have not yet encountered? This is a confusing question, the answer to which takes some insight to comprehend. The source of the confusion is that we do not realize that we are mixing levels of thought. We need to recognize the existence of (at least) two levels of thought: the modeling level and the metamodeling level.

The modeling level is the level we are normally at while trying to engage in the process of modeling. In a broad sense, we can think of any thought process as a modeling process – our brain is trying to figure out what we observe (the objects), how what we observe behaves (the processes), and what to expect based on this comprehension. The metamodeling level is the level at which we are at when we reflect upon how modeling is done. A metamodel is the model of the modeling methodology. Once we realize this, the answer to the question posed at the beginning of this section becomes obvious. Entities (objects, processes and states) as well as the various links are objects in the OPM metamodel, since time is not relevant to their being – they exist as concepts, primarily in our minds, independent of the time dimension. At the modeling level, the entities are objects, processes and states. For example, the follwing specification, which appeared in Chapter 9, is an example of a meta-model specification: "**Homogeneous** and **Heterogeneous** are two values of the **Homogeneity** attribute of a fundamental structural relation. The **Homogeneity** value of **Characterization** is **Heterogeneous**, while the other three are **Homogeneous**."

Figure 10.19 is a meta-OPD of an OPM **Connector**. It specifies, for example, that **Structural Link** and **Procedural Link** are **Connectors**, that **Connector** exhibits **Source** and **Destination**, and many more details. As this meta-OPD shows, all the things are objects. This is typical of meta-models, where the declarative, structural part is the lion's share of the meta-specification.

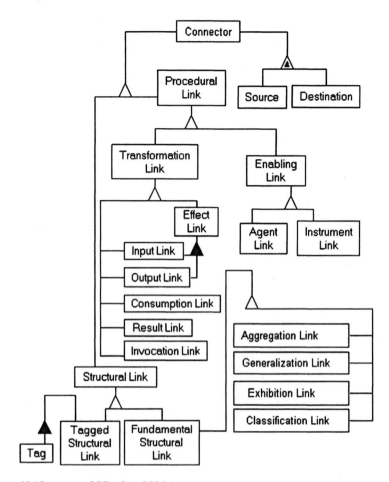

Figure 10.19. A meta-OPD of an OPM **Connector**

Processes in a meta-model specify behavior of the modeling method. Examples of processes at the OPM meta-model level are refining and abstracting, as well as transforming. These are types of behavior that characterize the modeling method itself, not the reality that OPM models. Referring to OPM as a system in its own right, it is frequently handy to specify OPM concepts using OPM itself. This is called reflective metamodeling.

> ***Reflective metamodeling*** *is modeling of a modeling method through itself.*

The unique reflective metamodeling ability of OPM is indicative of its expressive power. Throughout the book we have been switching back and forth from the modeling level to the metamodeling level without stating so explicitly. For example, Figure 4.3(b), which shows **Object** and **Process** as specialization of **Thing**, distinguished by the value of their **Perseverance** attribute, is a meta-OPD. As another example, while discussing **Feature** in Chapter 7, we noted that **Feature** exhibits several attributes. This discussion is at the OPM *metamodeling* level.

A *meta-OPD* is an OPD that applies reflective metamodeling, i.e., an OPD that specifies some portion of OPM. At this level, we apply reflective metamodeling to model OPM using OPM constructs. At the metamodeling level, we refer to OPM objects and processes, such as **Feature**, **Attribute**, **Operation**, **Value**, or **State**, just as we refer to "regular" objects and processes, such as **Machine** and **Manufacturing** at the modeling level. We even use the same typeface to express names of things at both levels. Thus, the sentence "We focus first on **Attribute**, and discuss its **Size**, **Objective**, and **Touch** attributes." implies that we are at the metamodel level, where **Attribute** is an object, and **Size**, **Objective**, and **Touch** are its attributes.

Summary

- A system is an object, which, in the eyes of some observer, carries out a particular function.
- Whether an object is a system is in the eyes of the beholder.
- Function describes the rationale behind the existence of a system, the intent for which is was built, the purpose for which it exists, the goal it serves, or the set of phenomena or behaviors it exhibits.
- An artificial system embodies a certain design concept and somehow benefits one or more people.
- Concept is the system architect's strategy for a system's architecture.
- Architecture is the combination of structure and behavior of a system that attains its function.
- Origin and essence are two important system attributes.
  - The values of the origin attribute are natural and artificial.
  - Systems whose origin is artificial are man-made. All other systems are natural.
  - The values of the essence attribute are physical and informatical.
  - Informatical objects are recorded on physical ones.
- Product is an artificial system that is produced by an entity with the intent of selling it to another entity.
  - Objects, systems and products constitute a specialization hierarchy, with objects being the most general and products, the most special.
- System's function and system's dynamic are two related but not equivalent terms. The same function can be achieved by different architectures.
- The function hierarchy is the hierarchy of functions in a system such that a set of lower level functions support or help achieve functions at higher levels.
- A function box in an OPD encloses a set of things that act to attain a certain function.

- A model is an abstraction of a system, aimed at understanding, communicating, explaining, or designing aspects of interest of that system.
- Natural languages, mathematical symbols and graphics are the main modeling paradigms.
- Reflective metamodeling is modeling of a method through itself.

Problems

1. Pick three systems and define their function.
2. For each system in Problem 1, describe the concept of the system's operation and explain how its architecture achieves the function. You may use prose or OPDs to do this.
3. Write OPL for the OPD in Figure 10.4. Compare and contrast the two systems.
4. Repeat Problem 3 for the system in Figure 10.3. Add a generalizing process to the OPD and OPL.
5. In Figure 10.7 add **Customer** as a specialization of **Beneficiary** and show how **Using** benefits the **Customer**. What is the relation between **Using** and **Bene-fiting**? Write the corresponding OPL paragraph.
6. Draw an OPD where **Good, Service**, and **Project** are added to the **Object-System-Product** hierarchy.
7. Provide two examples, each with two different systems that have the same goal and function but a different architecture.
8. Demonstrate the function hierarchy on a system of your choice. Draw an OPD to express this and provide the corresponding OPL.
9. Compare the role of modeling in systems architecture to the role of modeling in the architecture of buildings.
10. Show in an OPD or in free text how the function of a subsystem of filling a cavity in a soldier's tooth fits in the military's function hierarchy.
11. Model as much as you can from the following information (Encyclopedia.com, 2001):

    The atom is made from the proton, neutron, and electron. It turns out that protons and neutrons are made of varieties of a still smaller particle called the quark. At this time it appears that the two basic constituents of matter are the lepton and quark, of each of which there are believed to be six types. Each type of lepton and quark also has a corresponding antiparticle: a particle that has the same mass but opposite electrical charge and magnetic moment. Quarks appear to always be found in pairs or triplets with other quarks and anti-quarks-an isolated quark has never been found. The most familiar lepton is the electron; the other five leptons are the muon, the tau particle, and the three types of neutrino associated with each: the electron neutrino, the muon neutrino, and the tau neutrino. The six quarks have been whimsically named up, down, charm, strange, top (or truth), and bottom (or beauty); the top quark, which has a mass greater than an entire atom of gold, is about 35 times heavier than the next biggest quark and may be the heaviest particle nature has ever created. The quarks found in ordinary matter are the up and down quarks, from which protons and neutrons are made. A proton, for instance, consists of two up quarks and a down quark, and a neutron consists of two down quarks and an up quark. (Quarks have fractional charges of one third or two thirds of the basic charge of the electron or proton.)

12. Make a tabulated comparison of the advantages and shortcomings of the three modeling paradigms: natural language, mathematics, and graphics.

13. Write the OPL script of the OPD in Figure 10.8 and explain in your own words what these OPL sentences mean.

14. Draw an OPD and the corresponding OPL script of a flying system based on the following passage:

"A flying system can be an airplane, a glider, a Zeppelin, or a hot-air balloon. Although each is based on a different concept, all attain the function of carrying humans and goods in the air. While airplane and glider are heavier than air, and take advantage of Bernoulli's Law, Zeppelin and hot air balloon are lighter than air and take advantage of floatation. While airplane and Zeppelin are propelled, glider and hot-air balloon are not. Yet, all attain the function of carrying humans and goods in the air."

15. The OPD in Figure 10.20 specifies a portion of a document written for the US Federal Aviation Administration (RTCA Select Committee for Free Flight Implementation, 2000) on the future National Aviation System (NAS). Reverse engineer the requirements in that portion of the document by figuring them out from Figure 10.20. You may want to write the OPL paragraph first.

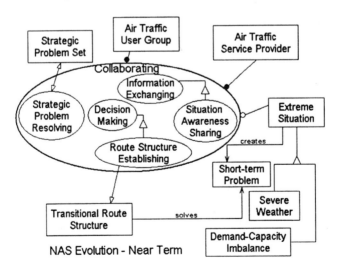

Figure 10.20. Problem 15

Chapter 11

System Lifecycle and Evolution

> *There is a perfectly killing little place that I want to show*
> *you, Anne. It wasn't built by a millionaire. It's the first*
> *place after you leave the park, and must have grown*
> *while Spofford Avenue was still a country road. It did*
> *grow – it wasn't built!*
>
> Anne of the Island by Lucy M. Montgomery

A complete methodology is not just an ontology and a set of notations. It prescribes a system whose function is supporting the lifecycle and evolution of an artificial system. There is confusion regarding the various terms that pertain to this issue. Most methods refer to the *development* of *software systems*, called the "Software Process", where there is a consensus that development entails analysis, design and implementation. OPM considers development as just one (albeit a central one) of three processes, the other two being initiating and deploying. This chapter introduces the architecture – the combination of structure and behavior through interacting objects – that comprises the objects and the processes that affect them, to attain the entire system function. We do this by way of presenting a set of OPDs and discussing their content using OPL. Since the architecture is aimed to be as general as possible, this chapter may seem abstract in first reading. The description of the **System Lifecycle and Evolution** system is probably initially quite vague. However, as we move forward and drill down to increase the level of detail of the system specification, the specification of the system's architecture gets less fuzzy and more concrete. This chapter discusses system lifecycle and applies OPM to describe the lifecycle and evolution of artificial systems.

11.1 System Lifecycle

A fundamental characteristic of systems is the lifecycles they exhibit. Knowledge about lifecycles of galaxies and planets is quite established. Examples of biological lifecycles and lifecycles of materials such as water and nitrogen in planet Earth abound. The life of a natural or an artificial system begins with its creation, construction, generation, or birth. As the system proceeds through its life, processes cause the system to undergo a series of state transitions. The processes that cause the transition from one state to another may also affect objects other than the system undergoing the evolution. Retreats and non-deterministic behavior can possibly accompany the progression along the series of states. This progression stops when the system reaches a terminal state – a state from which it cannot escape. At this point, the life of the system has come to its end. A new system of the same class is usually born at some point during the object's life. Next, we focus on lifecycle of artificial systems.

11.1.1 Lifecycle of Artificial Systems

The lifecycle of an artificial system typically starts with requirements. These are driven by factors such as the emergence of a new technology, the identification of a business opportunity or competition, or a need to solve a problem that has been deemed to harm the organization. Requirements are usually specified in unconstrained natural language prose. While rich in expressiveness, natural languages enable the posing of a list of requirements that may be contradictory or infeasible. The system architect must therefore analyze the requirements to fully understand the domain, the system and the environment. The analysis is the basis for design and implementation, which, in modern systems, are done in an iterative process of stepwise refinement and augmentation of an initial evolving prototype. This process continues until a full-fledged system, which meets the requirements, is in place, tested, assimilated and operational within its designated environment. If the system involves a significant hardware component of any kind, then the design must also accommodate for a proper, environment-friendly process for its disposal.

The main portion of the system's lifecycle is expected to be its deployment: usage and maintenance. Ideally, during this phase, the system provides the function it was designed to carry out. Frequently, the functionality of the system is partial, and as it is being utilized, new requirements arise, either due to new needs or omission of the functions from the requirements or from the architecture. It is up to the system architect to monitor these deficiencies and consider remedying them in the next generation of the product.

11.1.2 Software and Product Development Processes

Due to the dynamic nature of the environment, the functions the system is expected to fulfill, as expressed by the requirements, are highly likely to change between the time they were first formulated and the time the system becomes operational. As the time span of systems shortens, the rate of changes per time unit increases, causing system development to resemble "shooting a moving target." As Maciaszek (2001) noted:

> The literature on information systems management is full of examples of failed projects, exceeded deadlines and budgets, faulty solutions, unmaintainable systems...

The reasons for this well-documented fact are numerous and multifaceted, but many of them share in common lack of correspondence between requirements and system functionality. Strict configuration management must be exercised to minimize this effect and relate system capabilities to the requirements it is expected to meet.

Software developers often refer to the orderly development of software as *software engineering*, the *software process*, or shortly, the *process*. In OPM, the term "process" is reserved and has a very specific semantics of a thing that transforms an object. OPM refers to the entire lifecycle of systems as *system evolution*. The term system is more adequate than software systems, because software, as important and complex as it may be, is just one facet of the system and its development process. Evolution is a more adequate term than development, because development is only one of three major phases in a system's lifecycle, which are initiation, development and deployment.

These three processes and the objects that are involved might collectively be called the "OPM software process," or more generally, the "OPM system process," because it prescribes the development and lifecycle support of not just software systems but systems in general. In such systems, software may be a component, a subsystem, or a component in one or more subsystems. But since the OPM word **Process** is reserved for a more specific and more basic need, we'll stick with system evolution.

One process, which has been the subject of increasing attention and research in recent years, is the *product development process*. Since a product is a commercialized system, its development must take in account factors that affect its lifecycle cost while aiming at maximal functionality. It is here where factors such as regulations, competition, manufacturability, maintainability and supply-chain design become important considerations.

11.2 Systems Analysis and the Scientific Method

Systems analysis is the discipline that models a portion of reality within some domain in order to understand it, and, based on this understanding, possibly design new systems and/or improve existing ones. According to this definition, there is a continuum of purposes for the use of system analysis, which extends from theory to application. On one extreme of the spectrum lie the pure, basic sciences, whose primary goal is to enable human understanding. More specifically, science aims to satisfy human natural desire to know and have answers to questions of ever increasing depth and complexity regarding natural, social and artificial phenomena. The acquisition of knowledge has been made through generations of human endeavor to employ the scientific method. This approach advocates making and recording observations, analyzing them, building models in order to understand the laws of nature that govern the structure and behavior of systems and trying to make predictions that would prove or disprove the suggested model.

Figure 11.1 is an OPD of the scientific method. It explicitly shows how, given **Observations** of **Phenomena**, the **Research Community** generates **Hypothesis**, formulates a **Theory**, which is initially only "**conceived.**" **Experimentation** then determines whether **Prediction** and **Experimental Result** agree. If so, the **Theory** is **established**, else it is **refuted**, and a new **Theory** is formulated instead.

Through the understanding gained in this process, humans have become capable of explaining and possibly predicting phenomena in various domains. This category encompasses the natural sciences that include physics, chemistry, biology, and all the accompanying interdisciplinary sciences, with mathematics serving as the primary underlying modeling tool. Philosophy, which combines knowledge from natural and human sciences with mathematics, can also be considered a discipline on this side of the spectrum.

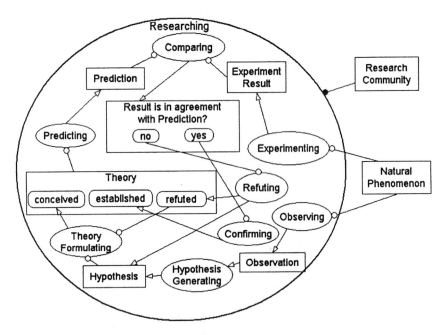

Figure 11.1. OPD of the scientific method

On the other extreme of the spectrum are all the domains of engineering, which are built on top of the valuable knowledge that humans engaged in the pure sciences have accumulated throughout the years. In-between these two extremes are a plethora of applied sciences that can be mapped along this theory-application orientation continuum. As part of an artificial system's lifecycle, where design typically follows analysis, analysis should handle the requirements of the domain or domains within which the system is developed. Ideally, the specifics and peculiarities of some solution, which is part of the design phase, ought to be disregarded as much as possible.

As Ruckelshaus (1990) has noted, although a scientific explanation may appear to be a model of rational order, one should not infer that the explanation was born from an orderly process. At the cutting edge of most scientific fields in particular, analysis itself, which ultimately yields that orderly reasoning, is chaotic and fiercely controversial. The same applies not only to science, but also to analysis and design of systems in the more general sense. As we noted in Chapter 9, real-life analysis and design of a complex system typically starts "middle-out" rather than the classical book recipe of "top-down" or "bottom-up." This is so because starting at either the top or the bottom mandates some prior knowledge of the architecture of the system under consideration, which is frequently not available from the outset. OPM recognizes this reality and provides analytical tools and mechanisms, such as consolidating, zooming and folding, discussed in Chapter 9, to manage "middle-out" beside the top-down and bottom up types of analysis.

11.3 Categorization vs. Interdisciplinarity

As system magnitude and complexity are constantly increasing, it becomes ever more difficult to identify a system that exists and operates within a single, specific scientific or engineering domain. Most non-trivial systems involve a host of such domains, and require a considerable amount of interdisciplinary knowledge, along with transfer and integration capabilities.

According to Bar-Yam (1997), for many years, professional specialization has led science to progressive isolation of individual disciplines. As Beer (1999) noted, circular causality has resulted in the undesirable phenomenon of what he call the *hardening of the categories*. Initiated primarily by librarians, who needed to classify books according to their main categories, university departments were founded along these disciplines. As they grew, they formed artificial walls among themselves and made it a point to hire new faculty that specialize in relatively narrow domains that are central to the discipline. As various taxonomies have emerged, it has become harder to integrate various science disciplines. However, a realization is starting to emerge amongst scientists and research funding agencies: separating science into various disciplines, with rigid borders among them, impedes the advancement of science. These are encouraging signals that ideas of General Systems Theory that theorists expressed at the first half of the 20th Century, as elaborated on in Chapter 14, are finding their way into science and engineering and are starting to materialize.

11.4 System Engineering and the Role of the System Architect

Engineering domains include traditional, well-established fields, such as mechanical, electrical and chemical engineering, as well as fields that are relatively younger, but already recognized and respected, such as computer and industrial engineering. The youngest and probably least recognized and established is the field of systems engineering. The system engineering discipline is based on system science, encompassing principles of General System Theory and other emerging theories. System science and engineering are based on the premise that there exists a set of principles, which are common to all systems. This emerging school of thought develops and utilizes tools for analysis, design, implementation and life-cycle support of systems, regardless of their domain. Engineering systems are man-made systems that exhibit significant cross-discipline components and require massive use of systems engineering. Leading institutions of higher education are just now starting to recognize the centrality and importance of engineering systems. MIT's Engineering Systems Division[1] is a prime example.

A system architect (also called "system integrator") is a senior systems engineer who is responsible for the development and lifecycle support of the architecture of large-scale systems that typically involve several engineering disciplines. As discussed in Chapter 10, the architecture prescribes the combination of the system's structure and behavior that attains its required functions under given constraints. To play the role of a systems architect, a systems engineer should be knowledgeable enough in the science and engineering fields that are pertinent to the system being studied and developed. To demonstrate the multidisciplinary nature of systems analysis and design and the need for a holistic system view, consider for example a large enterprise, engaged in the fabrication and distribution of microprocessors. Three major things (objects or processes) can be identified in the system: the product, the manufacturing and the organization. The products that the enterprise (the system) makes are based on electronics engineering, which, in turn, builds on semiconductor physics and material science. The manufacturing and the associated logistics are based on industrial engineering principles, which, in turn, require operations research that is based on mathematics. Organizational theories and human factors engineering address the structure and behavior of the organization as a whole. The considerations underlying its manufacturing are based on management and economy concepts. This familiarity with the domain of discourse should go hand in hand with a profound overall systems view, which enables the integration of the various system components at all levels.

11.5 An OPM Model of System Lifecycle Phases

The lifecycle of an artificial system typically includes initiation, development and deployment. In this section, we present an OPM model of a typical system's life-cycle and show how OPM can support system development throughout its entire lifecycle. The support starts right at the initiation stage, where requirements are gathered, through development (analysis, design, and implementation) all the way to deployment (usage, maintenance, disposal and initiating a new generation).

---

[1] http://esd.mit.edu

11.5.1 Top-Level Description of System Evolution

It may be useful to start with the generic system-environment OPD described in Figure 11.2. The system-environment OPD describes the system, its environment and the interacting process between them.

Figure 11.2. The generic system-environment OPD

Figure 11.3 is a top-level OPD of **System Lifecycle and Evolution** system, which is a variation of the generic system-environment OPD in Figure 11.2 that is geared towards artificial systems developed for a customer. Here, the system at the highest, most abstract level is called **System Lifecycle and Evolution.**

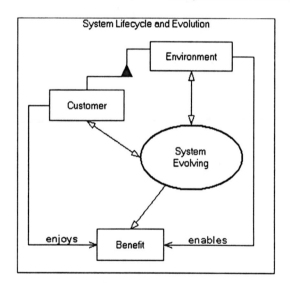

Figure 11.3. SD of **System Lifecycle and Evolution**

The **System** that is being developed and evolved is not visible at this most abstract level; it is an object inside the **System Evolving** process. The **Customer** is part of the **Environment,** which includes, among other things (not shown in the diagram), the **Customer**'s customers and suppliers, competitors, state-of-the-art technology, rules and regulations, and economic climate.

The dynamics of the **System Lifecycle and Evolution** system occurs within its top-level process, which is **System Evolving.** This process affects both the **Customer** and the **Environment,** and yields **Benefit,** which the **Environment ena-bles** and the **Customer enjoys.** Following the statement in the beginning of this section, the **System Evolving** process zooms in Figure 11.4 into three subprocesses: **Initiating, Developing**, and **Deploying. Initiating** yields the **System**, which fulfils the **Required Function Set**. To do this, it requires the **Environment**, the **Required Function Set**, and an **Integrated Systems Evolution Environment**, abbreviated **I SEE**, which implements OPM. **Customer** handles the **System Evolving** process, **elicits** the **Required Function Set, owns** the **System** and **enjoys** the **Benefit**, which the **Environment enables** and the **Deploying** process yields.

Developing, which **I SEE** handles, affects the **System** and yields the **Implementation**, which is part of the **System. Deploying** requires **I SEE** and the **Implementation**, and affects the **Required Function Set**. The effect is manifested in changing existing requirements and generating new ones as new experience is gained with the system while it is being deployed. We elaborate on the **Required Function Set** while discussing the **Deploying** process. When the **Required Function Set** is affected to the extent that the functionality of the **System** is no longer satisfactory, this new **Required Function Set** is again the instrument that triggers a new occurrence of **Initiating**. The beginning of a new generation of **System** completes a cycle in the life of the **System**. In the following sections we traverse the OPD set of **System Lifecycle and Evolution** system in a breadth-first fashion. This means that before drilling down to any subprocess at a refined level, we go over each of the processes at the abstract level.

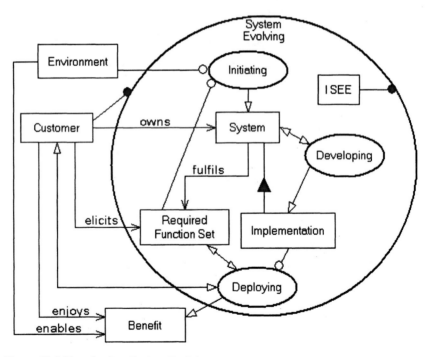

Figure 11.4. Zooming into **System Evolving**

11.5.2 Initiating the System

The birth of a system is a result of its **Initiating** process, which comprises three subprocesses: **Identifying**, **Conceiving** and **Initializing**. As Figure 11.5 shows, **Environment** exhibits the attribute **Need/Opportunity**: there are needs and/or opportunities in the **Environment** related to the **Customer**. **Customer** exhibits the attribute **Need/Opportunity Awareness**, with states **low** and **high**. **Identifying**, the first subprocess of **Initiating**, changes **Customer**'s **Need/Opportunity Awareness** state from being initially **low** to **high**, such that following the identification, the customer is aware of the needs and/or opportunities that the environment exhibits.

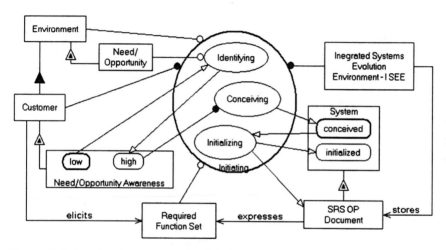

Figure 11.5. Zooming into the **Initiating** process

That the customer has identified the need and/or opportunity is denoted in the OPD by **Customer**'s **Need/Opportunity Awareness** attribute being in state **high**. **Customer** can now engage in the next subprocess, **Conceiving** the system. The **Conceiving** process yields a new **System**. As the OPD shows, the initial value of **System**'s state is **conceived**. At this state, the **System** enters the **Initializing** process. **Initializing** yields **SRS OP Document** (System Requirement Specification Object-Process document) as an attribute of the **System**, and transfers **System** into the **initialized** state. **I SEE**, which enables the **Initiation**, stores the **SRS OP Document**.

11.5.3 Developing the System

Developing is the central subprocess within the **System Evolving** process. In the OPD of Figure 11.6, **Developing** zooms into three processes, **Analyzing, Designing**, and **Implementing**, which are the major activities within **Developing**. Each of these processes yields part of the system. **Analyzing** yields **Analysis OP Document, Designing**, the **Design OP Document**, and **Implementing**, the **Implementation**.

11.5.4 Analyzing

Analyzing is the process within **Development** that pertains to probing into the requirements the system is supposed to meet, as expressed in the **SRS OP Document**, and translating them into a concise model of the pertinent domain and the system within it. Analyzing uses OPM right from the outset. As Figure 11.6 shows, the deliverable of **Analyzing** is the **Analysis OP Document**, which contains the analysis OPM system specification: the OPD set and its equivalent OPL script. This document fully conveys the knowledge gained by the **Architecting Team** whose members collaborate with domain experts on the side of the **Customer** in carrying out the analysis. The **Analysis OP Document** can contain the description of the system at its current state. It can contain more than one alternative for a proposed solution from the domain's perspective along with a scheme for evaluating the alternatives. This way, if a selected alternative turns out to be problematic during subsequent development phases, another one can be used.

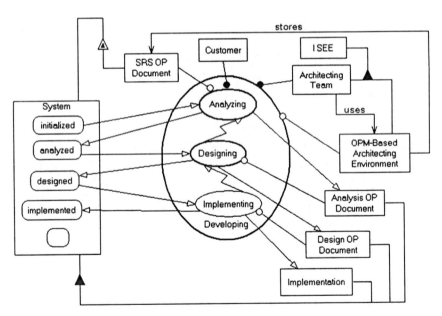

Figure 11.6. Zooming into the **Developing** process exposes its three main subprocesses: **Analyzing, Designing** and **Implementing**.

Unfolding **Integrated Systems Engineering Environment**, one can see that **I SEE**, consists of the **Architecting Team**, which is the agent for **Developing**, and the **OPM-based Architecting Environment**, which is the instrument for **Developing** that the **Architecting Team uses**. **Architecting Team** is the group of systems architects and integrators, analysts, designers, and developers, who, along with the domain experts, handle **Developing**. **OPM-Based Architecting Environment** is the software instrument the **Architecting Team** uses to apply OPM-based system development. When **I SEE** folds back, such that its two parts are concealed, of the two enabling links that run from the parts of **I SEE** to **Developing,** only the agent link remains. This is so because the agent link has precedence over the instrument link.

Analyzing, handled by the **Architecting Team** and the **Customer**, requires the **SRS OP Document** and an **initialized System**, i.e., a **System** which is at the state **initialized**. **Analyzing** changes **System** from **initialized** to **Analyzed**. **Analyzing** yields **Analysis OP Document**. The **Analysis OP Document**, like any OP document, is a set of inter-related OPDs and corresponding OPL script. This is the document that expresses the result of the analysis of the system in terms of names of objects, processes, states and tagged structural relations that reflect the problem domain and the system within it.

11.5.5 The Refining-Abstracting Cycles

In Chapter 9 we elaborated on the reality of middle-out analysis. We can now probe further into the observation that analysis and design mix top-down, bottom-up and middle-out modes or specifying systems. The middle-out mode results from refining-abstracting cycles. Each such cycle comprises a refining phase and an abstracting phase. These **Refining** and **Abstracting** subprocesses are depicted in Figure 11.7, where **Analyzing** is in-zoomed just for the purpose of explaining the Abstracting-Refining cycles. Note that in Figure 11.9, **Analyzing** is in-zoomed in a different manner, which is part of the **System Lifecycle and Evolution** system.

Using **Knowledge Sources**, the **Architecting Team** builds an initial **Analysis OP Document** through the process of **Refining**. This **Analysis OP Document** exhibits two attributes: **Completeness**, with states **incomplete** and **complete**, and **Clarity**, with states **unclear** and **clear**. The goal of the **Architecting Team** it to produce an **Analysis OP Document** that is both **complete** and **clear**. However, both **Refining** and **Abstracting** can potentially change both **Completeness** and **Clarity**.

Humans can "tolerate" and are willing to examine system schemes or graphic presentations that are neither too large nor too complex. Since the number of things in real-life systems is large, they cannot be effectively displayed in a single OPD. As more details about the system accumulate, **Analysis OP Document** eventually becomes **complete**, but at the same time **Clarity** is compromised. At some point, the accumulation of details in the OPD makes it too cluttered, rendering the state of the **Clarity** attribute **unclear**. With each occurrence of **Refining**, a **clear** specification may revert to **unclear**, calling for an action to fix this. **Abstracting** can potentially convert an **unclear Analysis OP Document** into a **clear** one. However, if

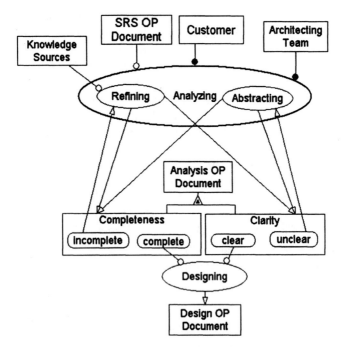

Figure 11.7. The **Refining Abstraction** cycles of **Analyzing**

Abstracting also omits details, it can render a **complete Analysis OP Document** into an **incomplete** one. Iterating through **Refining** and **Abstracting**, generating OPDs and OPL paragraphs along the way, should eventually lead to a state in which **Analysis OP Document** is both **complete** and **clear**, enabling **Designing** to start. Similar **Refining** and **Abstracting** cycles can be depicted for **Designing**, with **Completeness** and **Clarity** being the attributes of **Design OP Document**.

11.5.6 Designing

Once the **System** is **analyzed**, **Designing** can take place. According to Pahl and Beitz (1996), in systematic respects, designing is "the optimization of given objectives within partly conflicting constraints." While they focus on man-made physical systems, this definition of design is applicable to all artificial systems, including physical and informatical, or any combination thereof, such as reactive systems, which need to respond to stimuli from the environment, or embedded systems, which feature "embedded intelligence" inside operational hardware. **Designing** changes the **System** for **analyzed** to **designed**. It requires the **Analysis OP Document** and generates the **Design OP Document**.

As Figure 11.6 shows, **Designing** yields **Design OP Document**. The **Design OP Document** is an elaboration of the **Analysis OP Document**. It is also part of the **System**. **Designing** changes **System** from **analyzed** to **designed**. Once at state **designed**, the **System** can undergo **Implementing**. **Implementing** changes **System** from **designed** to **implemented**. **Implementing** requires **Design OP Document** and yields **Implementation**, which, like the **Analysis OP Document** and the **Design OP Document**, is also part of the **System**. Note that the **Implementation** does not have the "**OP Document**" suffix, because it is the manifestation of the **Design OP Document** – it is the end goal of the analysis and design processes.

11.5.7 The Waterfall Model vs. Iterative and Incremental Development

The classical Waterfall Model of information systems development advocated an orderly progression from analysis to design to implementation. However, as system developers have painfully learned since the introduction of the Waterfall Model in the 1970s, the major shortcoming of this model is its linearity. Linearity means that the basic model does not allow backtracking for correcting errors detected during the analysis, design and implementation. To remedy this, the iterative nature of the real development process, as opposed to the traditional idealized waterfall model, is manifested in loops that allow for backtracking, shown by the two invocation links in Figure 11.6.

One of the invocation links goes from **Designing** back to **Analyzing**, completing the analysis-design loop. **Analyzing** may fail to complete successfully because it does not meet requirements specified in the **SRS OP Document**. While engaged in **Designing**, the **Architecting Team** may find out that the analysis is infeasible or otherwise impractical, calling for another iteration of **Analyzing**. Backtracking to **Analyzing** enables **I SEE** to rewrite part(s) of the **Analysis OP Document** to meet the requirements not yet met.

The other invocation links goes from **Implementing** back to **Designing**, completing the design-implementation loop. **Designing** may fail due to faults in the **Analysis OP Document**. **Implementing** may consequently prove to be too difficult, costly or inefficient, which spells "back to the drawing board," i.e., repeat **Designing** in order to rewriting at least part of the **Design OP Document**. Documenting the intent or idea behind major design decisions can help manage these iterations and accelerate arrival at a good solution. A similar backtracking loop can exist in practice from **Implementation** back to **Analyzing**, bypassing **Designing** altogether. These loops can be repeated until the system's requirements are satisfied or the **Customer** adjusts the **Requirement Set**, expressed in the **SRS OP Document**.

While a formal **SRS OP Document** may exist, the **Requirement Set** may seem in reality like a moving target because the environment, technology, and/or customer expectations keep changing as the system is being developed. The dynamic nature of this process is not the only source of the complexity of the systems evolving process. Another source is the application of incremental development, which advocates the addition of manageable chunks of work unto a working piece of the system. Employing such incremental development approach, the above specification becomes more involved. A large development project is broken down into work packages, each of which undergoes a similar (albeit possibly simplified) process, and then the chunks are combined and/or build on top of each other.

11.5.8 Deploying the System

Deploying is the last major subprocess in the **System Evolving** process. It is within this process that the **System** is put to use and the **Customer** expects to enjoy its **Benefit**. Zooming into **Deploying** in Figure 11.8 exposes its **Assimilating, Using & Maintaining, Functionality Evaluating**, and **Terminating** subprocesses. **Required Function Set** consists of and generalizes **Current Function Set** and **Desired Function Set**. These two parts are at the same time also specializations of **Required Function Set**.[2] The developed **System**, which fulfils the **Required Set**, consists, among other things (as denoted by the small bar along the aggregation link next to **System**) of the **Implementation**. The **Implementation** is also zoomed in to show its two parts: **Assimilator** and **Maintenance Manager**.

Assimilator is the instrument for carrying out the **Assimilating** process, in which the **System** is embedded within its **Environment**. The **Environment** can be the organization within which the system would operate, or, more generally, any super-system within which **System** is embedded as a component. The assimilation is planned and carried out as a project in its own right. The **Assimilator** has provisions for laying out the assimilation project and for tracking and managing its execution, with appropriate feedback mechanism. **Assimilating** changes **System** from **implemented** to **assimilated.**

Once **assimilated**, the **System** starts to be used and maintained as a matter of routine through the **Using & Maintaining** process. Normally, this process is the longest of all the processes (phases) of the system lifecycle, and hence also the most expensive. This is where the **Customer** enjoys **Benefit**. More precisely, **Customer** is expected to reap the benefits of the system it invested for. **Using & Maintaining** affects both the **System** and the **Environment**. It yields **Benefit**, from which **Customer** enjoys – this is the purpose of the whole effort of initiating, developing, and deploying the system.

The **Maintenance Manager** is the instrument that allows the user to take care of the routine tasks while operating and using the **System**. It is responsible for monitoring, recording, and assisting in solving hardware, software, human interface, usability, and any other kind of problem detected during the operation of the **System**. More specifically, for the information technology portion of the system, the **Maintenance Manager** controls tasks that include storage management, net-

---

[2] A structural relationship like this, where the root object is concurrently the source of both aggregation and generalization, is termed a fractal relation. As shown in Figure 11.8, it is symbolized by a triangle whose left half is black and right half is white, symbolizing the concurrent existence of these two fundamental structural relations – aggregation and generalization. It is expanded on in Chapter 13.

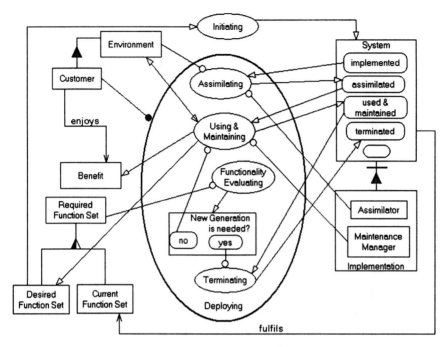

Figure 11.8. Zooming into **Deploying** exposes its **Assimilating, Using & Maintaining,** and **Terminating** subprocesses.

work and communication management, up time, backup and recovery, security, and privacy. Each such task is assigned to the corresponding sub-manager. Thus, the **Storage Manager** is the instrument for **Storing**, the **Network & Communication Manager** is the instrument for network and communication. **Robustness Manager** is the instrument for keeping the system up and managing its backup and recovery, and **Security & Privacy Manager** is the instrument for managing the system's security and privacy issues.

The **Requirement Manager**, which is another part of the **Maintenance Manager**, is charged with the unique role of collecting and organizing new requirements for the next generation of the **System**, as they arise during the **Using & Maintaining** process. These new requirements are recorded as part of the **Desired Function Set**. By changing requirements of existing functions and adding new ones, the **Requirement Managing** process affects the current **Required Function Set** by adding desired functions to it. When the **Required Function Set** becomes substantially different than the one fulfilled by the current generation of **System**, the functionality (performance, or quality of service) of the current **System** is growing more and more unsatisfactory. Functionality Evaluating determined whether **New Generation** is **needed**. A **yes** answer for this Boolean object is a condition for the **Terminating** process of the **System**, which affects the state of **System** from **used & maintained** to **terminated**.[3] **Terminating** is the last process of **Deploying**, which, in turn, is the last process of the entire **System Evolving** process.

We have finished traversing all the OPDs at the system (top) level, and at the second level, where we examined the **Initiating**, **Developing** and **Deploying** processes and touched upon their subprocesses. Continuing our breadth-first traversal of the OPD set, we now focus on two subprocesses of **Developing**: **Analyzing and Designing.**

11.6 Zooming into Analyzing

The first major process within **Developing** is **Analyzing**. As Figure 11.9 shows, **Analyzing** zooms into **Requirements Managing**, **Domain-Oriented System Specifying**, **Analysis Evaluating** and **Analysis Approving**. **Requirements Managing**, in turn, zooms into **Requirements Gathering** and **Requirements Analysis**. There are two possible triggers for **Requirements Analyzing**: **Required Function Set** is at state **gathered** or **Analysis** is not **meeting Requirements**. Since the Boolean operator here is OR (rather than XOR), both can be true at the same time. In OPL, the sentence reads: **Requirements Analyzing** occurs if **Required Function Set** is **gathered** or **Analysis** is not **meeting Requirements**.

Requirements Managing is the first subprocess within **Analyzing**. It is followed by **Domain Oriented System Specifying**, which is the core of the **Analyzing** process. **Analysis Evaluating** and **Analysis Approving** come at the tail end of **Analyzing**. **Requirements Managing** zooms into **Requirements Gathering** and **Requirements Analyzing**. **Requirements Gathering** takes **System** out of its **initialized** state and generates **Required Function Set** at its **gathered** state. **Requirements Analyzing**, which is performed next, changes **Required Function Set** from **gathered** to **analyzed**. **Requirements Analyzing** yields **skeletal Analysis OP Document** and **Evaluation Metrics Set**.

The **analyzed Required Function Set** is needed to start the **Domain-Oriented System Specifying** process. The result of this process is a significant elaboration of the **skeletal Analysis OP Document**, generated earlier by the **Requirement Analyzing** process. In OPL, **Domain-Oriented System Specifying** changes **Analysis OP Document** from **skeletal** to **elaborated**. **Domain-Oriented System Specifying** can also occur if **Analysis** is not **meeting Requirements**.

Using the **Evaluation Metrics Set** and the **analyzed Required Function Set**, the elaborated **Analysis OP Document** undergoes **Analysis Evaluating**. **Analysis Evaluating** determines whether **Analysis** is **meeting Requirements**. **Analysis Approving** occurs if **Analysis** is **meeting Requirements**. **Analysis Approving** changes **System** to **analyzed**. Also, **Analysis Approving** changes **Analysis OP**

---

[3] Due to the increasing importance of environmental considerations, disposal has become a major component as the terminal phase in the lifecycle of a physical product. Proper disposal is less relevant if the product is an information system. However, if the information system is embedded within a physical system, one can envision part of the software that will take care of disposal (recycling or environmental-friendly self-destruction).

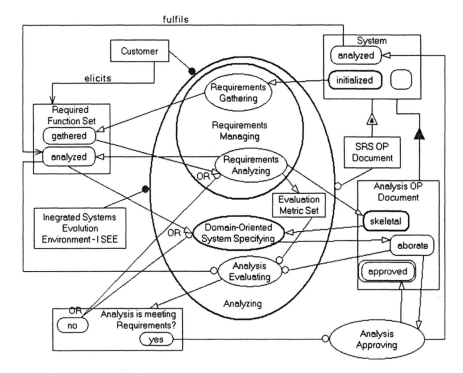

Figure 11.9. Zooming into **Analyzing**

Document from **elaborated** to **approved**. The **analyzed System** fulfils the **analyzed Required Function Set**. **Requirements Analyzing** or **Domain-Oriented System Specifying** occur if **Analysis** is not **meeting Requirements**.

  Analyzing determines whether **Analysis** is **meeting Requirements**. This is manifested by the result link from **Analyzing** to the Boolean object named "**Analysis is meeting Requirements?**" If so, i.e., if the Boolean object is in state yes, **Analysis Approving** occurs. In OPL, **Analysis Approving** occurs if **Analysis** is **meeting Requirements**. Furthermore, **Analysis Approving** changes **System** to **analyzed**. In other words, the state of the **System** is now **analyzed**.

  Domain-Oriented System Specifying is the major subprocess within **Analyzing**. Zooming in Figure 11.10 further into this process exposes its subprocesses. These are **Current Practice Specifying, Strategy Devising, Alternative Option Constructing,** and **Alternative Selecting. Current Practice Specifying** yields **Current Practice OP Document**. This is documentation of what is out there, be it the existing legacy system or the current state of affairs without any system in place to support the **Required Function Set. Strategy Devising** yields **Strategic Analysis** – an Object-Process model of the domain-specific factors that need to be considered to attain the system's goals and function set, the environment, the users, possible competition, etc. **Architecture Constructing** uses this documentation as background material along with the **gathered Required Function Set** to generate

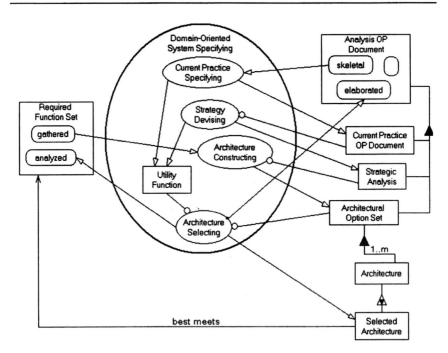

Figure 11.10. Zooming into Domain-Oriented System Specifying

the **Architectural Option Set**. This is a set of one or more **Architectures**, each being an Object-Process model of the system from the domain expert's viewpoint. The architecture specifies the detailed combination of structure and behavior of the alternative such that the function set is satisfied.

Current Practice Specifying and **Strategy Devising** yield **Utility Function**. **Utility Function** sets criteria and corresponding weights to various factors to be considered while selecting among alternative architectures. Using the **Utility Function,** the **Architecture Selecting** process yields the **Selected Architecture**, which is **Architecture,** and changes **Required Function Set** to **analyzed**. The **Selected Architecture best meets** the **Required Function Set**.

11.7 Zooming into Designing and Implementing

The basic objective of the **Designing** process is to flesh out the selected **Architecture** agreed upon during **Analyzing** by adding implementation details, such that the **Design OP Document** constitutes a complete blueprint for implementing the system.

Zooming into **Designing** in Figure 11.11, we see that it consists of **Strategic Designing** and **Tactical Designing. Strategic Designing** requires the **Strategic Analysis** part of the **Strategic Analysis OP Document**. It entails high-level issues that include, among other things and in generic terms, **Technology Selecting,**

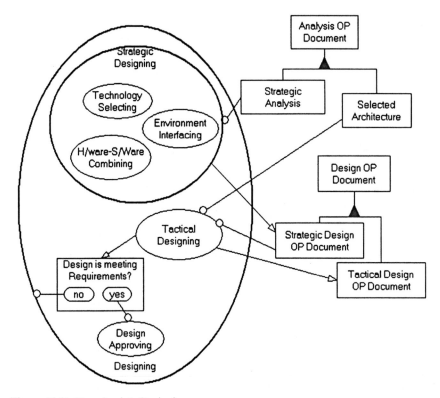

Figure 11.11. Zooming into Designing

Environment Interfacing, and **Hardware-Software Combining**. Depending on the nature of the system being developed, this would specialize into the appropriate selection of software and/or hardware development environment, and the tools with which the program logic, persistent data and graphic user interface will be developed. At this point, the required computational and communication resources and their topology need to be figured out to meet such attributes as load, security, privacy, up time, and reliability. The result of **Strategic Designing** is **Strategic Design OP Document**. Along with the **Selected Architecture**, which is the second part of the **Strategic Analysis OP Document**, it is the basis for the **Tactical Designing**, which deals with the details of the solution and yields the **Tactical Analysis OP Document**. This is a major elaboration of the **Selected Architecture**, in which all the smallest details are ultimately specified using OPM.

Designing includes **Strategic Design** and **Tactical Design**. In the **Strategic Design**, major design decisions are made regarding selection among solution alternatives that might have been the outcome of the **Analyzing** process, and optimal implementation facilities. These include the choice of development and target environments (which are not necessarily the same), the system configuration in terms of hardware, software and middleware, database and knowledge base organization, and human-machine interfaces.

Based on these strategic design decisions, the **Analysis OP Document** is augmented and refined as needed. The deliverable of **Designing** is the **Design OP Document**. Analogous to its **Analysis OP Document** predecessor, the **Design OP Document** contains the design OPD set and its equivalent OPL script, reflecting the newly designed system and all the fine details that pertain to its implementation.

11.8 From Design to Implementation

Implementing the system entails actually constructing or assembling its hardware and software components, so they are ready to be put to work. In other words, all the elements of the system should be ready for assimilation in the target environment. Three major components of the software are the program logic (expressed as code in some programming language), database schema, and user interface.

When the design is complete, the part of the OPL script that deals with software can be automatically translated into implementation. The implementation includes the combination of automatically generated program logic, i.e., actual code in a target language, with the application's database schema, which can also be derived automatically from the OPM specification.

The human interface of the system, which is an important component of the implementation beside the program and the schema, can be constructed semi-automatically. The automatic part concerns the derivation of the interaction hierarchy, which is extracted by identifying the agents and the various levels of processes in which they engage. The non-automatic part is the crafting of the actual graphics to endow it with the particular look-and-feel of a good, dedicated graphic user interface.

Summary

- Systems typically exhibit lifecycles.
- Systems analysis is the discipline that models a portion of reality within some domain in order to understand it, and, based on this understanding, possibly design new systems and/or improve existing ones.
- The "OPM system process" prescribes the development and lifecycle support of not just software systems, but systems in general, where software may be a component or a subsystem.
- We have laid out a complete OPM specification of a system for supporting the lifecycle of generic systems, products, and projects.
- Not only is the specification provided exclusively in terms of OPM, it also stipulates how an OPM-based support environment, the **Integrated Systems Evolution Environment** (**I SEE**), constitutes a basis for automated support throughout the life of the system.

- At the top level, system evolution comprises initiating the system, developing it, and deploying it.
    - Development includes analysis, design and implementation.
    - Using and Maintaining is the major process within Deploying. While the system is used, new requirements are recorded and used as input for the next system generation.
- Since OPM is a self-contained, comprehensive systems paradigm, it does not require the incorporation of auxiliary models, diagram types, or methods to enable the evolution of man-made systems.

Problems

1. Write the OPL paragraph for the OPD of the scientific method in Figure 11.1 and critique it regarding its accuracy and level of detail.
2. Augment the top-level OPD of **System Lifecycle and Evolution** system in Figure 11.3 by unfolding the **Environment**. Include the **Customer**'s customers, i.e., those entities to whom the customer of the developed system sells good or services, as well as **Customer**'s suppliers and competitors, state-of-the-art technology, rules and regulations, economic climate, and any other object and/or process you deem appropriate at this level of abstraction.
3. **System Evolving** zooms into three subprocesses: **Initiating**, **Developing**, and **Deploying**. Translate Figure 11.4 to OPL, identify the cycle, and discuss in free language how a complete cycle in the life of a system is represented.
4. **Initiating** yields the **System**, which **fulfils** the **Required Function Set**. How does the **Required Function Set** align with the definition of a system as an object that exhibits function?
5. Based on the description about the **Abstracting** and **Refining** cycles depicted in Figure 11.7, add the details in the text that are not in the OPD.
6. The OPD in Figure 11.7, which explains the refining-abstracting cycles, is not part of the OPD set that runs throughout this chapter. Integrate it into the OPD in Figure 11.9.
7. **System** can be at a number of states that are distributes across several OPDs. Construct a new OPD that shows all the states of **System** and the processes that change it from one state to the other. Make sure to include the process that closes the lifecycle loop.
8. Design an OPD and the corresponding OPL paragraph that zooms into **Architecture Selecting** in **Figure 11.10**.
9. Augment the OPD in Figure 11.11 by linking the inner processes within **Strategic Designing** to corresponding new objects and link these objects to the appropriate object(s) in the OPD.
10. Write the OPL paragraph that reflects the OPD in Figure 11.1. Criticize the model and try to add details to it.

Chapter 12
States and Values

> *The Caterpillar ... got down off the mushroom and crawled away into the grass merely remarking as it went, "One side will make you grow taller, and the other side will make you grow shorter."*
>
> Alice in Wonderland by Lewis Carroll, 1899

Along with objects and processes, states are an important entity in OPM. If objects and processes are the building blocks of OPM, and links are the mortar, states can be considered as the finish of the house: the paint job, the furniture, the architectural elements. States make the use of OPM a lot smoother and more expressive. They allow an object to change while retaining its identity; they provide for a wide range of interactions between objects and processes; their context sensitivity significantly enriches OPM's articulation power. We have been using states since the first chapter of this book. We have seen that values are generalizations of states. This chapter focuses on value and states' relations to other entities in OPM, as well as introducing a few more uses for states.

12.1 State-specified Objects and Links

It is sometimes desirable to specify not just the object that a process generates, but also the particular state at which that object is generated as a result of the occurrence of a process. Likewise, one may wish to specify not just what object a process consumes, but also the particular state that the object needs to be at when the process consumes it. State-specified objects are used for this purpose. To specify that an object is at a particular state, OPL simply puts the state name in front of the name of the object that owns that state. Figure 12.1 shows **sunny Sky, cloudy Sky, foggy Sky, human Pilot**, and **automatic Pilot** as examples of such *state-specified* objects.

Figure 12.1. Sunny Sky, cloudy Sky, foggy Sky, human Pilot, and **automatic Pilot** are examples of state-specified objects.

At the metamodeling level, Figure 5.8(a) indicates that when **Constructing** is done, **Object** is generated at state **existent**. The state-specified **existent Object** is used in the following OPL sentence:

> **Constructing** yields **existent Object.**
> *(State-specified result sentence)*

Similarly, Figure 5.8(b) shows that **Consuming** can happen if and only if **Object** is at state **existent**. The corresponding sentence is:

> **Constructing** consumes **existent Object.**
> *(State-specified consumption sentence)*

We call the link in Figure 5.8(a) *state-specified result link* and the one in Figure 5.8(b) *state-specified consumption link*. Figure 12.2 demonstrates the use of these two links. The link from **approved Raw Material** to **Manufacturing** is a state-specified consumption link, while the link from **Manufacturing** to **manufactured Product** is a state-specified result link.

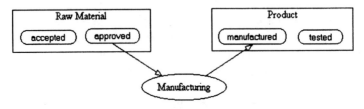

Figure 12.2. State-specified consumption and result links

The corresponding OPL paragraph follows.

> **Raw Material** can be **accepted** or **approved.**
> **Product** can be **manufactured** or **tested.**
> **Manufacturing** consumes **approved Raw Material.**
> **Manufacturing** yields **manufactured Product.**

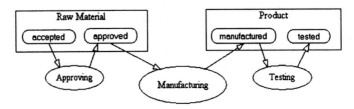

Figure 12.3. **Approving** and **Testing** are state-level processes, while **Manufacturing** is an object-level process.

The last two OPL sentences can be joined into one:

> **Manufacturing** consumes **approved Raw Material** and yields **manufactured Product.**
> *(State-specified transformation sentence)*

As we have seen, a change in the object's state is expressed by an OPL change sentence. The following OPL sentences demonstrate the dynamic aspect of the OPD in Figure 12.3:

> **Approving** changes **Raw Material** from **accepted** to **approved.**
> **Manufacturing** consumes **approved Raw Material** and yields **manufactured Product.**
> **Testing** changes **Product** from **manufactured** to **tested.**

12.1.1 Initial, Ultimate and Default States

It is often convenient or desirable to denote what an object is initially, and what it is ultimately. This is especially useful for objects that exhibit a lifecycle. The symbols for initial and ultimate states are shown in Figure 12.4, and their sentences below. Note that the two types do not have to be in conjunction; either one can be specified alone.

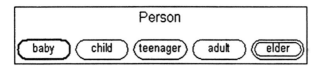

Figure 12.4. Initial and ultimate states demonstrated

> **Person** can be **baby, child, teenager, adult,** or **elder.**
> **Person** is initially **baby** and ultimately **elder.**
> *(Initial and ultimate state specification sentence)*

Another useful state type is default. As Figure 12.5 person is healthy by default; a product is operational by default; a door is closed by default.

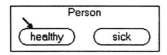

Figure 12.5. Default state demonstrated: **Person** is **healthy** by default.

> **Person** can be **healthy** or **sick**.
> **Person** is **healthy** by default.
> *(Default state specification sentence)*

12.1.2 The Transformation Attribute of a Process

By definition, a process transforms an object. The transformation can result in either a change in the existence of the object, i.e., the process generates or consumes an object, or just affect an object by changing its state. We define a meta-attribute of **Process** called **Transformation Extent**.

> *Transformation extent is a process attribute that determines the extent of the transformation the process has on the object it transforms.*

Transformation Extent has two values: **Existence** and **Effect**.

> *An **existence-impacting process** is a process, the value of the transformation extent attriburte of which is **Existence**.*

An existence-impacting process is a process that consumes or generates at least one object. As an example, **Manufacturing** is an example of an existence-impacting process, since it consumes an object and generates another object, as shown in both Figure 12.2 and Figure 12.3.

> *A **state-impacting process** is a process, the value of the transformation extent attriburte of which is **Effect**.*

A state-impacting process is a process that changes the state of at least one object, but does not generate or consume any other object. As an example, **Approving** and **Testing** in Figure 12.3 are state-impacting processes. They affect **Raw Material** and **Product**, respectively, but do not consume or generate any object. Recall that we distinguish a change in state from a change in object identity by how significant the change is. Existence-impacting processes are more profound than effect-impacting ones, because the transformations they cause are more significant. Hence, a process that generates or consumes objects and also affects one or more other objects is an existence-impacting process.

12.1.3 Object as a Role Player for State

States are sometimes related to objects in the system. For example, **Cash** can exhibit the status attribute **Owner**, whose states are **customer** and **bank**. **Customer** and **Bank** are also objects in their own right in the same system, but at the same time they are also states of Owner. How do we denote the fact that these are the same entities playing different roles, once as objects and once as states? The solution is presented in Figure 12.6 and in the corresponding OPL paragraph below.

A role-playing sentence enumerates the object(s) whose names are identical with names of states in the OPD. The only difference is that state names start with a lower-case letter. Each object is linked to its identical name state with a role link – a white triangle (like the Gen-Spec symbol) whose tip and base are linked with the object and state, respectively. Note that the state enumeration sentence "**Owner** can be **bank** or **customer**" is not required when the role-playing sentence is present. This is yet another example of context sensitive symbols. This relation is different than Generalization, since it does not connect two objects. It is also distinguishable from qualification, since the source and destination are reversed between the two: state to object in qualification, object to state in role-playing.

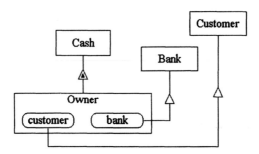

Figure 12.6. The objects **Bank** and **Customer** play the role of **Owner**'s states **bank** and **customer**.

Cash exhibits **Owner**.
Bank and **Customer** play the role of **Owners** for **Cash**.
   (Role-playing sentence)

The OPD in Figure 12.7 and its OPL paragraph below it show that object *instances* can also play a similar role. In this example Karen and David play the role of Owner as *values* rather than as states. The sentence **Karen** and **David** play the role of **Owner** values for **Dog** is a *value* role-playing sentence and it contains the reserved word value or values. Another example to demonstrate this relation type is the **Car-Driver Complex**; **New York** and **Boston** are real objects, playing the role of values of **Location**.

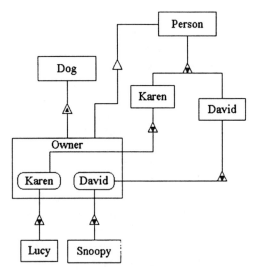

Figure 12.7. The instances **Karen** and **David** play the role of the values **Karen** and **David** as **Owners** of **Dog**, while **Lucy** and **Snoopy** are the **Dog** instances they own, respectively.

> **Person** is an **Owner.**
> **Karen** and **David** are instances of **Person.**
> **Dog** exhibits **Owner.**
> **Karen** and **David** play the role of **Owner** values for **Dog.**
>   *(Value role-playing sentence)*
> **Lucy** is an instance of **Dog,** the **Owner** of which is **Karen.**
> **Snoopy** is an instance of **Dog,** the **Owner** of which is **David.**
>   *(Instance qualification sentences)*

12.1.4 State Maintaining Processes

In Chapter 4 we noted that there is a family of processes that maintain the status of an object rather than change it. Members of this state maintaining process family include holding, maintaining, keeping, staying, waiting, prolonging, delaying, occupying, persisting, including, containing, continuing, supporting, and remaining. The semantics of these verbs is leaving the state of the object in its status quo. We noted that many of these verbs can be considered as working against some "force" which would otherwise change some object. As an example, consider the OPD in Figure 12.8.

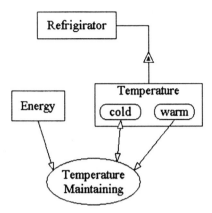

Figure 12.8. Refrigerator as a **Temperature Maintaining** system

> **Refrigerator** exhibits **Temperature**, which can be **cold** or **warm**.
> **Temperature Maintaining** consumes **Energy**.
> **Temperature Maintaining** changes **Temperature** from **warm** to **cold**.
> **Temperature Maintaining** maintains **cold Temperature**.
> *(State-maintaining sentence)*

A new type of OPL sentence, the state-maintaining sentence, reflects the OPD construct in which a process is linked to a state via a *state-maintaining link*. The state-maintaining link is a bi-directional link, which looks like the "regular" effect link, except that its context is different – rather than connecting a process to an object, it connects a process to a state of an object. The semantics of the link is determined by its source-destination combination.

12.1.5 Sentences and Phrases of States and Values

To see the differences between state and value OPL sentences and between state and value OPL phrases, consider the OPDs in Figure 12.9, which show two pairs of identical objects. The object on the left in each pair has one state, while that on the right, shows the state as a value. The only difference is that as states, **married** and **blue** are non-capitalized, while as values, **Married** and **Blue** are capitalized. The OPL sentences for the two objects in Figure 12.9(a) are:

> **Person** is **married**.
> *(State specification sentence)*
> Value of **Person** is **Married**.
> *(Value specification sentence)*

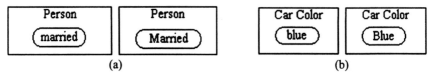

Figure 12.9. Pairs of objects with state on the left object and value on the right object

Since only one state and one value are depicted within the object, the sentences are state/value *specification* (rather than enumeration) sentences. The first sentence refers to the *state* **married** of **Person**, while the second refers to the *value* **Married** of **Person**. Obviously, the OPL sentence "**Person is married**" makes more sense than "Value of **Person** is **Married**." The corresponding OPL phrases for the two objects in Figure 12.9(a) are "**married Person**" and "**Person** with value **Married**." As with the OPL sentences, the former, which is a *state phrase* sounds better than the latter, which is a *value phrase*. The analogous OPL sentences for Figure 12.9(b) are similar:

> **Car Color** is **blue**.
> Value of **Car Color** is **Blue**.

The corresponding phrases are "**blue Car Color**" and "**Car Color** with value **Blue**." Here, too, **blue** sounds better as a state than as a value. Figure 12.10(a) shows **Water Temperature** as another such pair. The OPL sentences are:

> **Water Temperature** is **hot**.
> Value of **Water Temperature** is **Hot**.

Again, the state **hot** is preferred over the value **Hot**, as is the state phrase "**hot Water Temperature**" compared with "**Water Temperature** with value **Hot**." Figure 12.10(b) shows **Water Temperature** with the value **80 degrees Celsius**. Since it is a numeral, it can only be a value. The value sentence is:

> Value of **Water Temperature** is **80 degrees Celsius**.
> *(Value sentence)*

The value phrase is "**Water Temperature** with value **80 degrees Celsius**." If **80 degrees Celsius** were a state (which it is not), the state sentence would have been "**Water Temperature** is **80 degrees Celsius**" and the state phrase would have been "**80 degrees Celsius Water Temperature**."

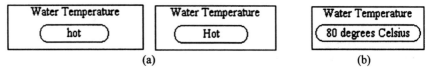

Figure 12.10. Water Temperature described in three subtly different ways

Summarizing these examples, we see that the terms value of an object's attribute and state of an object's status are intimately close and frequently equivalent. The state sentence and state phrase are more succinct and sound better than their value sentence and value phrase counterparts. However, in cases when the value is not a state, notably when it is a numeral, the value sentence and phrase are acceptable, as in the **Temperature** example above.

Reflecting back on the example in Figure 5.6(b), **New York** and **Boston** can be considered as values of **Location**, or as states of **Location**. Considering **New York** and **Boston** as states, the OPL paragraph is:

Car-Driver Complex exhibits **Location**.
Location can be **New York** or **Boston**.
  (State enumeration sentence)
Moving changes **Location** from **New York** to **Boston**.

If we think of **New York** and **Boston** as *values* rather than states of **Location**, the state enumeration sentence in the OPL paragraph above changes to:

Values of Location are **New York** and **Boston**.
  (Value enumeration sentence)

Consider the object **Car-Driver Complex** at a given point in time. As an *attribute* of **Car-Driver Complex**, **Location** is assigned a particular *value*. As a *status* of **Car-Driver Complex**, **Location** is at a particular *state*.

At any given point in time, the attribute has some value, which is drawn from the attribute's domain (set of legal possible values). Status is a specialization of attribute. This observation raises a question regarding when should an attribute be modeled as a status? This question is discussed in the next subsection. If we view **Location** as a status of **Car-Driver Complex** rather than as its attribute, then **New York** and **Boston** are states of **Location** rather than values of **Location**. Analogously, in the **Lamp** example of Figure 5.1, we can model **Status** of **Lamp** as the attribute, called **Operational Situation**, in which case **Off** and **On** would be values rather than states. This option, however, is less natural than **Lamp**'s **Status** with the **on** and **off** states, because the instantaneous operation of pushing a button, the **Lighting** process, provides for easy switching between the two.

Let us examine several other examples. **Marital Status** is an *attribute* of a **Person**. For the sake of simplicity, suppose its only *values* are **Single**, **Married** and **Divorced**, as Figure 12.11(a) shows. As Figure 12.11(b) shows, **Marital Status** can also be considered a *status* of **Person**, with the corresponding *states* **single**, **married** and **divorced**. The OPL script of Figure 12.11(a) follows.

Person exhibits **Marital Status** with values **Single**, **Married**, and **Divorced**.
Wedding changes **Marital Status** from either **Single** or **Divorced** to **Married**.
  (XOR source change sentence)
Divorcing changes **Marital Status** from **Married** to **Divorced**.

The OPL script of Figure 12.11(b) is different in the first sentence:

> **Person** exhibits **Marital Status,** which can be **single, married,** or **divorced.**

The state enumerated exhibition sentence from Figure 12.11(b) reads more naturally than the value enumerated exhibition OPL sentence from Figure 12.11(a), further supporting the case for using states. Next, we show that making **Marital Status** an implicit status improves the system if states are used, as it becomes simpler and more natural.

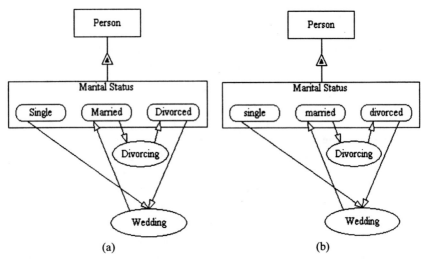

Figure 12.11. The effects of **Wedding** and **Divorcing** on a **Person**'s **Marital Status** as (a) values; (b) states. Can you spot the difference?

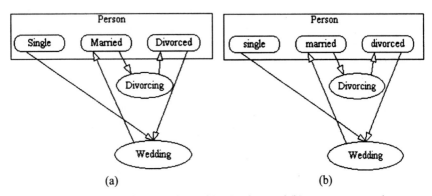

Figure 12.12. The OPDs of Figure 12.11 with (a) values and (b) states suppressed

The OPL script of Figure 12.12(a) is:

Values of **Person** are **Single, Married,** and **Divorced.**
Wedding changes **Person** from either **Single** or **Divorced** to **Married.**
Divorcing changes **Person** from **Married** to **Divorced.**

As in Figure 12.11, the OPL script of Figure 12.12(b) is different than that of Figure 12.12(a) only in the first sentence:

Person can be **single, married,** or **divorced.**

We see again that in this system states are better suited to describe the situation than values. The sentence "**Person** can be **single, married,** or **divorced.**" is more natural than "Values of **Person** are **Single, Married,** and **Divorced.**" We also see that when states are used, implicit status is preferred over explicit status. The sentence "**Person** can be **single, married,** or **divorced.**" is more natural than "**Person** exhibits **Marital Status,** which can be **single, married,** or **divorced.**" Suppressing the explicit mention of **Marital Status** caters to the natural language shortcut, which assumes human prior knowledge that **single, married,** or **divorced,** are states of **Marital Status.**

Considering another example, the **Color** of a **Car** is definitely an attribute of the **Car,** not a state. While it is possible to change the car's color through the process of **Painting,** we would not normally think of it as a change in the **Car**'s state but rather as a change of value of the **Car**'s **Color** attribute. This is so because painting a car is not as quick and easy a process as turning a lamp on or off.

Whether or not the **Car**'s **Transmission** is engaged is definitely a state of **Gearbox,** not its value. **Engaging** and **Disengaging** of **Transmission** can be done frequently and easily. Hence, **engaged** and **disengaged** are naturally modeled as states (not values) of the **Transmission** of **Car.** Likewise, (but unlike the color of a car) the **Color** of a **Traffic Light** is a status, not just an attribute. Being a status, at any point in time, **Traffic Light** is at some state, no just some value. This is so because traffic light colors are intimately associated with the passage of time and change periodically and automatically.

As another example, the status of a **VCR** can have one of the states **stop, fast forward, rewind, play, record,** or **pause.** Here too, we have no problem deciding that these are states – instances of the **VCR**'s status attribute. Conversely, when a **Salary** of an **Employee** is changes from **$80,000** to **$90,000,** these numbers are definitely values of **Salary,** not its states, because numerals are defined as values even if they can change easily.

12.1.6 Single Value Sentence

When an object can be at just one state, this state is considered a value. Let us examine the OPD in Figure 12.13 and its OPL paragraph below.

Empire State Building exhibits **Location.**
Location of **Empire State Building** is **New York.**
   (Single value sentence)

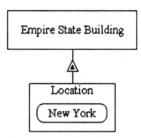

Figure 12.13. New York is a single value of **Location** of **Empire State Building.**

A single value sentence expresses the name of the attribute. Using the reserved word is, it relates the attribute, **Location**, to the exhibitor, **Empire State Building**. The reserved word is, which follows **Empire State Building**, relates the attribute to its single value, **New York**. The two OPL sentences above can be joined into the following one:

Empire State Building exhibits **Location,** the value of which is **New York.**
   (Value-specified exhibition sentence)

12.2 Telling States Apart from Values

A fine difference exists between state and value. To gain insight into this matter, consider the following example. The OPD in Figure 12.14(a) shows that **Size** is an attribute of the object **Mattress**. The OPD in Figure 12.14(b) shows **twin, full, queen** and **king** as states of the attribute **Size** of **Mattress**, which means that **Size** is a status, a specialization of attribute.

The OPL sentence that is equivalent to Figure 12.14(a) is:

Mattress exhibits **Size.**

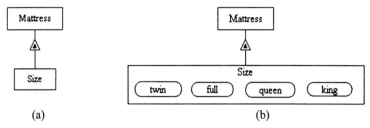

(a)   (b)

Figure 12.14. (a) **Size** as a status of **Mattress** (b) The four states of **Size** specified

The OPL sentence that is equivalent Figure 12.14(b) is:

> **Mattress** exhibits **Size**.
> **Size** can be **twin, full, queen,** or **king**.
> *(State enumeration sentence)*

These sentences make perfect sense. Recall, however, that states are intimately linked with the notion of time and are relatively easily amenable to change. For example, the three colors of a stoplight are states, while possible colors of a car are enumerated values. Stoplight colors change periodically and automatically, while car colors are changeable only with a considerable amount of effort. The states **twin, full, queen** and **king** of **Size** of **Mattress** do not fit this concept of state, as they are not related to time, neither are they easily changeable. Rather than being states of **Size**, they should be considered *values* of **Size**. As Figure 12.15 shows, to differentiate values from states, the name of a value starts with a capital letter, while that of a state starts with a lower-case letter.

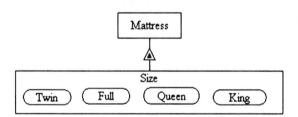

Figure 12.15. Expressing values as capitalized names within the rountangles

Size of **Mattress** is hence assigned one of the values **Twin, Full, Queen,** and **King**. The OPL sentence that is equivalent to Figure 12.15 is:

> **Mattress** exhibits **Size**.
> Values of **Size** are **Twin, Full, Queen,** and **King**.
> *(Value enumeration sentence)*

Generalizing from the examples we have seen, we model a collection of situations, at which an object can be, as a set of *states* (values of the *status* attribute) if it is easy or natural to switch from one situation to another. We model a collection of situations as *values* of an *attribute* if the situations are relatively fixed, unrelated to time, or if they are quantities. More specifically, the following guidelines are applicable for distinguishing between values and states:

(1) States are strongly associated with time.
(2) States denote phases in the object's lifecycle.
(3) States are easily changeable.
(4) States are expressed by phrases, not numbers.
(5) States are discrete and their number is small but greater than one.

Even with these guidelines, the decision regarding whether to model an attribute value as a state depends to a great extent on the modeler's intuition and may prove difficult. The good news is that most of the time this is not really a problem, because the two concepts are so closely related, that no damage can be caused by modeling values as states or vice versa.

12.3 Metamodeling the Attributes of Value and Their States

We have seen that values can be textually enumerated, but numerical values are at least as prevalent as textual ones. As we see below, at the metamodeling level, **Textual** and **Numeric** are two possible values of some attribute of **Value**.

> *Value is the concrete amount, quantity or specification of an attribute.*

Values are more general that states, as they do not require association with time. Each attribute has a set of possible, permissible, or legal values. Values can be textual, numeric or symbolic. They can be discrete or continuous, and their number can be finite or infinite. **Textual, numeric** and **symbolic** are states of the meta-attribute **Expression** of **Value**. **Discrete** and **continuous** are states of the meta-attribute **Granularity**. Finally, **Finite** and **infinite** are states of the meta-attribute **Finiteness**.

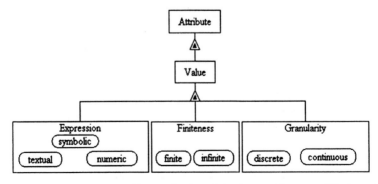

Figure 12.16. The three meta-attributes of **Value** of **Attribute** with two qualified specializations: **Enumerated Textual Value** and **Floating Point Value**

Figure 12.16 is a meta-OPD that presents the three meta-attributes of **Value** of **Attribute: Expression, Finiteness,** and **Granularity.** The OPL paragraph of Figure 12.16 follows.

Attribute exhibits **Value.**
Value exhibits **Expression, Finiteness,** and **Granularity.**
Expression can be **symbolic, textual,** or **numeric.**
Finiteness can be **finite** or **infinite.**
Granularity can be **discrete** or **continuous.**

Textual values are usually discrete and finite, because there is a finite number of words in natural language to describe variations in values. The attribute **Color,** for example, can be assigned a limited number of values in natural languages, ranging between **Violet** and **Red.** Physically, though, color is continuous, as it is the wavelength of the visible spectrum of light emitted from or reflected by the "colored" body. While modern computers boast over a million colors, only a tiny fraction of them have names. Some of these names cover a rather wide range of red-green-blue combinations and borderline colors are often subjective.

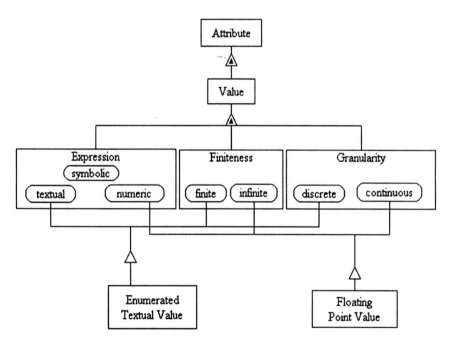

Figure 12.17. The three meta-attributes of **Value** of **Attribute** with two qualified specializations: **Enumerated Textual Value** and **Floating Point Value**

Various combinations of **Expression, Finiteness,** and **Granularity** values give rise to a variety of types of **Value.** For example, as Figure 12.17 shows, **Enumerated Textual Value** is a **Value** that combines **textual Expression** with **finite Finiteness** and with **discrete Granularity.** Another **Value** type in Figure 12.17 is **Float-**

ing Point Value, which exhibits numeric Expression, infinite Finiteness, and continuous Granularity. These two OPL qualification sentences are:

> Enumerated Textual Value is a Value, the Expression of which is textual, the Finiteness of which is finite, and the Granularity of which is discrete.
> Floating Point Value is a Value, the Expression of which is numeric, the Finiteness of which is infinite and the Granularity of which is continuous.

Examples of enumerated textual values are cold, warm and hot, which are values of the attribute Temperature of Water. An example of floating point values is the set of values 2.9, 29.8 and 88.7, which is another set of values of the attribute Temperature of Water, where the Units attribute of the Temperature is degrees Celsius.

12.3.1 Numeric and Symbolic Values

Numerals are by default values, not states. A symbolic value is symbolic in the mathematical sense, as well as in the computer programming sense. In the mathematical sense, standing alone, symbolic values, such as *A*, *Bzs*, *X*, *Uw* or *SPSK*, have no semantic meaning. They can abstract lower-level sets of symbols that may ultimately take on numeric values, and may or may not be restricted in range and/or type (natural, integers, rational, real, complex...).

In the programming sense, a symbolic value is like a variable or a memory location that can potentially store information about an object at any level of complexity, and ultimately a number, or, more precisely, a binary item (bit). Here, the symbol stands for a particular type or class of things (character, pointer, integer, floating-point, long, double precision, etc.) either by explicit declaration or by default. Each type has different storage requirements and is handled differently by the operating system.

12.3.2 Mapping Object States onto Attribute Values

Many of the quantitative attributes can have two alternative sets of values: textual and numeric. The OPD in Figure 12.18 and the corresponding OPL paragraph in Frame 25 demonstrate matching object states to attribute values.

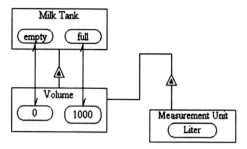

Figure 12.18. The states **empty** and **full** of **Milk Tank** are matched with the values **0 Liter** and **1000 Liter** of **Volume**.

Milk Tank can be **empty** or **full.**
Milk Tank exhibits **Volume** with values **0 Liter** and **1000 Liter.**
Empty Milk Tank and **Volume** with value **0 Liter** are equivalent.
Full Milk Tank and **Volume** with value **1000 Liter** are equivalent.
(State-value equivalence sentences)

Frame 25. The OPL paragraph of Figure 12.18

As this example demonstrates, values can be mapped onto states. In the OPD in Figure 12.18 and its corresponding OPL paragraph in Frame 25, the null bidirectional structural link expresses the equivalence between a state of **Tank** and a value of its **Volume** attribute. The reserved phrase that goes with the null bidirectional structural link, "are equivalent," is used in the two equivalence sentences in Frame 25.

While in Figure 12.18 and Frame 25 we specified exact **Volume** values, it is possible to specify a range of values of an attribute, such as **Temperature** of **Water**, and the ranges that correspond to the various **States Of Matter** of **Water**. Figure 12.19 and its corresponding OPL paragraph in Frame 26 show how ranges are expressed in OPL.

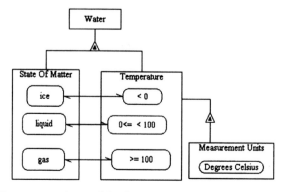

Figure 12.19. The correspondence of the three **Water**'s **States Of Matter** and its **Temperature** value range

---

Water exhibits **State Of Matter** and **Temperature**.

State Of Matter can be **ice, liquid,** and **gas**.

Temperature is measured in **degrees Celsius**.

Values of **Temperature** are less than **0 degrees Celsius,** equal to or greater than **0 degrees Celsius** and less than **100 degrees Celsius,** and greater than or equal to **100 degrees Celsius**.

Ice State Of Matter and **Temperature** with value less than **0 degrees Celsius** are equivalent.

Liquid State Of Matter and **Temperature** with value equal to or greater than **0 degrees Celsius** and less than **100 degrees Celsius** are equivalent.

Gas State Of Matter and **Temperature** with value greater than or equal to **100 degrees Celsius** are equivalent.

Frame 26. The OPL paragraph of Figure 12.20

The correspondence of exact ranges of values and inexact human "subjective" judgments, as expressed by states, is one of the fundamentals of fuzzy logic (Pedrycz and Zadeh, 1995), and this example can point the way to representing fuzzy logic principles effectively with OPM.

As another example, consider the following sentence pairs:

"The water is frozen."   "The water temperature is about 0°C."
"The water is chilling."   "The water temperature is about 4°C."
"The water is cold."   "The water temperature is about 10°C."
"The water is lukewarm."   "The water temperature is about 15°C."
"The water is warm."   "The water temperature is about 30°C."
"The water is hot."   "The water temperature is about 60°C."
"The water is boiling."   "The water temperature is about 100°C."

The first sentence in each pair is a textual statement about the qualitative "feel" of the water, which is related to its energy. The second sentence in each pair is a corresponding approximate numeric statement about the water's temperature.

12.4 Compound States and State Space

In Chapter 5, we have defined state as a situation at which the object can exist for some positive time duration. The examples we have seen so far are of states that are *atomic*, i.e., states that are not combined of other states. However, there are many instances of *compound* states.

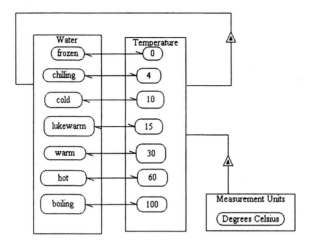

Figure 12.20. Matching object states to attribute values

Water can be **frozen, chilling, cold, lukewarm, warm, hot,** and **boiling.**
Water exhibits **Temperature.**
Temperature is measured in **degrees Celsius.**
Values of **Temperature** are **0, 4, 10, 15, 30, 60,** and **100.**
Frozen Water and **Temperature** with value **0 degrees Celsius** are
    equivalent.
Chilling Water and **Temperature** with value **4 degrees Celsius** are
    equivalent.
Cold Water and **Temperature** with value **10 degrees Celsius** are
    equivalent.
Lukewarm Water and **Temperature** with value **15 degrees Celsius** are
    equivalent.
Warm Water and **Temperature** with value **30 degrees Celsius** are
    equivalent.
Hot Water and **Temperature** with value **60 degrees Celsius** are
    equivalent.
Boiling Water and **Temperature** with value **100 degrees Celsius** are
    equivalent.

Frame 27. The OPL paragraph of Figure 12.20

---

*An **atomic state** is a state that is not combined of other states.*
*A **compound state** is a state that combines at least two other states.*

One attribute of **Car** in Figure 12.21 is **Location**, with values **New York** and
Boston. Another attribute is **Car**'s status **Drivability** (its ability to move on the
road), with states **operational** and **broken**.

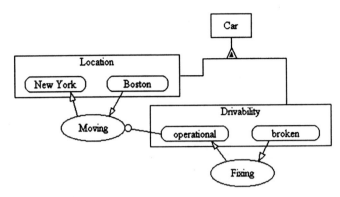

Figure 12.21. The **Location-Drivability** state space of **Car**

Since **Location** and **Drivability** are independent of each other, **Car** can be in one of four states: **operational** in **New York**, **operational** in **Boston**, **broken** in **New York**, and **broken** in **Boston**. The state space is actually a state/value space, because, as we have seen, states and values are very close terms. The Cartesian product of **Location** values and **Drivability** states, constitutes the *state space* of **Car**.

Moving on to a more complex example, **Airport** in Figure 12.22 exhibits four attributes: **Weather Conditions, Tower Services,** and **Radar Coverage**, with 2 states each, and **Pilot Familiarity** with 3 states. The Cartesian product of the sets of states of each attribute enumerates the object's state space. For our **Airport** example, this Cartesian product is {**Weather Conditions**} × {**Tower Services**} × {**Weather Conditions**} × {**Radar Coverage**} = {**fair, hazardous**} × {**available, unavailable**} × {**poor, fair, excellent**} × {**nonexistent, existent**} Thus, in the example of Figure 12.22, there are 2×2×2×3 = 24 such combinations. Each combination is a point, or a (compound) state in the state space of **Airport**. One of the 24 states (or points of the state space) of **Airport** for example is {**fair, available, poor, nonexistent**}. It is obtained by listing the first state in each of the four attributes of **Airport** above. Each such point can be the precondition for some process. Figure 12.22 shows preconditions for two processes. One is the precondition for **Permission Granting** and the other, for **Permission Denying**. The precondition for **Permission Granting** is expressed in the following instrument sentence:

> **Permission Granting** occurs if **Weather Condition** is **fair, Tower Service** is **available,** and **Radar Coverage** is **existent.**

This sentence does not mention **Pilot Familiarity**, which is therefore considered a "wild card." In other words, as long as the **Weather Conditions** are **fair, Tower Services** are **available,** and **Radar Coverage** is **existent**, the state or value of **Pilot Familiarity** does not matter. This precondition, then, accounts for three states of **Airport**, which is the set of points in the state space is {**fair, available, poor, existent**}, {**fair, available, fair, existent**}, and {**fair, available, excellent, existent**}.

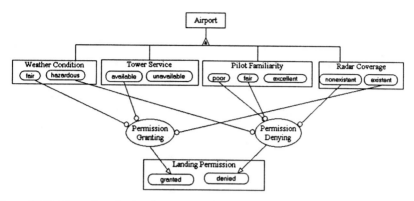

Figure 12.22. Examples of points in the state space of **Airport** that are required for **Permission Granting** and **Permission Denying**

The **Permission Denying** precondition is expressed in the following instrument sentence:

> **Permission Denying** occurs if **Weather Condition** is **hazardous**, **Pilot Familiarity** is either **poor** or **fair**, and **Radar Coverage** is **nonexistent**.

Here, **Tower Services** are the "wild card," which means that as long as **Weather Conditions** are hazardous, **Radar Coverage** is **nonexistent**, and **Pilot Familiarity** is either **poor** or **fair**, it does not matter whether **Tower Services** are **available** or **unavailable**. The condition "either **poor** or **fair** **Pilot Familiarity**" is expressed in the OPD by the fact that the two condition links from **poor Pilot Familiarity** and from **fair Pilot Familiarity** terminate at the same point on the **Permission Denying** ellipse. Recall that the semantics of procedural links terminating at the same point is logical XOR.

The number of compound states of an object is the size of its state/value space, i.e., the number of points in that space. It is the product of the sizes (number of states/values) of the individual attributes and/or parts of that object. In the **Airport** example, the size of the state space is $2^3 \cdot 3 = 24$. As noted, for three of these 24 compound states (state-space points) landing permission should be granted, for other four states landing permission should be denied, and for the remaining states this is undecided and could be left for the judgement of the airspace controller.

In general, if a thing has n attributes, each having v_i values, then the size f the state/value space is:

$$S = \prod_{i=1}^{n} v_i$$

In this context, each attribute can also be referred to as a *dimension*, analogous to the way that vectors can serve as dimensions (e.g., the three orthogonal vectors called X, Y and Z that span a 3-dimensional Cartesian point space).

12.4.1 The Attribute Feasibility Matrix

Not each point in the state/value space of an object is feasible. For example, examining Figure 12.16 for the value space of **Value** of **Attribute**, we realize that not each **expression-finiteness-granularity** of the $3 \times 2 \times 2 = 12$ possible combinations is feasible. For example, the combination in which the **Expression** is **textual** and the **Finiteness** is **infinite** is not feasible, regardless of the **Granularity** value, because the number of words is not infinite. To keep track of which points in the value space are feasible, we use the *attribute feasibility matrix*.

> The **attribute feasibility matrix** *of an object is a matrix, which denotes the feasible points in the attribute value space of some object.*

The attribute feasibility matrix of **Value** from Figure 12.16, presented in Table 13, lists the value space of **Value** and the names of the specialized **Value** combinations. In general, an attribute feasibility matrix of an object with $i = 1 \dots n$ attributes $A_1, A_2, \dots A_n$, each having v_i values, is a $v_1 \times v_2 \times \dots v_n$ n-dimensional matrix, whose elements are the members of the value space. The name of each feasible element is listed in the corresponding matrix cell and a dash implies an infeasible value combination.

Table 13. The attribute feasibility matrix of **Value**

Finiteness	Granularity	Expression		
		Textual	Symbolic	Numeric
Finite	Discrete	Text-enumerated	Symbol-enumerated	Integer-enumerated
	Continuous	-	Symbol-floating-enumerated	Floating-enumerated
Infinite	Discrete	-	-	Integer
	Continuous	-	-	Floating-point

12.4.2 Logical Compound States

So far, we have dealt with compound states that result from combinations of other states, but no logical expressions (i.e., expressions involving AND, OR or XOR) were involved. However, such logical compound states do exist.

> A *logical compound state* is a logical expression whose arguments are states.

The OPDs in Figure 12.23 and in Figure 12.25 are examples of logical compound states. **Table Lamp** consists of a **Switch** and a **Power Plug**. It is **dark** if either the **Switch** is **off** or the **Power Plug** is **disconnected** (see Figure 12.23) and it is **lit** when both **Switch** is **on** and **Power Plug** is **connected** (see Figure 12.24).

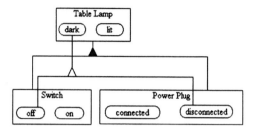

Figure 12.23. The compound state **dark** of **Table Lamp** symbolizes a logical OR

The corresponding OPL paragraph is listed in Frame 28.

Table Lamp consists of **Switch** and **Power Plug**.
Switch can be **off** or **on**.
Power Plug can be **connected** or **disconnected**.
Table Lamp can be **dark** or **lit**.
Table Lamp is **dark** when **Switch** is **off** or **Power Plug** is **disconnected**.
　(Compound-or state sentence)

Frame 28. The OPL paragraph of Figure 12.23

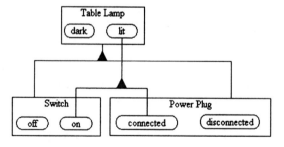

Figure 12.24. The compound state **lit** of **Table Lamp** symbolizes a logical AND

The states **dark** and **lit** are logical compound states. Note that when the black and white triangles are attached to states, they are respectively interpreted as logical AND, and logical OR, rather than as aggregation and generalization, which is their interpretation when they are attached to objects or processes. When objects are linked by the gen-spec triangle symbol and their states are also linked with this white triangle symbol, as in Figure 8.14, the relation between the states is one of overriding.

This is yet another example of graphic symbol overloading. The meaning of the symbol is inferable from its context.[1] Overloading these symbols should not confuse but rather enhance the readability of the OPD. There is a certain extent of analogy between aggregation and logical AND (both denoted by the black triangle) as well as between generalization and logical OR (both denoted by the white triangle).

Table 14. Logical compound states

Name and Sentence	Symbol
Logical AND compound state **A** is **a** when **B** is **b** and **C** is **c**.	
Logical OR compound state **A** is **a** when **B** is **b** or **C** is **c**.	

As the OPL sentences of Figure 12.23 shows, natural language uses "and" for both *aggregation* of objects into a whole and for logical AND of a compound condition. The two following OPL sentences demonstrate this:

> **Table Lamp** consists of **Switch** and **Power Plug**.
> **Table Lamp** is **lit** when **Switch** is **on** and **Power Plug** is **connected**.
> *(Compound-and state sentence)*

The and in the first sentence is an "aggregation AND," while the and in the second sentence is a "logical AND." Likewise, English uses "or" for both the generalization of objects into a generic object, as in "**Lamp** can be **Table Lamp** or **Street Lamp**." ("generalization OR") and for logical OR of a compound condition, as in the OPL sentence "**Table Lamp** is **dark** when **Switch** is **off** or **Power Plug** is **disconnected**."

---

[1] An example of another context-sensitive graphic overloading is the white lollipop, which is an instrument link when it links an object to a process and a condition link when it links a state to a process.

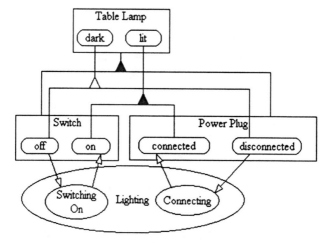

Figure 12.25. Lighting and its effects on **Switch** and **Power Plug**

Table 14 shows the logical AND and Logical OR compound states.[2]

In Figure 12.25, the process **Lighting** is added. Its effects are shown in the OPL sentences below, which are added on to the OPL paragraph of Frame 28.

> **Lighting** zooms into **Switching On** Process and **Connecting.**
> *(In-diagram in-zooming sentence)*
> **Switching On** Process changes **Switch** from **off** to **on.**
> **Connecting** changes **Power Plug** from **disconnected** to **connected.**

Summary

- State is a situation an object can be at, at a given point in time.
- Table 15 provides a summary of the various links that are related to states and their respective OPL sentences.
- State-specified objects are obtained by preceding the state name to the object name.
- Objects can have initial, ultimate and default states.
- An object can play a role of a state, if necessary, by linking the object as a source, via the gen-spec triangle, to the state.

---

[2] The relation between the object owning the compound state and the objects owning the lower-level (atomic or compound) states can only be aggregation or exhibition. Recall from Chapter 8 that when the relation is specialization, the "compound" state is really a generalized state.

Table 15. State-related links

Link Name	OPD, OPL sentence	Description
Condition	 **Processing** occurs if **Object** is **state 1**.	Object is an instrument. It must be at a specific state in order for the process to occur.
Agent Condition	 **Object** must be **state 2** for **Processing** to occur.	Object is an agent. It must be at a specific state in order for the process to occur.
Qualification	 **Qualified Object** is **Object**, the **Attribute** of which is **state 1**.	Qualified Object is a type of Object. It must be at a particular state of Object's Attribute.
Instance Qualification	 **Qualified Object** is an instance of **Object**, the **Attribute** of which is **state 1**.	Qualified Object is an instance of class Object. It must be at a particular state of Object's Attribute.
Role Playing	 **Role Player A** and **Role Player B** play the role of **Attribute** for **Object**.	Role Player A and B are objects, which are also states of another object, called Object.
State-specified Consumption	 **Processing** consumes **state Object**.	Process consumes object only if it is at a certain state.
State-specified Result	 **Processing** yields **state Object**.	Process creates object at a certain state.
State Maintaining	 **Maintaining** maintains **state Object**.	Process maintains object at a constant state.

- States and values are very similar concepts. States are easily changed as part of the object's cycles. Values are either numeric, or hardly changeable.
  - Textual states start with a lowercase letter while textual values start with an uppercase letter.
  - Expression, finiteness and granularity are three meta-attributes of values.
- States of an object can be related to numeric values of its attribute via the null bidirectional structural link.
- A compound state combines at least two other states.
  - The state space of a system is the Cartesian product of the states of the objects in the system.
  - Logical compound states use the black and white triangles, which symbolize AND and OR, respectively.

Problems

1. A car can be driven if it has fuel, the battery is charged, and the car keys are found. Use a compound state in an OPD to specify these conditions.
2. Enumerate the state space of the car in the previous question: (1) as a list, (2) in a table.
3. Draw OPDs and write the OPL sentences of three objects having initial and ultimate states.
4. Draw OPDs and write the OPL sentences of three objects having a default state.
5. The **Car-Driver Complex** requires not only the states of car enumerated in problem 1, but also a sober, awake, and licensed driver.
   a. Incorporate these requirements in an OPD of the Car-Driver Complex.
   b. Write the OPL paragraph of this OPD.
   c. Suggest an alternative way to display the OPD of (a).
6. An ordered set of values of **Size** of some object can be **Miniature, Tiny, Small, Medium, Big, Large, Extra Large, Great, Giant,** and **Colossal.** Alternatively, one can have a range of numbers of some specified measurement unit (e.g., meters for length or kilograms for mass) that indicate a more accurate and "objective" specification of the same **Size** attribute. Pick up an object with two sets of attributes, one qualitative and the other quantitative. Use the textual values of **Size** above and map them in an OPD to numeric ranges of your choice.
7. Using OPD, specify a system with at least one existence-impacting process and one state-impacting process. Write the corresponding OPL script.
8. Provide an example of an OPD for each of the patterns in Table 15 and write its OPL paragraph.

Chapter 13

Advanced OPM Concepts

When we strive to understand, we expect that this knowledge ... will enable us to gain additional control.

Y. Bar-Yam (1997)

The notation is the limitation. J. G. Long (2001)

In this chapter we present new advanced OPM concepts and elaborate on material presented in earlier chapters. These issues include real-time, metamodeling, scope, and structural relations.

13.1 Real-Time Issues

Real-time systems are systems that are required to react almost instantaneously to stimuli arriving from the environment. Processes and timing are central in such systems. This section discusses issues related to OPM analysis and design of real-time systems and other advanced process-related concepts.

13.1.1 Sequential vs. Parallel Process Execution

While OPDs are drawn such that the (implicit) timeline is directed from the top of the OPD to its bottom, the order of execution is dictated by the flow of control, or thread of execution, which is the sequence of alternating objects (or object states) and processes.

Figure 13.1 shows how the execution thread uniquely determines the execution order. **Breaking** a **whole Egg** enables its **Separating**. After those two processes are done, however, the next three processes can occur simultaneously; **Mixing, Whipping**, and **Disposing** of the **Shell** do not interfere with each other. There are two main execution threads: from **Yolk** to **Mixing**, through **Dough** to **Baking** and resulting with the **untopped Cake**, and from **White** of **Egg** through **Whipping** to the **Topping** Object. **Coating**, however, cannot begin until both threads are finished; it needs both **untopped Cake** and **Topping** Object, and the final result is a **topped Cake**. Yum!

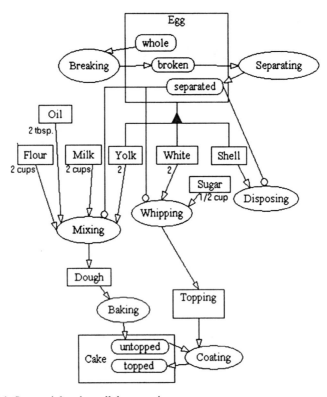

Figure 13.1. Sequential and parallel processing

Figure 13.1 shows how the execution thread uniquely determines the execution order. **Breaking** a **whole Egg** enables its **Separating**. After those two processes are done, however, the next three processes can occur simultaneously; **Mixing**, **Whipping**, and **Disposing** of the **Shell** do not interfere with each other. There are two main execution threads: from **Yolk** to **Mixing**, through **Dough** to **Baking** and resulting with the **untopped Cake**, and from **White** of **Egg** through **Whipping** to the **Topping** Object[1]. **Coating**, however, cannot begin until both threads are finished; it needs both **untopped Cake** and **Topping** Object, and the final result is a **topped Cake**. Yum!

13.1.2 Process Synchronization

Examining Figure 13.1 closely, a question arises as to how does **Coating** "know" when it can start? Or, more importantly, how does an object know it should be generated when two processes lead to it? For example, a new **Car** that has just been manufactured needs to undergo both **Fueling** and **Cleaning** before it is considered

---

[1]   The reserved word Object has been added to the object Topping since it ends with the ing suffix, which is reserved for processes. See Section 4.3.9.

shippable. The problem is that **Fueling** and **Cleaning** are not synchronized. Moreover, their execution time may be distributed over a range so they may take a different amount of time each time they are executed. These two processes are therefore unlikely to terminate at the same time. Any one of the two can terminate first and the amount of time between the termination of one and the other is not known. This necessitates a synchronization mechanism that will take care of generating **C** when **Bing** and **Ding** are over, regardless of the fact that they are not synchronized. The solution is shown in Figure 13.2. The fact that **Bing** is done is captured by the informatical object (also called "flag") **Bing Done**, and the same goes for **Ding**. The process **C Generating** requires that both **Bing Done** and **Ding Done** exist. Note how the result link from **C Generating** is linked to the contour of **C** from the inside. This legal option makes it explicit that the synchronization mechanism is contained inside **C**.

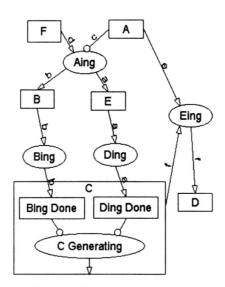

Figure 13.2. Synchronizing **Bing** and **Ding**

13.1.3 Events

An important and useful paradigm in information systems engineering is the Event-Condition-Action (ECA) flow of control mechanism. ECA follows the rule "*On Event if Condition then Action*," namely, if an event occurs, and an associated condition is fulfilled at the time of the event occurrence, then the associated action is triggered. In OPM terminology, Action is an OPM process. We need to define the two other concepts in the ECA mechanism, namely Event and Condition.

13.1.4 Chronon and Event

Data models may represent a timeline by a sequence of consecutive time intervals of identical duration. These intervals are termed chronons. Chronon, used also in quantum mechanics (Farias and Recami, 2001), is like a "time atom," a non-decomposable time interval of some fixed, minimal duration. A data model will typically leave the particular chronon duration unspecified, to be fixed later by the individual applications, within the restrictions posed by the implementation of the data model. Clocking instruments invariably report the occurrence of events in terms of time intervals, not time "points." Therefore, events, even so-called "instantaneous" events, can best be measured as having occurred during a time interval. With this in mind, we will use the term "point in time" as a synonym for chronon.

> *Event* is a significant happening in the system that takes place during a particular chronon.

To demonstrate the relativity of the chronon duration, consider the following extreme examples. In geology, an earthquake that lasts days can be an event, while in nuclear physics, an event, such as transition of an electron from one energy level to the other, spans time in the order of pico-seconds. Moving to more mundane examples, in an elevator system, pushing a button or arriving at some floor can be considered events. Touchdown of an airplane is an event that marks an important milestone in the landing process. In an accounting system, end-of-year is an event, because accounting is done on a yearly basis. The Big Bang is probably the most significant event in the Universe system.

Examples of generic events include the beginning or end of a process or the entrance or exit of an object to or from some state, a milestone during a process, or a process that is short enough to be considered of negligible length. Events of particular interest for modeling systems are those that trigger processes, called triggering events.

> *Triggering event* is an event that triggers some process in the system.

13.1.5 Basic Triggering Event Types

Beginning or end of a process, the birth or consumption of an object, or its transition from one state or to another state, are all events – significant points in the time of the system. Two attributes are related to an **Event**. One is the involved **Entity**: **Process** and **State** are two entities that are related to time, so they are the values of the **Entity** attribute. The other attribute is **Timing**, the time of occurrence, such as **Beginning** or **End** of a process or a state. For states, it may be more intuitive to talk about *entrance* into a state as the beginning of the state and *leaving* the state as its end.

The Cartesian product of the two values of timing (beginning and end) and two values of entities (process and state) yields four basic event types. Table 16 displays the four basic triggering event types.

Table 16. The four basic triggering event types

		Timing	
		Beginning	**End**
Entity	**State**	*State entrance event*	*State exit event*
	Process	*Process start event*	*Process termination event*

A state entrance event is an event marking the entrance of an object into a particular state. Similarly, a process start event is an event marking the start of a particular process.

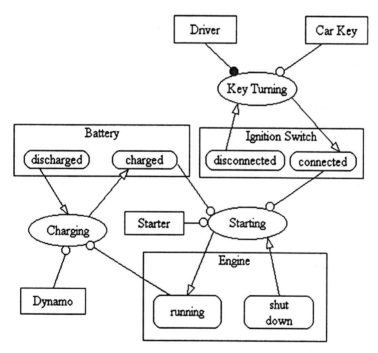

Figure 13.3. Use of triggering events and guarding conditions in a **Car Starting System**

A state exit event is an event marking the exit of an object from a particular state, and a process termination event is an event marking the end of a particular process. Each entrance or exit event can trigger a process in the system. These are the basic triggering events. A process can be triggered, for example, by an event defined as *"the point in time when 1/8 of the fuel tank is full"*. Other types of events can be a change in the value of some object (a data-driven event), an external stimulus (an external event), or an event caused by the arrival of some predetermined point in time (a clock-driven event). All of these event types, and others, can be modeled using OPM. The timing of triggering a process can be based on any one of these events.

Figure 13.3 shows the application of state entrance event. There are two condition links to **Starting**, one from **connected** and one from **charged**. In addition, **Starter** is an instrument for **Starting** and **Engine** must be in its **shut down** state. The precondition set therefore consists of four elementary conditions, repeated below.

The preprocess object set of **Starting** consists of all the objects or specific states from which procedural links lead into **Starting**. **Starter, charged Battery, connected Ignition Switch** and **shut down Engine** constitute the preprocess object set of **Starting**. This preprocess object set defines the precondition set: in order for **Starting** to take place, **Starter** need to exist, **Battery** needs to exists in its **charged** state, **Ignition Switch** needs to exist in its **connected** state and **Engine** needs to be **shut down**. In OPL:

> **Starting** requires **Starter, charged Battery,** and **connected Ignition Switch.**
> **Starting** changes **Engine** from **shut down** to **running.**

If **Battery** is not **charged**, **Ignition Switch** is not **connected**, or **Engine** is not **shut down**, **Starting** will not occur. Entrance of **Ignition Switch** into state **connected** results from **Key Turning**, which changes **Ignition Switch** from **disconnected** to **connected**. Note that **Charging** can occur only when **Engine** is **running**, and this state is entered as a result of the occurrence of **Starting**; yet **Starting** depends on a **charged Battery**. This realistically models the problem of a **discharged Battery**...

13.2 Process and State Duration

Processes take time to execute, and states are situations at which objects exist for some period of time. To be able to specify the time it takes to execute a process or to be in a state, both processes and states exhibit an implicit attribute called **Duration**. **Duration** is the amount of time it takes for a process to execute or for an object to be at a state.

Figure 13.4 is a meta-OPD specifying **Duration** as an attribute of both **Processing** and **state** of **Object**. **Processing** consists of an optional **Minimal Duration** and an optional **Maximal Duration** (Peleg and Dori, 1998). **Duration** exhibits Unit,[2] which can be millisecond (**ms**), second (**s**), minute (**min**), hour (**h**), day (**d**), or year (**y**). The default is **s**, so if no unit is specified, it is seconds. **Duration** also exhibits optional **Distribution**, which can be any probability distribution with the parameters that are associated with it.[3]

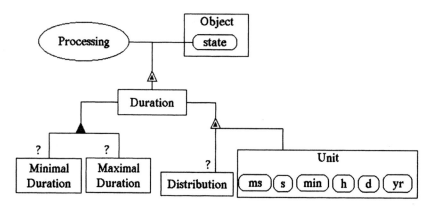

Figure 13.4. The implicit **Duration** attribute of **Processing** and **state**

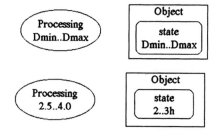

Figure 13.5. The conventions of denoting minimal and maximal duration on process and state

Figure 13.5 shows the conventions of denoting minimal and maximal duration of processes and states. Like participation constraints on objects, the minimal and maximal duration constraints are separated by two dots: **Dmin..Dmax**. The pair of numbers is recorded below the process or state name, followed by the **Measurement Unit**, if it is different than **second**. The range **Dmin..Dmax** should be interpreted as a closed range, i.e., **[Dmin..Dmax]**, where the lower and upper bounds are included. Parentheses around the range indicate that the range is open, so the bounds are not included. Hence, **(Dmin..Dmax)** means that **Dmin** and **Dmax** are not in the range.

---

[2] **Unit** is shorthand for the **Measurement Unit** reserved object, discussed in Section 7.7.2.

[3] The parameters of the probability distribution function are its attributes and the parameter values are the attribute values.

13.3 Processing states

At any given point in time, any object with states is at one of its states. A process affects an object by moving it from one of its states to another state. The change of state of an object is not instantaneous, because any process takes a positive amount of time to execute. Therefore, between exiting the input state and entering the output state, the object must be at some state, which is neither the input nor the output state, but at a *processing state*.

> *Processing state is an implicit state of an object, which already exited the input state but has not entered the output state.*

As soon as the process start event occurs, the object is in an implicit state called the processing state. The object is no longer at its input (preprocess) state. However, as long as the process executes, the object upon which the process is acting is not in its output (postprocess) state either. It is only when the process terminates that the object leaves its processing state and enters its output state. The name of the processing state is identical to the name of the process that operates on the object, except that its first letter is not capitalized, as is the convention for state names. The duration of the object being in the **processing** state is equal to the duration of the process called **Processing**.

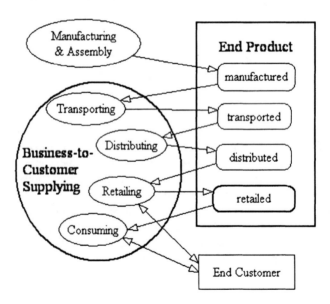

Figure 13.6. While **Retailing** takes place, **Product** is in the implicit processing state called **retailing**.

The significance of the processing state depends on the relative duration of the process. For example, the transition time of **Ignition Switch** in Figure 13.3 being started is insignificant, because the duration of the **Starting** process is very short with respect to other significant processes in the car starting system. Hence, the duration from exiting the **shut down** state of **Engine** to the **running** state of **Engine** is negligibly short with respect to other processes in the system. **Charging**, on the other hand, takes a significant amount of time before the state of **Battery** changes from **discharged** to **charged**. In this example, we define a third state called **charging**, between **discharged** and **charged**. By default, this state is implicit, i.e., we do not draw it as a state nor is it explicitly expressed in the corresponding OPL sentence. If, however, we need to refer to this state we would make it explicit in the OPD and OPL. The duration of **Battery** being in the **charging** state is equal to the duration of the **Charging** process.[4]

As another example, consider the OPD in Figure 13.6 and the following OPL sentence:

> **Retailing** changes **End Product** from **distributed** to **retailed**.

From the time **Retailing** starts (the **Retailing** start event) until the time it ends (the **Retailing** end event), **End Product** is already out of its **distributed** state but not yet at its **retailed** state. Rather, it is in the implicit processing state called **retailing**.

If we were to draw this state explicitly, we would place it between **distributed** and **retailed**. In a similar way, we can think of **transporting** and **distributing** as additional implicit processing states. At the more concrete level, when we zoom into **Retailing**, as done in Figure 13.7, we see what happens to **Product** while it is in its implicit **retailing** state between **distributed** and **retailed**. We can be more specific and note that as soon as **Selecting & Adding** started, **End Product** exits the state **distributed**, and that it enters its **retailed** state as soon as **Credit Card Processing** ends. In-between these two events (points in time) **End Product** is in the **retailing** state. Drilling into the details of the **Retailing** process, we note that **Virtual Cart** has two explicit states, **empty** and **full**. Here, too, a third, implicit

---

[4] Note that the processing state name ends with "ing". However, since it is lowercase it is still distinguishable from processes.

processing state, called **selecting & adding** exists as long as the **Selecting & Adding** process is in progress.

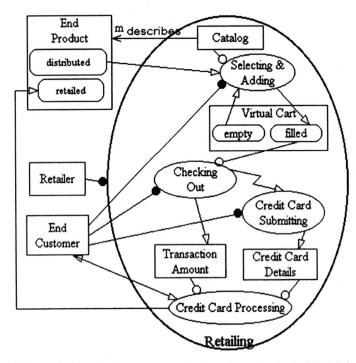

Figure 13.7. Zooming into **Retailing** we see what happens to **Product** while it is in the implicit **retailing** state, which is situated between **distributed** and **retailed**.

13.4 Probability in Procedural Relations

OPM provides a mechanism to express probability events. The probabilistic OPDs in Figure 13.8, Figure 13.9, and Figure 13.10 and the corresponding OPL sentences next to them show how probability can be applied as a readily available extension of the deterministic OPM. The ability to model stochastic processes enables OPM to carry out simulations of real occurrences.

The probability notation can be differentiated from path labels because it is always in conjunction with a XOR or OR logical operator. If both a path label and a probability are necessary on the same link, they should be separated by a comma, with the path label listed first.

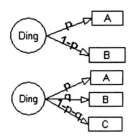

Ding yields either **A** with probability **p** or **B** with probability **1–p**.
(Probabilistic result OPL sentence)

Ding yields either **A** with probability **p**, or **B** with probability **q**, or **C** with probability **1–p–q**.

Figure 13.8. Probabilistic result OPD and OPL sentences with two (top) and three (bottom) possible outcomes

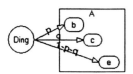

Ding yields either **b A** with probability **p**, or **c A** with probability **q**, or **e A** with probability **1–p–q**.
(Probabilistic multiple state specified result sentence)

Figure 13.9. Probabilistic multiple state-specified result OPD and OPL sentence

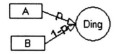

Ding consumes either **A** with probability **p**, or **B** with probability **1–p**.
(Probabilistic consumption sentence)

Figure 13.10. Probabilistic consumption OPD and OPL sentence

13.5 Scope and Name Disambiguation

In natural language, words, which are names of entities and relations, are understood from the context, or scope, within which humans speak or write them. The same word can be used in a variety of different circumstances and humans usually have no problem understanding the correct meaning from the context. In automated systems, such as programming language compilers, as well as OPM, special care need to be taken so that any ambiguity regarding the meaning or association of names is removed. To handle this name ambiguity problem, we define the fundamental directed acyclic graph. This is the basis for the definition of scope, which, in turn, is used to figure out what a name is connected to, using the reserved word of.

13.5.1 The Fundamental DAG

When drawn as links in an OPD, the four fundamental structural relations span a *fundamental directed acyclic graph* (fundamental DAG), whose nodes are objects and whose directed edges are the fundament structural links. The direction of the edges in this DAG is from the *ancestor* – the whole, exhibitor, generalization, or class – to the *descendant* – the part, feature, specialization, or instance, respectively.

> *A **fundamental directed acyclic graph** (DAG) of an object is the DAG spanned from that object, whose nodes are objects or object values and whose directed edges are the fundament structural links.*

Figure 13.11 is an example of a fundamental DAG, with nodes being the objects and the directed edges, the links between pairs of objects or object values. Their direction is from the tip of the triangle symbolizing the fundamental structural relation to the base of that triangle. The object or state connected to the tip is a *root*. Each root in the fundamental DAG defines a subset of the DAG, for which it is the root. This subset includes all the nodes that are reachable from the root object. For example, **Sink** is the root for **Kitchen Sink** and **Bathroom Sink**. The fundamental DAG is the basis for defining scope of an object.

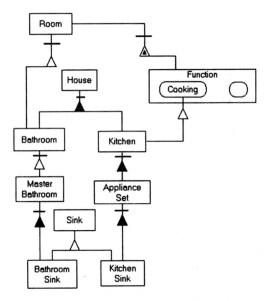

Figure 13.11. An example of a fundamental directed acyclic graph

13.5.2 Scope of an Object

Intuitively, the scope of an object can be thought of as the "extent" or "territory" of the object, the part of the system where it is recognized as the "master." The fundamental DAG provides for a formal rigorous definition.

> **Scope** *of an object is the subset of objects in a fundamental DAG for which that object is a root.*

In more simple terms, the scope is the extent to which the object is recognized. Objects well below Object A in hierarchy "look up" to Object A; they are in its scope. Many things, objects as well as processes, may appear many times in a system as part or features of things. A prominent example is **Name**. One should be able to use the same **Name** without being concerned about duplications and having to invent complicated names for each such attribute in order to avoid these duplicates. While most hierarchies are composed of objects, the scope definition applies to processes as well.

Scope is an important concept in solving name conflict problems, as the following example shows. Consider the attribute hierarchy in Figure 13.12, where two different attributes with the same name **Units** appear in the scope of **Person**. The corresponding "naive" OPL paragraph to the right of the OPD exposes the problem with this OPL paragraph: "unit" appears twice with two different sets of states.

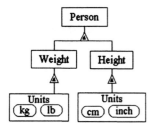

Person exhibits **Weight** and **Height**.
Weight exhibits **Units**.
Units can be **kg** or **lb**.
Height exhibits **Units**.
Units can be **cm** or **inch**.

Figure 13.12. Two different objects with the same attribute **Units** within the scope of **Person** expose the lack of ability to distinguish between **Units** of **Weight** and **Units** of **Height**.

The solution is to differentiate between the two **Units** attributes by adding the reserved word of and the immediate root:

Units of **Weight** can be **kg** or **lb**.
Units of **Height** can be **cm** or **inch**.

The word of associates the thing with a thing at a level immediately above it in the exhibition or aggregation hierarchy. By doing so, a name that is unique in the system is created: **Units** of **Weight** and **Units** of **Height** are now distinguishable as two different attributes.

13.6 The Reserved Words "of" and "which"

There are over 20 dictionary entries for the word "of." Following natural English, the OPM reserved word of is used to relate both a feature to its exhibitor and a part to its whole.[5] Thus, the word **Length** in the phrase **Length of Road** is an attribute of **Road**. In **Breaking of Glass**, **Breaking** is an operation of the object **Glass**. **Beauty of Country, Color of Car, Drying of Cloth,** and **Duration of Burning** are a few more examples for the use of the word of as a connector that references a feature to its exhibitor. The word "of" is used also to link things that are related by other structural links, notably Aggregation-Participation, as in "**Steering Wheel of Car**," which is derived from the OPL aggregation sentence **Car** consists of **Steering Wheel. Door of House** and **Tail of Dog** are other examples of how the word of relates a part to its whole. To see how the word of is used to define the scope of an object, consider the OPL phrase in Frame 29.

> **Car** consists of **4 Doors, 4 Wheels,** and additional parts.
> **Door** exhibits **Color.**
> **Color** exhibits **Brightness.**
> **Brightness** can be **dim** or **bright.**

Frame 29. A terse OPL paragraph, which does not use the word of

The scope of **Car** includes **Door, Wheel, Color,** and **Brightness**. The scope of **Door** is **Color** and **Brightness**, and the scope of **Color** is **Brightness**. **Brightness of Color of Door of Car** denotes the fact that **Brightness** is in the scope of **Door**. More verbosely, one could add the word of to emphasize the hierarchy of parts and features. The OPL paragraph in Frame 30 is a verbose version, which carries the same semantics as the OPL paragraph in Frame 29.

> **Car** consists of **4 Doors, 4 Wheels,** and additional parts.
> **Door of Car** exhibits **Color.**
> **Color of Door of Car** exhibits **Brightness.**
> **Brightness of Color of Door of Car** can be **dim** or **bright.**

Frame 30. A verbose OPL paragraph, which makes use of the word of

---

[5] Using the same word of to specify two types of relations, Aggregation-Participation and Exhibition-Characterization, may seem problematic due to potential ambiguity, but this is not really a problem. Both the OPD and the corresponding OPL paragraph specifically indicate which of these two relations exists between the two things. When an OPL sentence that makes use of the word of appears, the relationship between the two things listed before and after of are specified to be related via one of these two relations. See Figure 13.13 for example.

An "of" chain is a chain of things separated by the reserved word of. The objective of the of chain is to enable distinction between two things with the same name. The *length* of an of chain is the number of appearances of the word of in the chain. For example, the length of the of chain "**Units** of **Weight** of **Person**" is 2.

Brightness of **Color** of **Door** of **Car** is an example of an of chain. The further a thing is to the right in an of chain, the larger is its scope. Thus, in the above of chain, **Car** spans the **largest** scope, followed by **Door**, then **Color**, and finally **Brightness**. The word of is useful and not spurious in cases when the same name in the same scope is used for two different objects, as shown in Figure 13.12. Consider further the OPD Figure 13.13 and the corresponding OPL paragraph in Frame 31.

Car consists of **4 Doors, 4 Wheels,** and additional parts.
Car is parked in Garage.
Garage consists of **Door, 4 Walls,** and **Roof.**
Door exhibits **Color** and **Opening Mode,** as well as **Opening** and
  Closing.
Opening Mode can be **horizontal** or **vertical.**
Color exhibits **Brightness.**
Brightness can be **dim** or **bright.**

Frame 31. The OPL paragraph of Figure 13.13

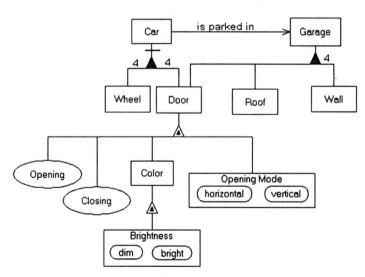

Figure 13.13. Opening Mode and **Brightness** are attributes of **Door** of both **Car** and **Garage.**

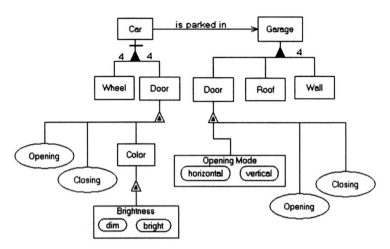

Figure 13.14. Opening Mode is only an attribute of **Door** of **Garage**, while **Brightness** is only an attribute of **Door** of **Car**.

This OPL paragraph specifies that **Door** is part of both **Car** and **Garage**. It also specifies that the operations of **Door** are **Opening** and **Closing** and that the attributes of **Door** are **Color** and **Opening Mode**.[6] This solution is not adequate if we are interested in colors of car doors only and in opening modes of garage doors only. The system architect should be able to specify that **Color** is an attribute of **Door** of **Car** only, while **Opening Mode** is an attribute of **Door** of **Garage** only. As it stands now, however, this is not possible, because the same object **Door** serves for both **Car** and **Garage**. One solution, presented in Figure 13.14, is to have two separate **Door** object classes, one for **Car** and one for **Garage**.

> **Door** of **Garage** exhibits **Opening Mode**, as well as **Opening** and
>   **Closing.**
> **Opening Mode** of **Door** of **Garage** can be **horizontal** or **vertical**.
> **Door** of **Car** exhibits **Color**, as well as **Opening** and **Closing**.
> **Color** of **Door** of **Car** exhibits **Brightness**.
> **Brightness** of **Color** of **Door** of **Car** can be **dim** or **bright**.

Frame 32. Sentences of the OPL paragraph of Figure 13.14 that use the reserved word of

The OPL sentences that are different than those in the OPL paragraph Frame 31 are listed in Frame 32. Here the word of is mandatory, because it enables distinguishing between **Door** as part of **Car** and **Door** as part of **Garage**. Now, **Door** of **Car** and **Door** of **Garage** are two different object classes. A possible simplification

---

[6] Note that a list that involves both attributes and operations in an exhibition sentence is separated into two lists. The first is the list of attributes, as in **Color** and **Opening Mode**, while the second is the list of operations, as in **Opening** and **Closing**. The two lists are linked with the reserved phrase as well as.

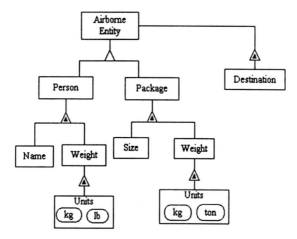

Figure 13.15. The attribute hierarchy of **Person** and of **Package** demonstrates that the reserved OPL word of needs to be applied recursively more than once to distinguish between **Units** of **Weight** of **Person** and **Units** of **Weight** of **Package**.

might be to call each **Door** a different name: **Car Door** and **Garage Door** would then be two different objects with two different names. However, this apparent simplification may prove to be a complication when we do need to use the word of, in which case we get awkward phrases like **"Car Door** of **Car"** and **"Garage Door** of **Garage."**

A single application of the reserved word of may not be sufficient, as the resulting object name may still not be unique. To see an example, consider the OPD in Figure 13.15 and it OPL in Frame 33.

Airborne Entity exhibits **Destination**.
Person and **Package** are **Airborne Entities**.
Person exhibits **Name** and **Weight**.
Package exhibits **Size** and **Weight**.
Weight of **Person** exhibits **Units**.
Weight of **Package** exhibits **Units**.
Units of **Weight** of **Person** can be **kg** or **lb**.
Units of **Weight** of **Package** can be **kg** or **ton**.

Frame 33. The OPL paragraph of Figure 13.15 demonstrates that the reserved OPL word **of** needs to be applied twice to distinguish between **Units of Weight of Person** and **Units of Weight of Package**.

Frame 33 shows an of chain applied in each of the last two sentences. Using of just once would not be sufficient, because in both cases we get the same phrase **Units** of **Weight**. Climbing one more step up the structure hierarchy in Figure 13.15 from **Weight** to **Person** and from **Weight** to **Package** and adding another link to

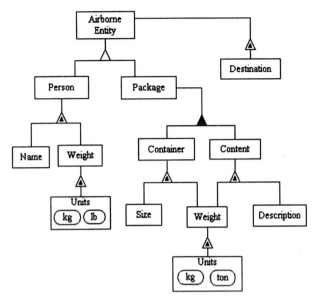

Figure 13.16. Units of **Weight** of **Content** of **Package** is an example of an "of chain" that combines attributes and parts.

the of chain resolves the ambiguity, as a distinction is achieved between **Units** of **Weight** of **Person** and **Units** of **Weight** of **Package**.

How long should the of chain be? Recall that the objective of the of chain is to enable distinction between two things with the same name. Two of chains should therefore be extended up the hierarchy until the first time in which the thing name at the end of one chain is different than the name at the end of the other chain.

As an example, consider the OPD in Figure 13.16. Here, **Units** of **Weight** of **Content** of **Package** is an example of length 3 of chain that combines attributes and parts. As another example, consider the OPL sentence "**Brightness** of **Color** of **Handle** of **Door** of **Store** of **Shopping Mall** is **dim**." The length of the "of chain" in this sentence is 5, and it combines the attributes **Brightness** and **Color** with the parts **Handle**, **Door**, and **Store**.

13.6.1 The Reserved Word "of" and the Dot Operator

In some database query and OO programming languages, the OPL reserved word of is implemented by the dot operator – a dot that separates the higher-level object from it lower-level subordinate. For example, in SQL statements, the dot identifies columns of tables, as in the following query:

```
SELECT company.employee.emp_id
FROM company
```

In OPM terms, the semantics of `company.employee.emp_id` is **"emp_id of employee of company."** Note that the order of objects is reversed, so the semantics of the dot operator is actually an inverse of the reserved word of. In Java, the dot is used to identify Java methods and fields. It is an operator that accesses fields, as in `company.employee.salary` and invokes methods, as in `company.employee. compute_salary()` or accesses the fields of a Java class or object.

13.6.2 Using "of" with Tagged Structural Relations

Tagged structural relations can also be associated with the reserved word of. Examining Figure 13.17, which is elaborated from Chapter 2, we see that while some tagged structural relations give rise to of chains that make sense, others do not. For example, **owns** is a tagged structural relation, for which the use of the word of makes sense, as in **Cash Card** of **Customer** and **Account** of **Customer**. However, **Account** of **Cash Card**, **Account** of **Bank**, **Account** of **Cash**, and **Cash** of **Cash Card** do not make sense, and should not be used. **Account** of **Bank**, for example, implies that the **Account** belongs to the **Bank**, while the OPD specifies that the **Bank** only **holds** the **Account**, while the **Customer owns** it. In the next subsection, we see a more general solution that makes use of the reserved word which.

13.6.3 The Reserved Word "which"

Like the reserved word of, which is also an OPL reserved word that is useful for a variety of purposes. We have already encountered it in a qualification sentence, such as **Car** is a **Vehicle**, the **Medium** of which is **Land**. The reserved phrase ", which" can be used to join two OPL sentences, such as **Tomato** is a **Vegetable** and **Vegetable** is a **Plant**, which is joined to the single sentence **Tomato** is a **Vegetable, which** is a **Plant**.

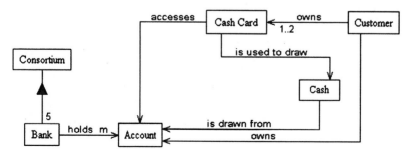

Figure 13.17. Elaboration of the OPD in Figure 2.3

Consider the pair of OPL sentences:

> **Product** is a **System**.
> **System** is an **Object**.

Using the reserved word which, these two sentences are merged into the following one:

> **Product** is a **System**, which is an **Object**.

As another example from Figure 13.17, **Customer owns Cash Card** and **Cash Card accesses Account** can be joined into **Customer owns Cash Card**, which **accesses Account**. Any two OPL sentences in which the last word in the first sentence is an object, which is also the first word in the second sentence, can be joined in this way. However, this is not the only way two sentences can be joined with the phrase, "which." Consider another pair of sentences from Figure 13.17: **Bank holds Account** and **Customer owns Account**, which are combined into: **Bank holds Account**, which **Customer owns**. Here, the common object, is not the last in the first sentence and first in the second, but rather second in both. Examining Figure 13.17, we see that in the first sentence pair, **Cash Card** is the destination of the tagged structural relation (**owns**) in the first sentence and the source of the tagged structural relation (**accesses**) in the second sentence. In the second sentence pair, **Account** is the destination of both structural relations, **holds** and **owns**. Indeed, looking at **Account**, we see that both the **holds** and **owns** tagged structural relations point at it. Looking at **Cash Card**, we see that while **owns** points at it, **accesses** originates from it. The difference between the two pairs explains the difference in the structure of the joint sentences. An example from Figure 13.17 of a third type of a pair of OPL sentences that can be joined by which is **Cash Card accesses Account** and **Cash Card is used to draw Cash**, which is joined to **Cash Card**, which **is used to draw Cash, accesses Account**. This sentence could also just as easily be written as "**Cash Card**, which **accesses Account, is used to draw Cash**."

Figure 13.16 contains three generic OPDs that summarize the three different cases, along with the corresponding "which" sentences resulting from joining the two sentences in each pair.[7] Note that due to symmetry, two sentences are possible in both case (a) and case (c) and only one in case (b). Since **B** is the common object, the pattern "... **B**, which ..." is common to all the five possible sentences.

The reserved word which can also be used to combine sentences that express fundamental structural relations. For example, consider the sentences "**B** consists of **A** and **C**." and "**B** exhibits **D, E,** and **F**." Since B is the common object and it the

---

[7]   Recall that a null tag structural link (a tag with nothing written in it), such as those in Figure 13.16, defaults to the tag relates, which can be substituted by any other tag.

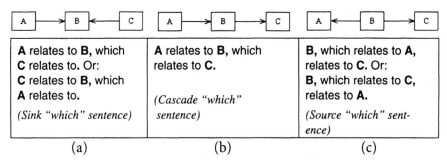

A relates to **B**, which **C** relates to. Or: **C** relates to **B**, which **A** relates to. *(Sink "which" sentence)*	**A** relates to **B**, which relates to **C**. *(Cascade "which" sentence)*	**B**, which relates to **A**, relates to **C**. Or: **B**, which relates to **C**, relates to **A**. *(Source "which" sentence)*
(a)	(b)	(c)

Figure 13.18. The three cases in which two structural links can be joined using which. (a) **B** is sink, yielding a sink which sentence. (b) **B** is cascaded, yielding a cascade which sentence. (c) **B** is source, yielding a source which sentence.

source for both the aggregation and the exhibition, to join these two sentence we apply the source "which" sentence, which yields two possible joint sentences:

> **A**, which consists of **B** and **C**, exhibits **D**, **E**, and **F**.
> **A**, which exhibits **D**, **E**, and **F**, consists of **B** and **C**.

The word of cannot be used to link a specialization to its generalizing thing, nor can it relate an instance to its class. Instead, OPL uses the reserved phrase "which is a" to relate a specialization to its generalization. For example, since **Car** is a **Vehicle**, the OPL way of relating **Car** and **Vehicle** is demonstrated in the sentence: **Car**, which is a **Vehicle**, exhibits **Racing**.

The reserved word which can be used more than once is a single sentence. For example, consider the pair of sentences below:

> **Lamp** can be **off** or **on**.
> **Lamp** exhibits **Energy Source**, which can be **electricity**, **gas**, or **oil**.

These two sentences can be joined in the following one:

> **Lamp**, which can be **off** or **on**, exhibits **Energy Source**, which can be **electricity**, **gas**, or **oil**.

The reserved words of and which provide for some interesting shortcuts and combinations of sentences. However, the importance of these options are not so much in the ability to generate complex and sophisticated OPL sentences, as it is in the ability to parse natural language specification into a set of consistent OPL sentences. In other words, this insight provides OPM with the ability to automatically convert certain types of sentences in unconstrained prose of systems requirement specification into a more formal specification that is expressed in terms of simpler OPL sentences. This ability, if developed properly, is highly valuable as a preliminary pass on a requirement specification document.

13.6.4 Operation: A Process Without Side Effect

Having defined scope, we are in a position to delineate the difference between a process and operation as its specialization. We have encountered operation as a feature; now, looking back, we can define it from a different angle. In OO, all processes are operations. In OPM, processes can stand alone, but they do not have to. As we have seen, an operation is a process that a thing exhibits, such that the thing "owns" that process. How can we tell when a process is an operation of some object? OPM provides the ability to model processes that can stand alone, beside objects rather than inside them. Still, OPM recognizes the existence of operations, namely processes that are owned by objects.

Examining the OPD in Figure 4.5(b), we see that **Manufacturing** cannot be an operation of **Machine**. If we allow that, why wouldn't we assign **Manufacturing** as an operation of **Operator**, or of **Raw Material**, or of **Work Order**, or of **Production Report**, or of **Product**? Each of these objects is involved in some way in the **Manufacturing** process, but none of them encompasses **Manufacturing** in its entirety. It therefore would not make sense to arbitrarily assign any one of these involved objects as the "owner" of the **Manufacturing** process. The **Factory**, however, is an adequate object to own **Manufacturing**. **Factory** consists of the objects **Operator, Machine, Raw Material, Product, Work Order** and **Production Report**. These objects are the scope of **Factory**. All the objects on which **Manufacturing** has any effect are inside **Factory**. **Manufacturing** does not affect any object that is outside the scope of **Factory**. In other words, **Manufacturing** is a process that has no *side effect* outside the scope of **Factory**. Therefore, **Manufacturing** is an *operation* of **Factory**. In general, we say that a process has a side effect on an object if the process transforms (affects, generates, or consumes) one or more objects that is not in the scope of the object.

> A process has a **side effect** on an object if the process transforms one or more objects outside the scope of that object.

An operation is a specialization of a process in the following sense.

> An **operation** of an object is a process that has no side effect beyond the scope of that object.

13.7 Structure-Related Issues

This section discusses a few issues related to structure: transitivity strength, Hamiltonian distance and the fractal relation.

13.7.1 Transitivity Strength

Recall that we have defined a transitive structural relation $\mathfrak{R}$ as a structural relation, for which if $A\mathfrak{R}B$ and $B\mathfrak{R}C$ then $A\mathfrak{R}C$. Careful examination of fork relations, such as the Aggregation-Participation structural relation, reveals a difference in semantics. Although $\mathfrak{R}$ is transitive, there is some difference in meaning between the combination of the two statements $A\mathfrak{R}B$ and $B\mathfrak{R}C$ on one hand, and the statement $A\mathfrak{R}C$ on the other hand. The meaning of **A** consists of **B** and **B** consists of **C** compared with **A** consists of **C** is not quite the same.

To see this semantic difference, let us examine the Aggregation-Participation hierarchy in Figure 7.9. Applying transitivity a number of times, one OPL sentence obtained from the OPD of Figure 7.9 is "**Road Vehicle** consists of **Block.**" Although this is true, it would not be quite accurate to say that a road vehicle consists of a block without providing further information about the nature of this composition. One should be told that the aggregation is indirect, i.e., that there is at least one intermediate level in-between the two. Specifically, we would like to know what is the intermediate object(s) mediating between **Road Vehicle** and **Block**. It is much more informative to specify that a **Vehicle** consists of a **Power Train** which, in turn, consists of an **Internal Combustion Engine**, which, in its own turn, consists of a **Block**.

The reason for this semantic difference is the hierarchy that the Aggregation-Participation relation induces. Due to this hierarchy, the relation between **A** and **C** is indirect, while between **A** and **B** and between **B** and **C** it is direct. A direct relation means that the two things are related by a structural link with no intermediate thing in-between them. Examples of transitive structural relations are "**is co-planar with**" for planar shapes or "**is collinear with**" for line segments. If **A**, **B** and **C** are planar shapes, "**A is co-planar with B**" and "**B is co-planar with C**" then "**A is co-planar with C**" has exactly the same semantics as the two previous sentences. This is so as there is no hierarchy among all the shapes in the same plane. We call such transitivity a **Strong Transitivity**. In general, since there is no hierarchy, all the things that relate to each other with strong transitivity are at the same level.

Weak Transitivity, on the other hand, occurs with transitive relations that are hierarchical. Hierarchical transitive relations give rise to indirect links between two things. In such relations, a chain of two or more transitive structural links with the same label relates two things. In this case, there is at least one intermediate thing in-between the related things. An indirect link occurs with fork relations that are transitive, because forks give rise to hierarchy. The Aggregation-Participation relation is an instance of a relation whose **Transitivity** is **Weak**.

> *Strength is an attribute of a positive transitivity of a structural relation, which denotes whether or not the result of applying transitivity yields a sentence with exactly the same semantics as the source sentences.*

Weak Transitivity is the opposite of **Strong Transitivity** discussed above, which is the default value.

> *A transitive structural relation is* **strong** *if the semantics of the sentence resulting from applying transitivity is identical with the semantics of the source sentences.*
> *A transitive structural relation is* **weak** *if the semantics of the sentence resulting from applying transitivity is not exactly the same as the semantics of the source sentences.*

The value "strong" is the default of the strength attribute. This follows the *default attribute value convention*, according to which the default value of an attribute is the adjective derived from the name of that attribute (which is a noun).

13.7.2 Hamiltonian Distance

Hamiltonian distance is a term borrowed from Graph Theory that measures how far apart two objects related by a weak transitive structural relation are from the viewpoint of that relation. In other words, it is a measure of the extent of indirectness of the relation.

> *The* **Hamiltonian distance** *between two objects, related by a weak transitive structural relation, is the number of hierarchy levels separating these two things.*

Zero Hamiltonian distance means that the two things are directly related, i.e., there is a direct structural link from one thing to the other. For example, in the OPD of Figure 7.9, the Hamiltonian distance between **Driving Unit** and **Steering Wheel** is 0, as these two objects are directly linked. The Hamiltonian distance between **Road Vehicle** and **Cylinder** is 3, because to get from one to the other we need to traverse through three intermediate hierarchy levels – those containing **Power Train, Internal Combustion Engine** and **Block**, in this order.

The Hamiltonian distance is relevant only for *positive, weak* transitivity. It has no meaning for negative or strong transitive structural relations. For any two things related by a strong transitive structural relation it should make no difference how many times the transitive law was applied until that relation was established.

13.7.3 The Fractal Relation

Sometimes, *two* fundamental structural relationships exist between a root object and a set of lower-level, more concrete objects. The most common pair of fundamental structural relationships that go hand in hand is Aggregation-Participation and Generalization-Specialization. Their combination gives rise to the fractal relation.

> A *fractal relation* is a relation between an ancestor object and a set of one or more descendent objects that combines Aggregation-Participation with Generalization-Specialization.

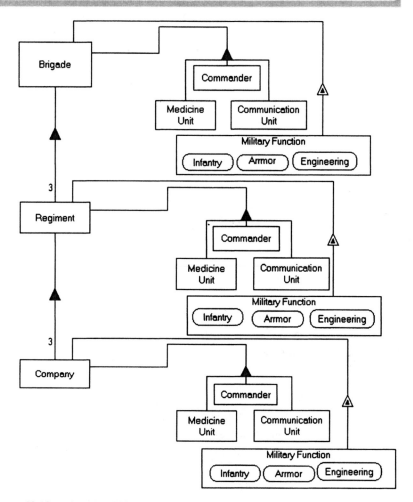

Figure 13.19. A three-level hierarchy structure of military units

The OPD in Figure 13.19 represents a three-level hierarchy structure of military units: A **Brigade**, which consists of 3 **Regiments**, and a **Regiment**, which consists of 3 **Companies**. In addition, each one of those units consists of a **Commander**, a **Medicine Unit** and a **Communication Unit**, and exhibits **Military Function**, which can take on value of **Infantry, Armor**, or **Engineering**. Applying the fractal relation between **Brigade** and **Regiment** and between **Regiment** and **Company** expresses the structure in Figure 13.19 more compactly and more elegantly. The corresponding OPL script follows.

> **Brigade** consists of **3 Regiments,** which inherit from **Brigade.**
> **Regiment** consists of **3 Companies,** which inherit from **Regiment.**
> *(Fractal sentences)*

Note that the phrase "which inherits from" is used rather than "is a," which would normally be used for the gen-spec relation, because otherwise we would get sentences such as "**Regiment** is a **Brigade.**" We wish to avoid such sentences, because although they may be correct methodically, they are counter-intuitive: a regiment is part of a brigade, and saying it *is* one is quite confusing.

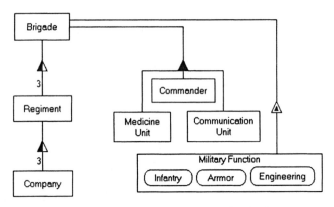

Figure 13.20. Applying the fractal relation to the OPD in Figure 13.19

13.7.4 Covariance and Contravariance

Covariance and Contravariance, shown in Figure 13.21, are opposite ideas. They refer to a situation (Figure 13.21(a)) in which **Object 1** is related to **Object A** and **Object 2** is related to **Object B**. **Object 2** is a specialization of **Object 1**. Covariance (Figure 13.21(b)) claims that **Object B** is then a specialization of **Object A**. Contravariance means that Object A is a specialization of Object B (Figure 13.21(c)).

A debate in the OO community has concerned the question of whether covariance or contravariance best model real-life situations. A well-known example that supports the covariant case relates herbivores and cows. Herbivores eat plants. Cows are herbivores. Grass is a plant. Cows eat grass but not other plants. Figure 13.22 is the OPD that expresses this, along with the corresponding OPL paragraph.

Eating of **Cow** is a specialization of **Eating** of **Herbivore**[8], and it overrides it. While **Eating** of **Herbivore** consumes **Plant**, **Eating** of **Herbivore** consumes **Grass**, which is a specialization of **Plant**. This example indeed supports the case for covariance. An example of contravariance would be if instead of **Grass** we would have

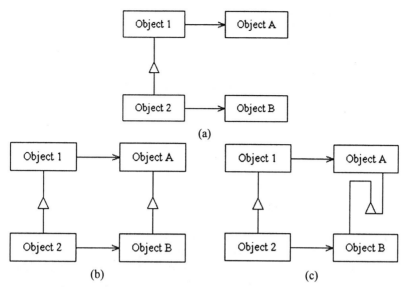

Figure 13.21. Covariance and contravariance expressed in OPDs: (a) Given. (b) Covariance. (c) Contravariance

> **Herbivore** exhibits **Eating**.
> **Eating** of **Herbivore** consumes **Plant**.
> **Eating** of **Herbivore** affects **Herbivore**.
> **Cow** is a **Herbivore**.
> **Cow** exhibits **Eating**.
> **Grass** is a **Plant**.
> **Eating** of **Cow** consumes **Grass**.
> **Eating** of **Cow** affects **Cow**.

Organism, since **Organism** generalizes **Plant**. OPM can model both covariance and contravariance, and both seem to exist, although covariance is probably more common.

13.8 OPM Metamodeling Issues

This section takes us to the metamodeling level. We first define a metamodel for **Thing**. We then examine relationships between object, attribute, and status, as well as between instance, value and state and model them in a specialization-specification hierarchy. We proceed with a metamodel that specifies how features are incor-

---

8   Note that the reserved word of in the phrase **Eating** of **Herbivore** relates the operation **Eating** to its owner **Herbivore**, and that it is required since **Eating** appears twice in the same OPD, so we need to specify which **Eating** is being referred to. See Section 13.6 for more about the reserved word of.

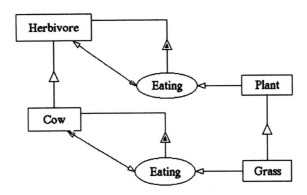

Figure 13.22. A covariance example

porated within the framework of things, objects, processes, values and states. Zooming into the generic **Processing**, we establish a pattern, which can serve as a basis for automating code generation, as it specifies behavior common to all processes. The section ends with a presentation of the OPM construct hierarchy.

13.8.1 A Metamodel of Thing

We now have sufficient knowledge to make an attempt at constructing a metamodel of **Thing**. The OPD in Figure 13.23 and the OPL paragraph below specify this. We list the OPL sentences and add some explanations in-between.

> **Thing** exhibits **Perseverance.**

Recall that **Perseverance** is an attribute that characterizes **Thing** and provides for distinguishing between an object and a process, as expressed below.

> **Perseverance** can be **static** or **dynamic.**
> **Object** is a **Thing**, the **Perseverance** of which is **static.**
> **Process** is a **Thing**, the **Perseverance** of which is **dynamic.**

The value of **Perseverance** determines whether the **Thing** is an **Object** or a **Process**. When the value of **Perseverance** is **static**, the **Thing** is an **Object** and when it is **dynamic**, the **Thing** is a **Process**. Note that at the metamodeling level we think of process as an existing concept, therefore we model **Process** as an object, and it is not called **Processing**.

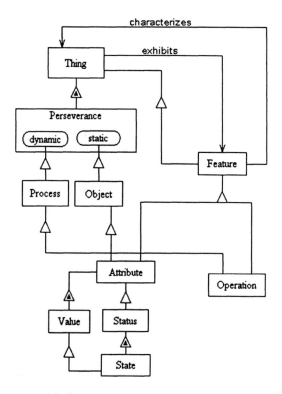

Figure 13.23. A metamodel of **Thing**

Two tagged structural links exist between **Thing** and **Feature**:

> **Thing exhibits Feature.**
> **Feature characterizes Thing.**

This pair of structure OPL sentences specifies the bidirectional relations between a **Thing** and its **Feature**. Note the difference between the sentence "**Thing** exhibits **Perseverance.**" and the pair of sentences "**Thing exhibits Feature.**" and "**Feature characterizes Thing.**" The first sentence is an exhibition sentence. The reserved OPL word exhibits (in non-bold) is the textual equivalent of the triangular Exhibition-Characterization symbol that connects **Thing** with **Perseverance.** The two other sentences are tagged structural sentences. The non-reserved OPL words **exhibits** and **characterizes** (in bold) are the tags of the tagged structural relation that connects **Thing** with **Feature** in both directions.

> **Feature** is a **Thing.**
> **Attribute** and **Operation** are **Features.**
> **Attribute** is an **Object.**
> **Operation** is a **Process.**

Thing specializes into **Feature**, which, in turn, specializes into **Attribute** and **Operation**. Since specialization is transitive, **Attribute** and **Operation** are also **Things**. More specifically, **Attribute** is a specialization of **Object,** while **Operation** is a specialization of **Process.** The rest of the sentences deal with the relations among **Value, Status** and **State.**

Attribute exhibits **Value.**
Process changes Value.
Status is an **Attribute.**
Status exhibits **State.**
State is a **Value.**

Status is a specialization of **Attribute.** As we have seen in the specialization-specification hierarchy in Figure 13.24 and in Table 17, just as **Attribute** exhibits **Value,** so does **Status** exhibit **State.** Hence, **State** is to **Status** what **Value** is to **Attribute**; **State** is a specialization of **Value,** just as **Status** is a specialization of **Attribute.**

13.8.2 The Specialization-Specification Hierarchy

Table 17 shows the relationships among six concepts – **Object, Attribute, Status, Instance, Value** and **State** – as a two dimensional hierarchy. The two dimensions are **specification** and **specialization. Status** is a specialization of **Attribute,** which, in turn, is a specialization of **Object,** while **State, Value** and **Instance** are the specified (identified) versions of **Status, Attribute** and **Object,** respectively.

Table 17. The Specialization-Specification hierarchy

Specification
⟶

Object	Instance
Attribute	Value
Status	State

Specialization ↓

To exemplify these relationships, let us consider again the object **Car** – an artificial system whose function is moving people from one location to another. We have seen that since **Car** is designed to move, **Location** is one of its most important attributes. An attribute is defined as an object that describes another, higher level object. As an *object,* **Location** has *instances,* such as **New York** and **Boston.** As an *attribute,* **Location** has *values,* which are identical to the instances of the object **Location,** such as **New York** and **Boston.** Finally, as a *Status,* **Location** has *states,* which are again identical to the instances of the object **Location** and to the values of the attribute **Location**, such as **New York** and **Boston**. In other words, **New York** and **Boston** are at the same time *instances* of the *object* **Location**, *values* of the *attribute* **Location**, and *states* of the *status* **Location**. Looking at Section 12.2, we can even say that **New York** and **Boston** play the role of **Locations** for **Car**.

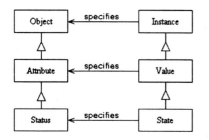

State is a Value.
State specifies Status.
Status is an Attribute.
Value specifies Attribute.
Value is an Instance.
Instance specifies Object.
Attribute is an Object.

Figure 13.24. Meta-OPD of the specialization-specification hierarchy

Figure 13.24 is a meta-OPM specification of the specialization-specification hierarchy that appears in Table 17. We see again the equivalence between object instances, attribute values and status states. When an object plays the role of an attribute, its instances play the role of attribute values, and when an attribute plays the role of a status, its values play the role of states.

13.8.3 A Refined Generic Processing Model

Elaborating our generic OPM model of **Processing,**[9] presented in Figure 8.27, the internal mechanism of how a process occurs can be specified in more refined generic terms. Figure 13.25 is an OPD in which **Processing** of Figure 8.27(b) was in-zoomed. The details are explicit in the following corresponding OPL script in Frame 34.

Processing from SD1 zooms in SD1.1 into Triggering, Executing,
   Terminating, and Exception Handling, as well as "Pre-Condition
   Set is met?" and "Post-Condition Set is met?"
SD1.1 Paragraph:
Agent handles Processing.
Processing requires Instrument.
Triggering determines whether Pre-Condition Set is met.
Executing occurs if Pre-Condition Set is met.
Executing consumes Consumee.
Executing affects Affectee.

---

[9] This generic *model* of **Processing** should not be confused with the *metamodel* of **Thing** in Figure 13.23, where **Process** is modeled as an object.

> **Executing** yields **Resultee.**
> **Executing** determines whether **Post-Condition Set is met.**
> **Terminating** occurs if **Post-Condition Set is met,** or if **Pre-Condition Set is not met.**
> **Exception Handling** occurs if **Post-Condition Set is not met**
> **Exception Handling** invokes **Termination.**

Frame 34. The OPL paragraph of Figure 13.25

    Executing is the "body" of the process **Processing.** It is within **Executing** that the process-specific occurrences take place, and therefore it is different from one **Processing** to another. The first subprocess of **Processing** is **Triggering.** It yields the Boolean object **"Pre-Condition Set is met?"** The idea here is similar to "Design by Contract" proposed by Meyer (1997), which has been implemented in Eiffel (Meyer, 1990). **Pre-Condition Set** is the set of conditions that must be met in order for the **Executing** subprocess to be enabled. **Pre-Condition Set** is obtained by examining all the procedural links incoming into **Executing.** These links may include enabling (agent and/or instrument), input, consumption, and effect link. For each such link, checking is carried out to verify that the object from which the link originates exists. For each *input* link, i.e., a link from a particular *state* to **Executing,** the check further validates that the object is at the state from which the input link originates. If this check is successful, **Executing** occurs. Otherwise, **Terminating** occurs and nothing else happens with **Processing.**

    Likewise, the Boolean object **"Post-Condition Set is met?"** is designed to ascertain that the set of conditions required in order for the (normal) **Terminating** subprocess to occur, are met. **Post-Condition Set** is obtained by examining all the procedural links outgoing from **Executing.** These links may include output, result, and effect links. For each such link, checking is carried out to verify that the object to which the link points exists. These objects constitute the **Postprocess Object Set** of **Processing,** defined in Section 5.3. For each *output* link, i.e., a link from **Executing** to a particular *state*, the check further validates that the object is at the required output state. If this check is successful, **Terminating** occurs. Otherwise, **Exception Handling** takes place, and only then, **Terminating** is invoked. **Exception Handling** is designed to respond to all conceivable abnormalities that can take place during **Executing,** which cause one or more conditions in the **Post-Condition Set** to be violated.

SD1.1

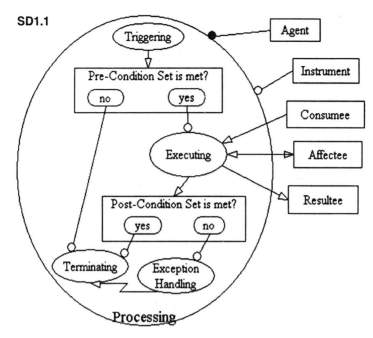

Figure 13.25. Zooming into **Processing** of Figure 8.27(b)

13.8.4 Time Exception Handling

Time Exception Handling is a specialization of **Exception Handling** that option-ally occurs when the actual **Duration** of **Processing** or of being at a state is below or above the limits.

Figure 13.26 and the OPL paragraph in Frame 35 specify the meta-OPM defini-tion of **Exception Handling** for time. Processing is assumed to have an internal implicit **Time Measuring** subprocess, which enables the generation of the **Dura-tion** object. Note the logical XOR relation between the conditions requiring that **D < Dmin** or **D > Dmax**, which is denoted by the tips of the two instrument links arriving at the same point along the **Exception Handling** process circumference. An analogous meta-OPM describes the time **Exception Handling** process that is triggered when a state is dwelled at less than **Dmin** or more than **Dmax**.

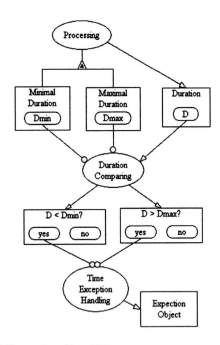

Figure 13.26. OPD of **Exception Handling**

> **Processing** exhibits **Minimal Duration** with value **Dmin** and **Maximal Duration** with value **Dmax**.
> **Processing** yields **Duration** with value **D**.
> **Duration Comparing** requires **Dmin** and **Dmax**.
> **Duration Comparing** consumes **D**.
> **Duration Comparing** determines whether **D** is less than **Dmin** and whether **D** is more than **Dmax**.
> **Time Exception Handling** occurs if either **D** is more than **Dmax** or **D** is less than **Dmin**.
> **Time Exception Handling** yields **Exception Object**.

Frame 35. The OPL paragraph of Figure 13.26

Figure 13.26 and the OPL paragraph in Frame 35 specify the meta-OPM definition of **Time Exception Handling**.

13.9 The OPM Construct Hierarchy

By now we have been exposed to the entire spectrum of OPM constructs that span the abstraction levels, starting at the single OPL word and ending at the entire OPM specification. In this section, we elaborate on Section 9.6 and provide definitions for the various OPM constructs and the hierarchical relationships among them.

An **OPL word** is just like a word in English. A collection of one or more OPL words make up an **OPL phrase,** the basic OPL construct, which is an ordered collection of OPL words that convey specific semantics. If an OPL word is an "atom," then an OPL phrase is a molecule.

OPL phrases can be *reserved* and *non-reserved.* Examples of reserved OPL phrases are "consists of" and "is a." Non-reserved OPL phrases are names of entities (objects, processes or states), structural relation tags or non-reserved participation constraints. A reserved OPL word, such as exhibits, is an OPL phrase that consists of one word. An OPD symbol is the graphic representation of an OPM entity (object, process, or state) or link (either structural or procedural).

An **OPD Phrase** is the diagrammatic or graphic equivalent of an **OPL Phrase.** For example, ▲ is the OPD phrase that is semantically equivalent to the OPL phrase "consists of." A pair of an **OPL Phrase** and a semantically equivalent **OPD Symbol** is an **OPM Phrase Pair.**

> An **OPL phrase** is an ordered collection of one or more English words.
> An **OPD phrase** is the graphic equivalent of an OPL phrase.
> An **OPM phrase pair** is a pair of semantically equivalent OPL and OPD phrases.

Next in the OPL hierarchy are the OPL and OPD sentences.

> An **OPL sentence** is an ordered collection of OPL phrases that expresses a piece of knowledge about the system.
> An **OPD sentence** is the graphic equivalent of an OPL sentence.
> An **OPM sentence pair** is a pair of semantically equivalent OPL and OPD sentences.

An OPL paragraph is the textual equivalent of an OPD.

> An **OPL paragraph** is a collection of OPL sentences.
> An **OPD** is the graphic equivalent of an OPL paragraph.
> An **OPM paragraph pair** is a pair of semantically equivalent OPL paragraph and OPD.

At the top of the OPM hierarchy are the OPL script and the OPD set.

> An **OPL script** is the collection of all the different OPL sentences in all the OPL paragraphs of the system.
> An **OPD set** is the collection of all the OPDs of the system.
> An **OPM specification** is a pair of the system's OPL script and OPD set.

The OPL script is the union of all the OPL sentences that appear in the OPL paragraphs derived from all the OPDs in the OPD set. For the sake of clarity, and to make each OPD self-contained, the same OPD sentence may appear in more than one OPD. Duplicate OPL sentences appear only once in the OPL script.

An *OPL construct* is a generalization of an OPL phrase, OPL sentence, OPL paragraph and OPL script.

An OPD construct is the graphic equivalent of the OPL construct.

An *OPD construct* is a generalization of an OPD symbol, OPD sentence, OPD, and OPD set.

An OPM construct pair is a generalization of an OPM phrase pair, sentence pair, paragraph, and OPM specification.

An *OPM construct pair* is a pair of semantically equivalent OPD and OPL constructs.

Table 18 lists the levels and modalities of OPM constructs. The four OPD-OPL pairs, as shown in the table, are *OPD set – OPL Script* at the *System* level, *OPD – OPL Paragraph* at the *System view* level, OPD Phrase – OPL Sentence at the *Related things* level, and finally, OPD Symbol – OPL Phrase at the *Symbol* level.

Table 18. Levels and modalities of OPM constructs

Modality / Level	OPD Construct	OPL Construct	OPM Construct Pair
System	OPD set	OPL script	OPM specification
System view	OPD	OPL paragraph	OPM paragraph pair
Related things	OPD sentence	OPL sentence	OPM sentence pair
Symbol	OPD phrase	OPL phrase	OPM phrase pair

Figure 13.27 is a meta-OPD that depicts the equivalence and structural relationships among system descriptors: **OPL Script** and **OPD set** specify **System** textually and graphically, respectively. **OPD set** and **OPL Script** are semantically equivalent. The **OPD set** consists of one to many **OPDs**, while the **OPL Script** consists of one to many **OPL Paragraphs**. This logic goes on all the way from the system level to the thing/relation level.

Summary

- The flow of control is the sequence of alternating objects (or object values/states) and processes.
- Event is a significant happening in the system that takes place during a particular chronon, or point in time.

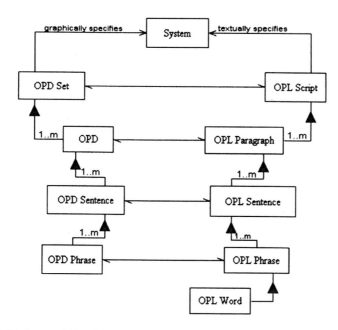

Figure 13.27. A meta-OPD of the OPM construct hierarchy

- A process or state duration can be denoted underneath the process or state name.
- A processing state is an in-between state; it is a situation in which the object already left the input state but has not yet reached the output state.
- Probability of events can be modeled
- The scope of a thing is the extent to which that thing is recognized. Any thing below it in the hierarchy is in its scope.
- The reserved word of is used to link an object with the one immediately above it in the structural hierarchy (mainly with aggregation and exhibition). It can be used to distinguish between same-name objects that are semantically different.
- The reserved word which is used to join two OPL structure sentences into one.
- An operation is a specialization of a process, which has no side effects. In other words, an operation is encapsulated within an object and all the effects are contained within that object.
- Transitivity strength is an attribute of a transitive relation: strong transitivity means that applying transitivity retains the exact semantic meaning of the two original relations.
  - Hamiltonian distance is the number of hierarchy levels separating two objects, related by a weak transitive structural relation.
- A fractal relation is a combination of generalization and aggregation relations.

Problems

1. Practice your which skills: find ten pairs of sentences that can be combined using which. Have at least one of each type (sink, cascade, source).
2. Write the OPL paragraph of the OPD in Figure 13.1 and explain how it determines the order of process execution. Then reorder the sentences in the best way, in your opinion.
3. Apply a synchronization pattern similar to that in Figure 13.2 to model the following requirement. A new **Car** that has just been manufactured needs to undergo both **Fueling** and **Cleaning** before it is considered at a state **shippable**, and the **Fueling** and **Cleaning** can be done in any order, but not simultaneously.
4. Based upon your real-life experience, and building on Figure 13.3, add a second system diagram in that OPD set, explaining the method to start up a car when the battery is discharged.
5. Find at least two systems in this book where a processing state would be helpful if explicitly stated. Explain.
6. Write the longest of chain you can think of. Write a short paragraph explaining whether using long of chains is counter-productive or not.
7. On Halloween night, in Wichita, Kansas, three witches went out to scare some children. The first witch said: "The Maddisons, which live on a farm, are out trick-or-treating tonight. I was watching them and they left at eight o'clock, by my watch." The second witch answered back: "The Arbeits, which are in town now, are watching a movie two hours long." The third witch retorted: "Oh yeah? Well, the Lukashev boy is walking around pretending to be a vampire." Assume the witches aren't lying and model the facts, which the witches mentioned.
8. Manufacturing can result in a product, which is sellable or defective. Sellable products are shipped, while defective ones are repaired. The outcome of the repair can be that the product is sellable or that it remains defective. After four unsuccessful repairing attempts the product is recycled.
   a. Specify the system through OPM.
   b. Add the fact that the probability of success in repairing a defective product is 0.9.
   c. Add the fact that the time to repair a defective product is between 15 and 20 minutes.
9. Specify the system described in Problem 8, given that if the time to repair a defective product is more than 20 minutes, than the product is recycled.
10. A human tester has been assigned to inspect the product in the system described in Problem 8. He uses a testing device that determines whether the product is grade A, B, C, or D. Grade A products are exported, Grade B products are shipped for local consumption, Grade C products are sent to repair and Grade D products are recycled right away. Add these facts in **SD1**.

11. Find in the book at least three examples in which two or more sentences can be joined into one using the reserved word which, as in:

> **Lamp,** which can be **off** or **on,** exhibits **Energy Source,** which can be **electricity, gas,** or **oil.**

12. Based on your solution to problem 11 extend the ruled for joining sentences using the reserved word which.
13. Pick up a paragraph of a technical specification and generate as many OPL sentences as you can from it.
14. Draw the OPD set of the OPL paragraph you generated in Problem 13.
15. **Challenge problem:** Using the experience gained from problems 13 and 14, as well as any other source of knowledge, sketch a preliminary outline of an automated algorithm to generate OPL sentences from a free prose text of a technical specification.

Chapter 14

Systems Theory

> *Cybernetics and Systems Science ... constitute a some-*
> *what fuzzily defined academic domain that touches virtu-*
> *ally all traditional disciplines, from mathematics,*
> *technology and biology to philosophy and the social*
> *sciences.* F. Heylighen (2001)

Systems theory encompasses the observations made over the past two and a half millennia about common features that characterize systems, regardless of their domain. As defined by Heylighen (2001), systems theory is the trans-disciplinary study of the abstract organization of phenomena, independent of their substance, type, or spatial or temporal scale of existence. It investigates both the principles common to all complex entities, and the (usually mathematical) models, which can be used to describe them. More specifically, it is related to the recently developing "sciences of complexity," including artificial intelligence, neural networks, dynamical systems, chaos, and complex adaptive systems. Building on the foundation of systems laid out in Chapters 10 and 11, this chapter discusses various system theories and their relationships with OPM. We begin with an introduction of the informatics hierarchy and General Systems Theory. We discuss the term beneficiary in the context of system, environment, and interface. Control and feedback are modeled in OPM terms. Classical physics and the quantum theory are discussed with an eye toward their effects on OPM. We define and demonstrate objectifying: converting a process occurrence to an object record. The informatics hierarchy is then defined and ontology are finally related to OPM.

14.1 The Informatics Hierarchy

We may think of a hierarchy of information-related concepts, which, for lack of a better name we will call the *informatics hierarchy*. The informatics hierarchy includes (from bottom to top) the concepts of data, information, knowledge, understanding, expertise, wisdom, and ingenuity.[1] The levels in this hierarchy are defined in terms of lower ones.

Data are elementary symbols and relations among them.

Information is the result of processing pieces of data of various types and volumes, and packing them in a compact and meaningful way.

Knowledge of a subject is usually the result of absorbing, assimilating, processing, and analyzing information.

---

[1] This level was suggested by Ed Crawley.

Understanding is the result of accumulated and mentally digested knowledge, which enables building a mental mode, specifying cause and effect, and predicting future phenomena based on the model.

Expertise is a high degree of understanding of some domain, obtained through deep familiarity and experience with the domain, which enables solving new problems, using good judgment and making smart decisions.

Wisdom is the ability to capitalize on expertise to find novel, original, creative solutions to problems and explanations for phenomena. It often involves the ability to pose questions, express opinion, and transfer expertise from one domain to another through generalizations and projections.

Ingenuity is the special spark or burst of wisdom that is very rare and leads to new insights, ways of though, or innovations, that change the way mankind views the universe and itself within it.

For lack of an existing term, each concept in the informatics hierarchy will be termed *informatic*. The definitions of the informatics in the hierarchy are not sharp, and the borderlines between any two adjacent concepts may not be clear-cut. Most people use a representative informatics element – usually information, and often data – to refer to any informatic. This semantic blur is a manifestation of the fuzziness built into natural languages.

Interestingly, except for the lowest item in the hierarchy, data, and understanding, the informatics exist in their singular form only. We cannot talk about "informations," "knowledges," "expertises," "wisdoms," or "ingenuities," indicating that these are indeed mentally challenging, intangible, elusive concepts that are not amenable to quantification.

At the bottom of the informatics hierarchy are *data*, a relatively easy to understand concept. Data are atomic units of information, each of which in itself may have only limited significance, if any. A set of related data elements might be processed to yield some more meaningful *information*. Pieces of information, when put together in the correct context, may yield human *knowledge*, which leads through analysis to *understanding* of the domain or an investigated portion of it. Understanding enables decision-making and provides the basis for design. *Expertise* employs lower level informatics to solve problems in existing systems and design new ones. *Wisdom* is the ability to transfer knowledge across domains and to generate new knowledge from observations and inferences. It often requires expertise in the field. Finally, *ingenuity* is a one-of-a-kind trait, endowed only to talented few. It is seeing beyond wisdom and finding bridges between seemingly unconnected ideas, and answering fundamental questions.

14.1.1 Computers Are Climbing the Informatics Hierarchy

The informatics lower in the hierarchy, data, information, and, to some extent, knowledge, are conventionally stored in computers, which can efficiently access and manipulate high volumes of data. As we climb the hierarchy, more human intellect and less computing power is involved. Knowledge, understanding, expertise, and ultimately wisdom and ingenuity require increasing level of human intelligence. As we climb the informatics hierarchy, the level of required (human or artificial) intelligence increases.

The roles that computers have played throughout the short history of their evolution reflect their gradual ascent through the informatics hierarchy. Early computers in the mid 1940's served as military and business data processing (DP) machines, operated by DP personnel employed in DP departments within large corporations. This was the application of the lowest informatics level. In the 1970's computers were "promoted" to serve at the next informatics level – the information level – as they became vehicles for handling information systems devised by information system engineers, in which database systems constitute the lower level of the hierarchy. Since the 1980's, computers advanced to a yet higher informatics level, as they have also begun to undertake roles as tools to support knowledge base infrastructure, upon which AI-induced expert systems and decision support systems at a variety of levels and domains are built. Computer-based robots of increasing intelligence, equipped with various sensing capabilities to interact with the environment and the simulated expertise to perform specific tasks are also emerging. In the context of the Informatics hierarchy, OPM is an effective tool for converting information into knowledge and help generate expertise.

14.1.2 Knowledge and Understanding

Knowledge and understanding are two adjacent and central levels in the informatics hierarchy. As Winograd and Flores (1978) have pointed out, the question of what it means to know, or "what can we know" is one of the most central issues of philosophy. This question proceeds to "How can we know" or how humans decide which of several possible explanations is most "correct." The rationalistic view accepts the existence of an objective reality, made up of things that bear properties and participate in relations. A cognitive being "gathers information" about those things and builds a "mental model," which can be in some respect correct (a faithful representation of reality) and in other respects incorrect. Knowledge is a storehouse of representations, which can be called upon for use in reasoning and which can be translated into language.

Mediating between the two extremes of rationalism vs. empiricism, Kant coined the phrase *objects are our way of knowing*. This view provides for a practical way of knowing in the absence of objective knowledge, yet allowing observation to count as supportive evidence in argumentation. Kant called it "a scandal of philosophy and of human reason in general" that over the thousands of years of Western culture, no answer to the question "How can I know whether anything outside of my subjective consciousness exist?" could be found. Heidegger (1962), on the other hand, argued that "the 'scandal of philosophy' is not that this proof has yet to be given but that *such proofs are expected and attempted again and again.*" (p. 249; emphasis in original). Heidegger and Nietzsche follow Hegel in supporting the argument approach and call for practical reasoning and common sense to arrive at a community consensus.

Two points in the philosophy of Heidegger (1962) are relevant to our discussion: First, our implicit beliefs and assumptions cannot all be made explicit. There is no neutral viewpoint from which we can see our beliefs as things, since we always operate within the framework they provide. Second, practical understanding is more fundamental than detached theoretical understanding. We have primary access to the world through practical involvement in what is *ready-to-hand* – the world in which we are.

14.2 Ontology

Ontology is defined as a branch of philosophy that deals with modeling the real world (Wand and Weber, 1989). Ontology discusses the nature and relations of being, or the kinds of existence (Ontology Markup Language, 2001). More specifically, ontology is the study of the *categories of things* that exist or may exist in some domain (Sowa, 2001). The product of such a study, called *ontology*, is a catalog of the types of things that are assumed to exist in a domain of interest from the perspective of a person who uses a specific language, for the purpose of talking about that domain.

The traditional goal of ontological inquiry is to discover those fundamental categories or kinds that define the objects of the world. So viewed, natural science provides an excellent example of ontological inquiry. For example, a goal of subatomic physics is to develop taxonomy of the most basic kinds of objects that exist within the physical world (e.g., protons, electrons, neutrons). Similarly, the biological sciences seek to categorize and describe the various kinds of living organisms that populate the planet. The natural and abstract worlds of pure science, however, do not exhaust the applicable domains of ontology. There are vast, human-designed and human-engineered systems such as manufacturing plants, businesses, military bases, and universities, in which ontological inquiry is just as relevant and equally important. In these human-created systems, ontological inquiry is primarily motivated by the need to understand, design, engineer, and manage such systems effectively. Consequently, it is useful to adapt the traditional techniques of ontological inquiry in the natural sciences to these domains as well (IDEF Family of Methods, 2001).

The types in the ontology represent the predicates, word senses, or concept and relation types of the language L when used to discuss topics in the domain D. An uninterpreted logic, such as predicate calculus, conceptual graphs, or Knowledge Interchange Format (2001), is ontologically neutral. It imposes no constraints on the subject matter or the way the subject may be characterized. By itself, logic says nothing about anything, but the combination of logic with ontology provides a language about the entities in the domain of interest and relationships among them.

An *informal ontology* may be specified by a catalog of types that are either undefined or defined only by statements in a natural language. In every domain, there are phenomena that the humans in that domain discriminate as (conceptual or physical) objects, processes, states, and relations. A *formal ontology* is specified by a collection of names for concept and relation types organized in a partial ordering by the Generalization-Specialization (also referred to as the type-subtype) relation. Formal ontologies are further distinguished by the way the subtypes are distinguished from their supertypes. An *axiomatized ontology* distinguishes subtypes by axioms and definitions stated in a formal language, such as logic; a *prototype-based ontology* distinguishes subtypes by a comparison with a typical member or prototype for each subtype. Examples of axiomatized ontologisms include formal theories in science and mathematics, the collections of rules and frames in an expert system, and specifications of conceptual schemas in languages like SQL. OPM concepts and their type ordering are well defined, hence OPM belongs to the family of axiomatized ontology.

The IDEF5 method (IDEF Family of Methods, 2001) is designed to assist in creating, modifying, and maintaining ontologies. Ontological analysis is accomplished by examining the vocabulary that is used to discuss the characteristic objects and processes that compose the domain, developing definitions of the basic terms in that vocabulary, and characterizing the logical connections among those terms. The product of this analysis, an ontology, is a domain vocabulary complete with a set of precise definitions, or axioms, that constrain the meanings of the terms sufficiently to enable consistent interpretation of the data that use that vocabulary.

14.3 General Systems Theory

General systems theory (GST) argues that however complex or diverse the world that we experience is, we will be able to describe it by concepts and principles that are common to all systems and are independent of their domain of discourse. Following this argument, uncovering and establishing those commonalties would enable us to analyze and solve problems in any domain, pertaining to any type of system.

14.3.1 A Brief History of General Systems Theory

The dawn of General Systems Theory (GST) can be traced back to Aristotle, who articulated the basic synergy principle: *The whole is more than the sum of the parts.* Galileo, who emphasized the analytic approach, replaced this synthesis-oriented view. The analytic approach, which is based on experimentation and introspection, opened the door for modern scientific analysis. Descartes developed the scientific method to be able to analyze complex phenomena by breaking them into elementary particles (which we now call objects) and processes. OPM rests on these same premises: objects and process.

In modern times, efforts to construct a unifying theory that tackles complex systems in the various domains of human activity – natural, social and engineering sciences – dates to the early 1920s. Lotka (1956) articulated the principles of what would become modern systems theory, and applied them to biological phenomena, such as the circulation of elements and growth of organisms. Defay (1929) and Schrödinger (1967) utilized thermodynamic principles to explore biological systems and made it clear that an organism is an open system that exchanges matter and energy with its environment in order to remain stable.

Bertalanffy (1968) established GST principles on the basis of ideas he developed in the 1930's and published in 1955. GST reconciles competing concepts of cybernetics and system dynamics. Just as in economics, Keynes defined "whole" as the entire economic system, in biology Darwin defined "whole" as a system of nature. In spite of the remoteness between economics and biology, they share common system principles. Understanding the need for communication among domain experts to increase the overall knowledge of the operation of the system, Bertalanffy (1975) listed the following aims of GST:

- There is a general tendency towards integration in the various sciences, natural and social;
- Such integration seems to be centered in a general theory of systems; and
- This theory may be an important means for aiming at exact theory in the non-physical fields of science.
- Developing unifying principles that run through the universe of the individual sciences, this theory brings us nearer to the goal of the unity of science, and may lead to a much-needed integration in scientific education.

Bertalanffy viewed GST as being comprised of three elements:

Mathematical Systems Theory: The description of the system is provided in terms of a set of measures that define the states and transformations of the system at various points in time. Formal mathematics, such as a set of differential equations or a graph-theoretical description are employed for this purpose. The system is precisely described from an *internal* aspect, using attributes such as stability, wholeness and sum, growth mechanisms, competition and finality. Externally, the system is described in "black box" terms of inputs and outputs – we do not know what goes on inside – and in control theory terms, such as feedback and goal. This mode is useful for thinking about the system and its environment.

System technology: Society and its use of technology have become so complex, that they are no longer amenable to traditional analysis. GST allows one to effectively cope with this complexity. Ecosystems, industrial complexes, education, urban and political environments, socioeconomic entities and a variety of organizations exhibit structure and behavior that lend them to analysis within the GST framework.

System philosophy: GST strives to be a fully articulated worldview that contrasts with the mechanistic framework of the traditional scientific approach. GST is a new paradigm, complete with ontology and epistemology, covering "real systems, conceptual systems and abstract systems." The coverage of real systems pertains to the scientific approach, while the conceptual and abstract ones may be new to traditional human thinking habits.

In 1956, Boulding (1956) identified the communication problems that can occur during systems integration: subsystem specialists (which we call domain experts, e.g., physicists, ecumenists, chemists, sociologists, etc.) have their own languages, but in order for successful integration to take place, all subsystem specialists must speak a common language, such as mathematics. The communication among specialists of various domains at some level in the hierarchy of systems discussed below contributes to the development of knowledge at higher levels.

Mathematics is so widely recognized for its generality and abstracting power, that it has been used in almost any domain of knowledge and human interest. Building on graph theory, OPM can be viewed as a field of mathematics for general systems modeling. Conversely, mathematics can be expressed in OPM terms, with variables being objects and operators, processes.

In their seminal work, Shannon and Weaver (1949) listed three stages in the development of scientific analysis: (1) Organized simplicity, which is the basis for classical mechanics and based on the assumption that the orderliness of the world is built up from simple units and relations; (2) unorganized complexity, the basis for statistical physics, which accounts for complexity arising from random occurrences; and (3) organized complexity, expressed by information theory, which accounts for complexity by identifying fundamental ordering relations. They claimed that the third stage is the model for science in the 20th century. In 1948, Wiener (1961) suggested that cybernetics draws on systems, information, and control theories. He analyzed feedback and goal-directed behavior and applied them to social, biological and mechanical systems. His envisaging of the central role of computers in industry and intellectual processes provided the impetus for systems dynamics theory and cognitive science.

Kerzner (1995) defined GST as "an approach that attempts to integrate and unify scientific information across many fields of knowledge." GST tries to solve problems by looking at the total picture, rather than through an analysis of the individual components. He goes on to note that an increasing number of scholars are recognizing the central role General Systems Theory should play in various, seemingly remote domains, such as project management.

14.3.2 The Hierarchy of System Levels

Boulding (1956) postulated that all areas of scientific interest could be categorized according to their level of development in the following universal hierarchy of system levels:

(1) The *level of frameworks* – the level of static structure, for example the anatomy of the universe;
(2) The *level of clockworks* – the level of simple dynamic systems with predetermined motion;

(3) The *thermostat level* – the level at which the system is self-regulating in maintaining equilibrium through control or cybernetic mechanisms;

(4) The *cell level* – the level of self-maintaining structure, at which life begins to differentiate from not-life;

(5) The *genetic-societal level*, which is typified by the plant and dominated by the empirical world of the botanist;

(6) The *animal system level*, which is characterized by mobility, teleological (purposeful) behavior and self-awareness;

(7) The *human level*, at which the human being is considered as a system with self-awareness and the ability to utilize language and symbolism;

(8) The *social system level*, which exhibits human organizations, complete with value systems, communication and education abilities, emotions and history recording capabilities; and finally

(9) *Transcendental systems level*, which feature the ultimate and absolutes, and the inescapable and unknowable things that nevertheless exhibit systemic structure and relationship.

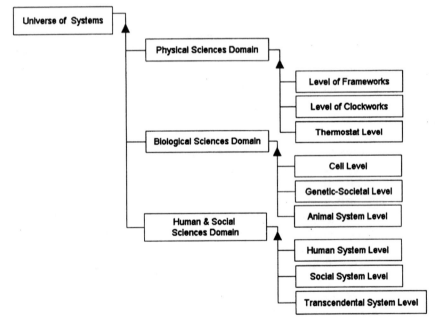

Figure 14.1. The aggregation hierarchy of levels of systems according to Boulding (1956)

Kerzner (1995) noted that although these nine levels appear somewhat vague, they split neatly into three groups of three. The first three concern the physical sciences, the next three include the biological sciences, and the last three include the human and social sciences. These levels and their groupings are depicted as an aggregation hierarchy in the OPD in Figure 14.1.

14.4 Autopoietic vs. Allopoietic Systems

Due to their marvelous functioning, complexity, and evolutionary nature, living organisms have provided ample inspiration for system researchers in their quest for common system characteristics and ideas for devising artificial systems. Maturana and Varela (1980) characterized a living organism as an "autopoietic" system,

> ... a network of processes of production (transformation and destruction) of components that produces the components that: (i) through their interaction and transformations continuously regenerate the network of processes (relations) that produced them; and (ii) constitute it (the system) as a concrete unity in the space in which they (the components) exist.

Processes in this definition are identical with OPM processes. Components are OPM objects. Autopoietic systems are contrasted with allopoietic ones, which exhibit a purpose other than maintenance of their own organization. In an allopoietic system, the product of operation of the system is different from the system itself. The majority of man-made systems are allopoietic. Encyclopedia of World Problems and Human Potential (1995) gives an example of a car, which exhibits a set of processes that differ from the set of processes that brought about its creation in the factory and the assembly plant. Cars move people and goods around, but they do not give birth to new cars...

While artificial systems differ from natural ones in that the former are mostly allopoietic whereas the latter are autopoietic, artificial systems feature a set of life-like attributes, which is reflected in the way they evolve over time. For example, one can trace the evolution of a vessel – a system for sailing in water – from early canoes to modern cruise ships, or of a writing system from a clay tablet to a color laser printer. The observation that natural systems can serve as a successful model for artificial ones has inspired many system researchers and developers to base artificial systems on such biological models as neural networks, evolution and genetics. For example, Winograd and Flores (1978) noted that "the most successful designs (of systems) are ... those that are 'in alignment' with the fundamental structure of that domain and that allow for modification and evolution."

14.5 Systems and Humans

GST stipulates that certain things are shared by all systems. Advocating "seeing things whole," Laszlo (1986) took up the philosophical aspects of GST that views the universe as an interconnected and interdependent field, continuous with itself. Within this universe, the human is included as an entity with knowledge, intentions and goals. These human aspects of GST serve to unite Aristotelian insights with contemporary theories of complexity, and these traits give rise to artificial systems that exhibit function. Indeed, OPM distinguishes between a natural and an artificial system by examining whether or not humans have built the system with some intent or goal in mind. To be understood, natural systems require analysis. To materialize the system's goal through functioning, artificial systems require in addition the elements of design and implementation.

Forrester (1961, 1973) has observed the generality of the feedback mechanism, noting that "the same principles of cause, effect and feedback underlying various weapon systems were applicable to explaining the dynamic behavior of governments, business systems, and human behavior." Regarding human and social systems management, Forrester (1975) suggest why understanding behavior of social systems is problematic:

> *Orderly processes are at work in the creation of human judgment and intuition, which frequently lead people to wrong decisions when faced with complex and highly interactive systems. Until we come to a much better understanding of social systems, we should expect that attempts to develop corrective programs will continue to disappoint us.*

In the early 1960's, practitioners started to realize that GST is applicable and useful for business management. Senge (1990) used Forrester's ideas to analyze a variety of institutes and the type of reflective thinking such institutes carry out. Johnson et al. (1967) related the corporate enterprise to a living cell – an open system that is influenced by its environment and influences it through a process we call interacting. Moore (1964) pointed out the dynamic nature of the business system and characterized it by the flow of resources. GST has thus provided the basis for what has become to be known interchangeably as systems management, project management or matrix management. This managerial approach cuts across many organizational disciplines, including finance, manufacturing, engineering, marketing and human resource management (Kerzner, 1995).

Due to its generic nature, OPM can be instrumental as a vehicle for analysis of organizations of all types for purposes of introspection, continuous improvements and new system development, as well as for self-study and understanding.

14.6 Systems Theory Characteristics

A common denominator to systems of all domains and areas of research and human thought becomes apparent as we discuss systems theory characteristics. These include the terms environment, beneficiary, time, goal, control and feedback. Many concepts are not well specified, and they call for accurate definitions. Some have already been discussed in previous chapters, while others are discussed below, and clearly defined and applied in OPM.

14.6.1 Previously Defined Characteristics

We have already encountered some concepts, which are common to all systems, no matter their domain. The ideas of time, change, goal, function, and others, are not domain-specific.

Time, a central variable in systems theory, provides a reference for the very idea of dynamics. *Change* is seen as a transformation of the system in time, which, nevertheless, conserves its identity. Generation, growth, state transition, consumption and decay are major types of change. We have defined OPM transformation and change in a concise and coherent way.

All systems seem to be actively organized in terms of their *function*. Natural, autopoietic systems strive solely to keep themselves running, while the function of an artificial system is derived from its *goal*. Goal-directed behavior characterizes the changes observed in the state of the system. We have seen how goal and function define systems in the context of OPM. Feedback, the mechanism that mediates between the goal and the behavior of the system, will be explained and modeled in OPM terms later on in this chapter.

14.6.2 System, Environment and Beneficiaries

An important difference that is not so well understood is the distinction between *system* and *environment*. For biological systems, this distinction is usually straight-forward, because an organism is adapted to its environment, over which it has no control, through millions of year of evolution. If the environment changes drastically and abruptly, the organism would not survive. For artificial systems, delineating the border between objects that are inside the system and those that surround it as part of the environment is more involved. It entails examining those objects and determining whether they are controllable by the system architect. Objects that are amenable to such control belong to the system; others are part of the environment. Examples of objects that would be considered as environment objects for most systems are the earth's atmosphere, political borders, earthquakes, the global weather system, legal limitations, language barriers, rules and regulations, gravity, the movement of the earth, high tide and low tide.

In addition to system and environment, artificial systems often involve the notion of *beneficiary*. Consider a space shuttle system as an example. The environment of the space shuttle is the earth, its atmosphere and outer space. The astronaut crew is part of the system, because without it the shuttle would be missing an important element that controls it. The same argumentation applies to a crew of a passenger aircraft. However, what about the passengers? Are they part of the system or the environment? Since the passengers are served *by* the system, almost by definition they are not part *of* the system. The aircraft and its crew do not need the passengers to fly the airplane. On the other hand, neither can the passengers be classified as part of the environment. One reason, for example, is that there is some level of control of the airplane system regarding how the passengers are seated and how they behave in normal and emergency situations. The passengers seem to be neither part of the system nor of the environment. They are the users and *beneficiaries* of the system. They pay for the service of being flown from location A to location B. Similarly, for a car, the driver is both a part of the system and a beneficiary user. A car driver plays two roles: one is as a part of the system, when she or he operates the car, and the other as the beneficiary – the one who benefits from being transported. This distinction makes more sense when we consider a taxi driver, who is part of the taxi-driver system, and the passenger that benefits from the service.

Generalizing these examples, we conclude that an artificial system may have, either directly or indirectly, a component, which is neither in the system nor in the environment, but somewhere in-between: the beneficiary. The beneficiary is an agent who ultimately enjoys the use of the artificial system.

14.6.3 Control and Feedback

A system needs an informatics component to be managed, or controlled.

> *Controlling is the process of using information to maintain attributes of a system within a desired range of values.*

The more intensive the system is on the informatics side, the more controllable it is. The concept of control is tightly linked with *feedback*. Forrester (1968) has established a quantitative theory of systems dynamics. His assertion is that within the system boundary, the basic building block is the feedback loop. Every decision is made within a feedback loop. Every decision controls actions, which, in turn, influence the factors triggering the decision. For example, if a budget slippage is detected, a decision is triggered to tighten the financial control over the subsequent phases of the project. The feedback loop, then, is a closed path that links a decision, a state of the system, information that is external to the system, and returns to the decision point. Meinhardt (1995) provides other examples:

An essential element of dynamic systems is a positive feedback that self-enhances the initial deviation from the mean. The avalanche is proverbial. Cities grow since they attract more people, and in the universe, a local accumulation of dust may attract more dust, eventually leading to the birth of a star. Earlier or later, self-enhancing processes evoke an antagonistic reaction. A collapsing stock market stimulates the purchase of shares at a low price, thereby stabilizing the market. The increasing noise, dirt, crime and traffic jams may discourage people from moving into a big city.

As the generic feedback mechanism in Figure 14.2 and the corresponding OPL script beneath it show, a **Controlled Object** exhibits at least one **Controlled Attribute** that needs to be kept at the **acceptable** state.

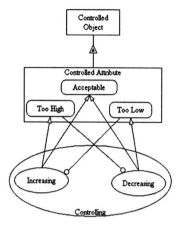

Figure 14.2. A generic feedback mechanism: **Controlling** takes care of bringing **Controlled Attribute** of **Controlled Object** from **too high** or **too low** back to the **acceptable** state.

> **Controlled Object** exhibits **Controlled Attribute**.
> **Controlled Attribute** can be **too high, acceptable,** or **too low.**
> **Controlling** zooms into **Increasing** and **Decreasing.**
> **Increasing** occurs if **Controlled Attribute** is **too low.**
> **Increasing** yields either **acceptable Controlled Attribute** or **too high Controlled Attribute.**
> **Decreasing** occurs if **Controlled Attribute** is **too high.**
> **Decreasing** yields either **acceptable Controlled Attribute** or **too low Controlled Attribute.**

As both the OPD and OPL paragraph show, if the **Controlled Attribute** is **too high**, a **Decreasing** process aims at bringing it back to the **acceptable** state, but it may overshoot and make it **too low,** in which case **Increasing** occurs. **Decreasing,** on the other hand, may overshoot below the **acceptable** level, which would again trigger **Increasing.**

Figure 14.3 builds on the basic feedback concept and shows a **Feedback System.** It exhibits **Functioning** (the process of providing the function for which the system exists), the **Environment** as an instrument to **Functioning,** and the **Beneficiary** as the object that **Functioning** affects. The **Controlling** process governs one or more **Controlled Attribute. Controlling** zooms into **Sensing,** which requires **Sensor** to determine the **Sensed State** of the **Controlled Attribute** through the **Sensing** process. A **too high** or **too low Sensed State** triggers **Increasing,** which requires the **Actuator** and brings the **Controlled Attribute** back to the **acceptable** state.

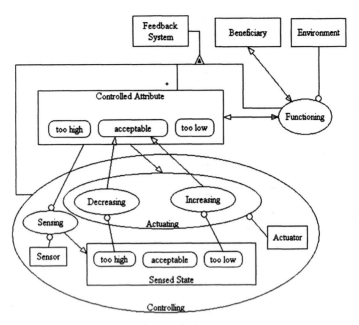

Figure 14.3. A generic model of a **Feedback System** contains the feedback mechanism.

Figure 14.4 depicts a yet more comprehensive specification of a **Controlled System**. The **Controlled Attribute** has a **Value** and exhibits **Acceptability**. Analogous to the three states before, **Acceptability** can be **below limit, within range**, and **above limit. Increasing** or **Decreasing**, collectively called **Affecting**, change the value of the **Controlled Attribute** such that its **Acceptability** changes from **below limit** or **above limit** to the desired **within range** state. The **Controlled System** consists of a **Controller**, which, in turn, consists of **Measuring Instrumentation** and an **Actuator**. The **Controller** stores the **Control Limits**, which can be set or changed by the human operator, and are compared with the actual value of the **Controlled Attribute** to determine the **Acceptability** state. **Measuring** zooms into **Activating, Gauging**, and **Recording. Activating** requires a **Clock** (which can be part of the system or external to it), and changes the status of the **Measuring Instrumentation** (which is the **Sensor** in Figure 14.3) from **idle** to **active**.

In natural systems, the "decision" is done by mechanisms that build on "laws of physics" to restore stability. For example, in a mammal's body, lack of oxygen can be detected due to increased effort. In response, the respiratory system, which is an actuator of the brain, the organism's controller, is ordered to increase the breathing rate such that more oxygen into the lungs. Bacteria or viruses can replicate, but a proliferating bacteria often triggers an immune response that neutralizes the bacteria. In artificial feedback systems, which exhibit functioning and controlling, there is at least one attribute that needs to be kept at the acceptable state. If it is too high, a decreasing process brings it back to the desired level, and if it is too low, increasing does the opposite.

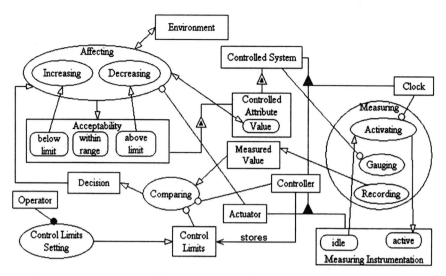

Figure 14.4. A detailed version of a generic **Controlled System**

14.7 Classical Physics vs. Quantum Theory

Since the 17th Century, it has been accepted that physics describes a physical realm that has an objective existence, which is independent of humans and their views, and that this reality is managed through precise mathematical and logical rules. The classical physics theories that were accepted prior to the introduction of the quantum and relativity theories – Newton's mechanics theory of the 17th Century and the electromagnetic and thermodynamic theories developed in the 19th Century – were based on several underlying assumptions or features that may be considered "intuitive." The main classical features, discussed below, are visualization, causality, locality, self-identity, and objectivity. The scientific method is modeled by OPM in Section 11.2. The success of the classical theories to accurately predict outcomes of observations and experiments made them a guiding model of investigation in various domains, from chemistry and biology to psychology and sociology.

At the end of the second decade of the 20th Century, quantum theory became accepted as the fundamental theory, through which physicists research the nature of matter. The quantum theory, which has had impressive successes in explaining natural phenomena in a plethora of domains that range from the sub-atomic to the cosmic scale, has shattered the train of thought that was based on the classical physical theories. The conceptual problems that the quantum and relativity theories evoked extend beyond the pure physical research and touch upon the most fundamental principles, upon which modern western scientific culture is based (Ben-Dov, 1997). In particular, quantum theory raises serious questions about General

Systems Theory, e.g., Bertalanffy's system philosophy discussed above. We discuss the five classical assumptions, how quantum theory possibly contradicts them, and how systems theory and OPM should relate to them.

14.7.1 Visualization

The classical physics theories can be visualized in terms of human daily experience. They can be described in unambiguous terms that humans can relate to with confidence. Quantum theory exhibits duality: in two different circumstances, the same physical thing is described by two seemingly contradictory modes. For example, a beam of electrons or light in certain experimental settings behaves as a flow of particles, while in other settings it exhibits characteristics of a wave. These particle and wave images seemingly contradict each other, as a particle is always located is a certain place in space, while a wave can be found at different locations at the same time. Since contradiction in nature is unacceptable, the "particle-wave" image has been suggested to describe the physical reality through a visual image. Yet, wrote Ben Dov (1997), nobody has succeeded in describing a clear picture of how it looks like.

To reconcile this problem, in 1927 Bohr suggested the complementarity idea, according to which we ought to accept the fact that different, seemingly contradictory descriptions of the world can nevertheless be all true. He claimed that this idea represents a general epistemological concept that is applicable to all domains of human knowledge, from physics and biology to research of humanity, society and morals. Although classical science advocates the use of the mechanistic approach alone, in biology, for example, it is often useful to apply the mechanistic and purposeful approaches interchangeably. The purposeful approach becomes more dominant as we move from primitive life forms, generated through Darwinian selection and survival of the fittest, to artificial systems. OPM's definition of system indeed includes the element of function, which is applicable to both natural and artificial systems.

In OPM, the distinction between natural and artificial systems is apparent of we look at the fundamental difference in their function. In artificial systems, the function is a result of a premeditated design, aimed at accomplishing the desired function, which is a materialization of the system's goal. In natural systems, the function is a result of "laws of nature." In artificial systems, human cognition, will, intent, and goal yield the system's function. In biology, the metaphor of purpose is often helpful in understanding why a certain biological system is "architected" the way it is. Complex biological systems can be explained in terms of goal, although they are the result of eons of evolution. For example, we can explain the fact that a flower looks like a bee by arguing that this look helps attract real bees that help the flower spread its pollen. Attraction is the function of the peculiar flower shape, but the evolution theory explains that it results from survival of the fittest rather than a-priori design.

OPM can specify both natural and artificial systems using the same ontology and conceptual framework and with the same degree of precision. For design of artificial systems, the designer can build on OPM analysis results as a reference baseline, and continue from this point while using the same set of concepts and symbols.

14.7.2 Causality

Classical physical theories ascribe a mandatory link between cause and effect: each event they describe is *determined with certainty* by the physical status that prevailed before the event happened. Beer (1999) has noted the *circular causality* phenomenon: science is expected to amass "the facts," and then to form hypotheses based on those facts. Nevertheless, the hypothesis is already covertly implicit in the selection of the facts, and further selection occurs while these hypotheses are elaborated.

Another aspect of causality is brought up in quantum theory. Even if all the starting conditions are fully known, the equations can only predict the *probability* with which a certain event will occur. For example, quantum theory cannot predict when exactly a given radioactive atom will disintegrate, only what the probability is that it will do so within the next hour. Given a large collection of such atoms, we can say, however, what fraction will disintegrate within an hour.

As long as the systems specified by OPM are not at the elementary particle level, it is possible to consider cause and effect in the classical sense. Bearing this restriction in mind, intuitive knowledge representation and reasoning leads us to believe that classical physics embraced with confidence can still be useful and convenient. This is true as long as the systems we model are not at the atomic or sub-atomic level and relativistic effects can be neglected. OPM handle stochastic processes through a variety of probabilistic sentences, as shown in Chapter 13.

14.7.3 Locality

The locality feature, which dates back to Aristotle, stipulates that nothing can act in a place where it is not found. Since the discovery of gravitation and electric/magnetic fields and electromagnetic radiation we know that this is not the case. Light-speed telecommunication and networking further diminish the significance of this feature.

14.7.4 Self-Identity

According to the self-identity feature, also found in Aristotle's theory, each object is made of a "materialistic cause" and a "formative cause." The materialistic cause provides the object with an identity of its own, which is different from the identities of other objects. The formative cause is the collection of the object's concrete features. One can change the features, but the identity of the object remains the same. Self-identity holds also in the classical mechanics, but not so much in thermodynamics, because it emphasizes energy, a feature of particles, which have their self-identity. In quantum mechanics the self-identity feature does not exist – one cannot talk about two electrons as being distinct from each other. Most artificial systems, in which practitioners are interested, do maintain self-identity of their constituent objects. Object-oriented approaches are based on the self-identity principle.

OPM distinguishes between an object that merely changes its state and one that loses its identity altogether while being transformed into an object that belongs to a different class. A distinction is also made between changing the values of the attributes and changing the attributes themselves. While it is clear that attribute value changes do not affect object identity, changing the set of features (attributes and/or operations) may be a reason to change the object's identity. Consider, for example, a land vehicle, which exhibits a rolling or driving operation. If this land vehicle is converted into a water vehicle, which exhibits sailing, it makes more sense to change the object's identity than to preserve it. This is due to a basic change in the function of the object, which is more significant than the fact that most of the physical parts of the land vehicle remain in the water vehicle.

14.7.5 Objectivity

The objectivity feature implies that the theory is able to describe the physical realm regardless of whether it is being inspected. Classical theories are formulated to describe a physical realm that exists without regard as to whether we view it or not. Aristotle argued that knowing needs to be tested in a dialectic reasoning type of debate. Reasoning allows for the conscious use of natural language as a way to communicate ideas and arguments. The Aristotelian method implicitly bypasses the problem of objective knowledge, namely that the knowledge cannot be separated from the individual observer or the consensus of a group.

We have argued that this view, which accepts the existence of an objective reality, is rooted in the metaphysical revolution of Galileo and Descartes, and grew out of a tradition going back to Plato and Aristotle. This view accepts the existence of the objective world of physical reality and the subjective mental world of an individual's thoughts and feelings. According to Heidegger (1962), the rationale rests on several taken-for-granted assumptions:

- We are inhabitants of a "real world" made up of objects bearing properties. Our actions take place in that world.
- There are "objective facts" about that world that do not depend on the interpretation (or even the presence) of any person.
- Perception is a process by which facts about the world are (sometimes inaccurately) registered in our thought and feelings.

- Thoughts and intentions about action can somehow cause physical (hence real-world) motion of our bodies.

Much of the philosophical efforts have been directed to draw the relations between the mental and the physical domains. Some have argued that we must understand all behaviors in terms of the physical world. Others advocated solipsism – the denial that we can establish the existence of an objective world at all, since our own mental world is the only thing of which we have immediate knowledge. Following the success of Galileo and Newton in explaining and predicting natural phenomena through experimentation, 17th Century philosophers at the dawn of Enlightening, like Bacon, Locke and Hume, supported empiricism as a way to discern religion from science by demanding experimental evidence. Representing rationalism, Descartes and Leibniz advocated axiomatic logic as a central to knowing what is reliable and rejected empiricism as being too close to the observer and hard to distinguish from dreams.

In quantum theory, according to Heisenberg's uncertainty principle, this does not hold: an electron that is being inspected behaves differently than if it is not. If we do not inspect it, it can be in one of several forms, but if we do, we must assume that the measurement causes exactly one option to be materialized. In most cases, however, we can assume that an observation does not alter the observed, but we should be aware that there are cases outside the quantum realm, where observation does affect the result. In particular, systems that involve humans are subject to this phenomenon. A well known example is the Hawthorn experiment, done in the early days of industrial engineering, which proved that when human workers were observed, they performed better than otherwise – the mere attention they were paid changed their productivity.

As Beer (1999) has noted, because any subject of our attention is limited by our senses, and because individuals differ in acuity of perception, any system is a subjective phenomenon. We cannot have an objective system, which means that no system is ever right or wrong. A system is a model that is more or less useful for some purpose. If that purpose is not defined, then there is no criterion of utility. Aiming at modeling systems in some domain, OPM is concerned with the practical type of understanding. If the resulting operational system functions satisfactorily from the beneficiary viewpoint, then OPM can be viewed as an effective tool for systems development.

14.8 Objectifying: Converting a Process into an Object

Natural languages convert a process into an object when there is a need to keep a record of the occurrence of the process. Once a process occurs, or is scheduled or expected to occur at some future time point, it often needs to be recorded. The record keeps the information that such a process has occurred, or is expected to occur. The record can also store the time interval of the occurrence, additional attributes, and the associated object transformations that the process caused. To explain this point, consider a **Car Crashing**, in which the car involved was damaged. By OPM definitions, this noun is a process, since it passes the process test, presented in Chapter 4, as we show below.

(1) Object dependability: there are at least two objects – **Car** and **Driver** – that must be present in order for **Car Crashing** to occur.

(2) Object transformation: **Car Crashing** affects **Car** by changing its state. The change is in the value of the attribute **Condition** of **Car**. The change is from **intact** to **damaged**. The driver may also be affected. A new object that may be created as a result of **Car Crashing** is **Claim**.

(3) Association with time: time is an important attribute of **Car Crashing**. Indeed, the exact time at which the accident took place is one of the first things any interested party (police, insurance company) asks about and records.

(4) Association with verb: the more commonly used term is **Car Accident**, but we adhere to the convention of using the appropriate gerund form for processes when possible. Therefore, we elected the name **Car Crashing**. They convey the same semantics.

Having shown all four criteria of the process test to be valid in this case, we have established that **Car Crashing** is a process. Consider, however, the accident from the point of view of the **Insurance Company**, to which a **Claim** has been filed for compensations due losses incurred by this **Car Accident**. From the **Insurance Company**'s viewpoint, the fact that the accident was a process at the time it took place is of little significance. What is significant is that the occurrence of the accident has become a fact that needs to be dealt with, just as the physical damage inflicted on the car that needs to be fixed. To handle the accident, the **Insurance Agent** would fill out a claim form, in which **Car Accident** and its details are recorded. These details typically include the date and time, location, circumstances, description and damage estimate, and they will be written on paper and/or in a computer database. Thus, it is the *record* of the accident, rather than the accident itself, which stands for the object **Car Accident**. To be accurate, we should call this object **Car Accident Record**. This informatical object represents a physical process. Informatical objects can also represent physical objects, not just processes. An example of an informatical object that represents a physical object is **Car Record** in a motor vehicle registry of some state.

As another example for a process turning into an informatical object, let us consider again the **Wedding** system from Chapter 1. **Marrying** is a social and legal process, carried out according to some religious or secular protocol. The preprocess object set of **Marrying** includes two instances of the object class **Person**, which play in this process the roles of **Bride** and **Groom**. It also includes an agent, whom we will call **Officer**. **Marrying** transforms the **Bride** and **Groom** by affecting them. The effect of **Marrying** is to change the attribute **Marital Status** of both the **Bride** and **Groom** from **non-married** (which generalizes single, bachelor, bachelorette, widower, widow, or divorced) to **married**. Another transformation the process **Marrying** causes is the generation of the new **Married Couple** object. Once **Marrying** is over, it can be viewed as an object, at which point **Marriage** seem to be a better word. By so doing, we actually refer to the recollection of the record of the marriage and the outcomes it brought about, rather than the pattern of behavior exhibited during the wedding. Put differently, **Marrying** is a process during its occurrence and an object thereafter. That object, which can be named **Wedding Object**, is the memory of the **Marrying** process – the fact that it took place along with additional information.

In general, **Objectifying** is the reserved process that converts a process into an informatical object. Processes become objects if we are interested in the outcome, or the *effect* of the process, rather than the details of the dynamics that the process exhibited when it occurred. A process becomes an object when a record of it is kept in some memory. Figure 14.5 shows the **Objectifying** process, invoked by **Processing**, yields **Processing Record**.

Figure 14.5. The **Objectifying** process generates **Processing Record** from **Processing**.

The record is itself an instance of an informatical object. In particular, the human recollection of a specific occurrence of some process is an informatical object. This includes personal events, such as a birth or a journey, historic events, such as elections or wars, and natural events, such as earthquakes or floods. The memory – the mental record – of these things, which, at the time of their occurrence were undoubtedly processes, is analogous to the record of processes of interest on a physical medium. In a case like this, the record of the occurrence of the process, along with all the pertinent information associated with it, becomes an (informatical) object. As such, this record can be inscribed in a human's mind, on paper, digitally, or by any other physical medium.

Summary

- The informatics hierarchy includes (from bottom to top) the concepts of data, information, knowledge, understanding, expertise, wisdom and Ingenuity.
- Ontology is a branch of philosophy or metaphysics that deals with modeling the real world and is concerned with the nature and relations of being.
- General Systems Theory argues that however complex or diverse the world is, there exists a set of concepts and principles that are common to all systems and are independent of their domain of discourse.
- Autopoietic systems are systems that keep themselves in constant existence by regenerating their components. In contrast, allopoietic systems are generated for a purpose other than their self-maintenance.
- Human factors are the most complicated and most challenging to model.

- Environment is the collection of objects surrounding the system and interacting with it such that the system has little to no effect on the environment.
- Beneficiary is the agent who benefits from the function of the system.
- Controlling is the process of using information to maintain attributes of a physical system within a desired range of values.
- Feedback is the mechanism many systems use to control attributes of interest within a certain range. Controlling is part of feedback.
- Quantum theory raises serious questions about five principles of General Systems Theory.
- Objectifying is the process of converting a process occurrence into a record object.

Problems

1. Specify the informatics hierarchy in OPM.
2. Use OPM to describe the changing roles of computers as they have been progressing since the early 1950's.
3. Model the various types of ontology with OPM.
4. Add attributes to the aggregation hierarchy of systems in the OPD in Figure 14.1. Decide which are inherent and which are emergent (you may wish to refer to Chapter 7). Which attributes are inherited to higher levels?
5. Pick one autopoietic and one allopoietic system. Specify them in OPM in a way that would clearly show the difference between them.
6. Pick an organization (company, school, university...) with which you are familiar.
    a. Model it using OPM.
    b. Discuss what elements of human nature in your system that are difficult to model with OPM.
    c. Can you think of any other method that would do a better job of modeling those elements of human nature in your system? Explain.
7. Give two examples of systems in which the user is the beneficiary and two systems in which the user is not the beneficiary of the system.
8. Apply the generic feedback OPM model to a feedback system with which you are familiar. Insert the feedback model into an OPM model of the system it is controlling.
9. Give three examples of process objectifying and explain their usefulness.
10. Controlled Systems.
    a. Write the OPL paragraphs for Figure 14.3 and Figure 14.4.
    b. Suggest a way to split Figure 14.4 into two equal-level system diagrams.

Chapter 15

Object-Oriented Modeling

*I've seen people give the customer a huge list of classes
and message traces and the like. ... The customer has no
idea what the developers are doing, and most of the benefit of writing requirements is lost.*   Ben Kovitz (1998)

Analysis is key to understanding a domain, a system within the domain, and the environment in which it operates. Complete, coherent, high-quality analysis is a prerequisite for the success of systems evolution. While General Systems Theory, discussed in Chapter 14, has attempted to tackle the difficult problem of exploring common underlying systems principles, it has not been concerned with prescribing *how* systems should be developed. In other words, GST is descriptive and not pre-scriptive. Being more pragmatic and realizing that computers do not tolerate a bit of ambiguity (pun intended), information systems methodologists have focused on developing and perfecting methods that prescribe practices for achieving quality information systems.

The object-oriented (OO) approach to the development of software systems development has become the prevailing paradigm during the last decade of the 20th Century. While OPM is applicable to systems in general, many of the ideas in OPM have evolved from OO principles and from previous approaches to developing information systems. This chapter reviews approaches to and methods for informa-tion systems development that inspired OPM in one way or another. Following a short survey of pre-OO approaches, it focuses on OO and discusses its advantages and shortcomings in conjunction with OPM.

15.1 The Evolution of System Analysis Methods

A variety of approaches to and paradigms of systems analysis have been suggested. As Wand and Weber (1989) have noted, constructs used for analyzing systems include data flows (De Marco, 1978), activities (Kung and Sølvberg, 1986), objects (Essink, 1986), linguistic objects (Verheijen and Van Bekkum, 1986) and entities (Chen, 1976). Early approaches have advocated the structured design (Gane and Sarson, 1979) or information-flow idea. The customary graphic representation of information-flow-oriented, approaches, or "functional decomposition", as they are called, is the data-flow diagram, or DFDs (De Marco, 1978). Other methods that evolved from it, such as ADISSA, designed by Shoval (1988) also emphasize the processes that occur within a system and the flow of data among these processes.

15.1.1 Data Flow Diagrams

DFDs have served to successfully model information systems long before the object-oriented approach started to gain popularity. DFDs are *process-oriented*: they focus on processes as the main building blocks of the model. "Objects" in DFDs are represented either as *external entities* or as *data stores*. A data store is the representation in the information system of a real world object. Processes in DFDs are capable of being "exploded" such that their inner content becomes visible. This was one idea that inspired the OPM zooming mechanism. A DFD *context diagram* is a top-level DFD that shows the entire system as a process and the surrounding external entities as the environment. As Maciaszek (2001) notes:

> *Although DFDs have been suppressed in UML by the use case diagrams, the context diagram is still a superior method of establishing the system boundary.*

A major drawback of DFDs, besides lacking the explicit concept of objects, is that they are implicitly *solution-driven*. A DFD-based model uses concepts such as *data flow* and *data store* that are rooted in the IT area. Such terminology tends to influence the analyst to prematurely consider design aspects that belong in the solution domain, before the problem domain has been fully considered and understood and before all the requirements and intricacies of the system have been analyzed.

15.1.2 Entity-Relationship Diagrams and Their Combination with DFD

Entity-Relationship Diagrams (ERDs) model the static-structural aspects of a system. ERDs became popular as a means of data modeling and are still in wide use, mostly for relational database design. An ERD is built from entities, which are analogous to objects and the structural relationships among them in OO approaches. Attributes can be attached to each entity, and participation constraints can be attached to each side of the relationship.

Prior to the introduction of object technology, combining DFDs to model the behavior of the system with ERDs to model the information, was a popular approach. Indeed, the combination of the two covers quite nicely the two major aspects of all systems, namely structure and behavior. The problem with this combination, however, is that these are two different approaches that do not naturally merge to provide a comprehensive insight of the system being analyzed and designed. The following citation of Davidson (1998) illustrates problems of teaching a combination of ERDs with DFDs.

> *I spend about the same amount of time on DFDs and E-R diagramming, and try to integrate the two by having students do the same problems / cases using both. My students frequently have the problem that, once I drill DFDs into their heads, their E-R diagrams look like DFDs. One semester, I reversed the process, and they got E-R diagramming down, but the DFDs looked like E-R diagrams!*

One would assume that the OO paradigm would solve the problem. However, this has not happened. It turns up that the structure-behavior gap, which existed in pre-OO methods, notably between ERDs and DFDs, has prevailed in the OO world. The gap may be less noticeable because various models are packaged within the OO method, but it does exist (Dori and Goodman, 1996).

15.1.3 The Object-Oriented Paradigm

As systems have become more complex, the DFD/ERD approach has gradually been turned down in favor of the currently accepted object paradigm, which places objects as the basic building blocks of the analysis. Since the late 1980's, the object-oriented (OO) paradigm has gained popularity as a means to carry out sound analysis and design rather than being merely a programming method.

A major difference between ERDs and many of the OO methods is that ERDs do not support inheritance, which is an important contribution of the object paradigm. The object model, the basic model in typical OO methods, which takes care of the system's static aspect, is essentially an ERD extended with inheritance. Shifting the emphasis from processes to objects has been a step in the right direction, because objects are the things that processes act on, and therefore they are indeed more fundamental than processes. However, an adverse side effect of this complete switch from processes to objects has been the suppression of the dynamic aspect of the system, which DFDs so nicely represented.

Being aware of this deficiency, OO methods added one or more dynamic models beside the basic object model. As the survey below shows, several OO techniques adopted DFDs as part of their method, applied as a separate analysis activity. For example, Shlaer and Mellor (1992) use Action DFD as a last step in an object-oriented analysis process, but the DFD is not integrated into the OO analysis. DFD is also used as the functional model of Object Modeling Technique (OMT) (Rumbaugh et al., 1991), and by Martin and Odell (1992) in their Object-Flow Diagram, discussed below. OO approaches typically combine the basic object model with a number of other models, each representing another aspect of the system.

15.2 Pre-UML Object-Oriented Methods

Evolving bottom-up from OO programming languages, OO analysis and design is based on the premise that *software* systems can be modeled in terms of *objects* that embed both data (as data members or attributes) and behavior (as operations or methods). Objects are most adequate for modeling the structure of the system, but structure is only one of three main aspects of any system, the other two being function and behavior. This section briefly surveys object-oriented analysis and design methods that were in use prior to the adoption of UML (Object Management Group, 2000) as a standard in 1997. While UML has replaced these methods, it is still worth studying them in some detail to gain insight into the evolution of the field. As this survey shows, each method employs at least three models, each handling another aspect of the system. A comprehensive table and a discussion conclude this section.

15.2.1 Object Modeling Technique

Object Modeling Technique (OMT), developed by Rumbaugh et al. (1991), combines the object model, the dynamic model and the functional model. These three models are used as part of the analysis procedure. The object model contains most of the declarative structure, the dynamic model specifies the high-level control strategy for the system, and the functional model captures functionality of objects that must be incorporated into methods. The objects comprising the system are identified and displayed in an object diagram. A state diagram is drawn for each object, and operations are shown in a DFD. In order to cope with the combinatorial explosion expected in a flat state diagram, OMT uses Statecharts – a broad extension of the conventional state machines and state diagrams introduced by Harel (1987; 1988; 1992). Statecharts expand conventional state diagrams with three elements: hierarchy, concurrency and communication. The dynamic and functional models handle system aspects beyond structure. The three models have different graphic representation, leaving the burden of integration among the various system aspects to the developer. Two of the three OMT models – the Object-Class model and Statecharts – have been "inherited" to UML, but DFD was not.

15.2.2 Object-Oriented Software Engineering

Object-Oriented Software Engineering (OOSE), developed by Jacobson et al. (1992) is a Use-Case driven approach to modeling behavior of systems that can be applied to business situations Jacobson et al. (1994). The Use-Case approach defines typical scenarios and examines them in view of the system's requirements and specifications. It has gained popularity as a complementary means to "pure" object-based analysis, as it potentially overcomes problems in addressing system dynamics directly by many object-oriented analysis methods. The Use-Case model has become part of UML. OOSE defines the following five models for the various states of software development:

- Requirements model: defines the limitations of the system and specifies its behavior. It consists of a Use-Case model, interface descriptions, and a problem domain model. The Use-Case model uses actors and use-cases. These concepts are an aid in defining what exists outside the system (actors) and what should be performed by the system (use-cases).
- Analysis model: at structuring the system independently of the actual implementation. At this stage, the objective is to capture information, behavior and presentation of the system. The object types used in the analysis model are entity objects, interface objects and control objects, while in most other OO

analysis methods, only one object type is used, typically entity objects. The introduction of control objects was motivated by the need to handle the dynamic part of a system.

- Design model: adapts the analysis model to an actual implementation environment. In the first phase of the design, each object in the analysis is mapped into a design block. In the second phase, the Interaction diagram defines the interaction between the objects, and use-cases play a major role in clarifying the requirements from each object.
- Implementation model: consists of the annotated source code. The basis for the implementation is the design model.
- Test model: developed for testing the system at different granularities. The designers test lower levels, such as object modules and blocks. Units at the subsystem level are then tested, followed by an integration test.

15.2.3 Object-Oriented Analysis and Object-Oriented Design

Object-Oriented Analysis, or OOA (Coad and Yourdon, 1991) consists of five layers: Subject, Class & Object, Structure, Attribute and Service. These correspond to the five major OOA activities: Finding Classes & Objects, identifying structures, defining subjects (groups of objects with related semantics), defining attributes, and defining services (methods) for each object class. A service chart, which is an enhanced flow chart, describes each service. Objects interact through message passing. The OOA approach prescribes a method of identifying the objects in the system.

Object-Oriented Design (OOD) (Coad and Yourdon, 1991A) comprises four components: Problem Domain Component, Human Interaction Component, Task Management Component and Data Management Component. Each component corresponds to a particular design activity, in which each one of the five analysis layers should be handled.

15.2.4 Object-Oriented Systems Analysis

Object-Oriented Systems Analysis, or OSA (Embley et al., 1992), defined as "an approach for capturing and organizing information pertinent to the design and implementation of a software system," is a "model driven approach," implying that there is no predetermined set of steps to carry out the analysis. Rather, the analysis comprises a collection of different models. The ability to work concurrently with these models is a basic concept in OSA. Such an approach is expected to be more suitable for system analysis than a method-driven approach, since analysis cannot usually be prescribed as a given set of steps.

OSA consists of three major parts, each using a different model. The Object-Relationship Model is used for modeling the objects in the system and the structural relations among them. The Object-Behavior Model introduces the concept of states. Every object in a system can be in one of many states. These states and the conditions for switching among them are defined. Each object found in the ORM is "exploded" into a set of states, conditions for changing these states, and the action(s) that should be performed as a result of each change. The Object-Interaction Model models the interaction among the objects.

15.2.5 Object-Oriented Analysis & Design

Martin and Odell (1992) divided the analysis into two parts: Object Structure Analysis (OSA) and Object Behavior Analysis (OBA). These two activities, performed together, form an integrated model of the system. The object-relationship diagram, essentially an ERD, is used for OSA. OBA handles system dynamics. The resulting states and transitions among them are drawn in state transition diagrams. For each object, a fence diagram (state transition diagram) is drawn to model the object life-cycle. Recognizing the difficulty involved in integrating these two separate analysis tools, the method offers yet another diagram – the Object-Flow diagram, which is similar to DFD, enables objects to flow from one activity to the next. The result of the analysis is four sets of diagrams: the event schema, the process dependency diagram, the object-flow diagram and the state-transition (fence) diagram.

Object-Oriented Design has two aspects: Object Structure Design and Object Behavior Design, which continue OSA and OBA, respectively, providing ways of mapping objects and events to OO programming language constructs.

15.2.6 Object Life-Cycles

Shlaer and Mellor (1992) propose the Object Life-Cycle approach as a three-step method for analyzing information systems:

- Information Models, similar to ERDs, abstract the world as a collection of objects, their attributes, and the relationships among them, based on policies, rules and physical laws.
- State Models formalize object lifecycle through state transition diagrams and the actions associated with the transitions. An object communication diagram models the interaction among different objects.
- Process Models are Action Data Flow Diagrams (similar to DFDs), representing processes and object data stores. A process is a fundamental unit of operation and an object data store corresponds to the data (attributes) of an object in the information model.

Recursive Design directly maps the analysis model into object-oriented design using Object-Oriented Design Language.

15.2.7 The Booch Method

Booch (1991) tried to incorporate the best from each OO method. Considering object classification as a central object-oriented analysis problem, a set of methods for *discovering* the classes and objects, is suggested. A collection of models for capturing two dimensions of the analysis and design is proposed: the logical/physical view and the static/dynamic view. The physical model describes the concrete software and hardware composition of the system, using Module diagram. This diagram shows the layout of the various classes among the physical files. A process diagram handles multiprocessor systems. The logical model is composed of static and dynamic parts. The static part includes two diagram types. *Class Diagrams* describe classes with their attributes and operations, and relations among them: inheritance, association, "has a," "using" and cardinality. *Object Diagrams* show objects and the messages passed among them, specifying scenarios that are part of the system's behavior. The dynamic part of the logical model adds two more diagrams. Based on Statecharts (Harel, 1987; 1988; 1992), *State Transition Diagrams* show the events that transition a class between states along with the associated actions. Based on event trace diagrams (Rumbaugh et al., 1991) and interaction diagrams (Jacobson et al., 1994), *Interaction Diagrams* trace the execution of a scenario specified in an object diagram. The notation in Booch's method is designed for both analysis and design.

15.2.8 MOSES

MOSES (Henderson-Sellers and Edwards, 1994) has a static model and a dynamic model. The basic concept in the static, O/C model, is O/C – a concept that encompasses both object and class. The dynamic part of MOSES uses two models: event models and Objectcharts (Coleman et al., 1992). Event models link O/Cs with messages that invoke published services. Messages are linked to relationships in the O/C model. Objectcharts include decomposable states and transitions, as well as and services, which are invoked by events that cause transitions.

15.2.9 The Fusion Method

The Fusion method (Coleman et al., 1994) has integrated and extended then-existing object-oriented approaches to handle analysis, design and implementation. In the analysis, an ERD-like *Object Model* describes the structure of the system using classes and relationships. Classes are modeled with attributes but without methods for communication among objects. Relationships include cardinality constraints, aggregation and specialization among classes, and relationship attributes. A System Object Model defines the parts of the modeled world that are part of the system and those that are external.

The *Interface Model* handles the behavior of the system in terms of events and state changes they cause. A scenario is a sequence of events flowing among agents (external to the system) and the system, for some purpose. Each scenario involves agents, the tasks they want the system to do, and the sequence of communications involved in getting it done. These scenarios help build the interface model. It includes the *Operation Model*, which declaratively specifies the system's behavior in terms of change of state and the events that are their output. The operation model is expressed as a series of schemata, each of which is assembled from structured text.

The *Lifecycle Model* defines the allowable sequences of interactions that a system may participate in over its lifetime. The Lifecycle is defined with a lifecycle expression, which includes various operators, such as | to denote "followed by" and * to denote "repeated." Both the Operation and the Lifecycle models use extensive text and complex expressions, but do not take advantage of graphical tools. A Data Dictionary is used throughout the development phases. The analysis results are input to the design, which is performed using four separate models: Object Interaction Graphs, Visibility Graphs, Class Descriptions, and Inheritance Graphs.

15.2.10 OPEN Modeling Language

Object-oriented Process, Environment and Notation, or OPEN for short (Henderson-Sellers and Graham, 1996; Henderson-Sellers et al., 1998; OPEN, 2001), encapsulates issues of business, quality, modeling and reuse within its lifecycle support for software development. OPEN's heart is a pair of two-dimensional matrices, which provide probabilistic links between Activities of the lifecycle and Tasks, which are the smallest units of work within OPEN. A second two-dimensional matrix then links the Task, which provides the statement of goals, to the Techniques, which provide the way the goal can be achieved, i.e., the "how." A metamodel and notation, collectively called OPEN Modeling Language, or OML (Firesmith et al., 1997), supports OPEN.

15.3 Unified Modeling Language – UML

Based on Booch (1991), Rumbaugh et al. (1991) and Jacobson et al. (1992), UML (Object Management Group, 2000) became the Object Management Group (OMG, 2001) standard for software systems development in 1997, and it has been dominating the "methods market" ever since. Prior to UML, there was no clearly leading OO modeling language. Most of them shared commonly accepted concepts with slight differences in expression and language. Required to choose from among these modeling languages, industry users were discouraged from entering the OO market and pushed for the adoption of a general-purpose standard. UML notation is a melding of graphical syntax from various sources, with some symbols removed a few others added. A major contribution of UML is that it has standardized naming and symbols. UML has one model that can be looked at using nine different views. Each view is represented using a different graphical diagram types The nine views are as follows (Object Management Group, 2000):

(1) Use-case diagram; (2) Class diagram; (3) Object diagram; Behavior diagrams, which include (4) State transition (Statechart) diagram and (5) Activity diagram; Interaction diagrams, which include (6) Sequence diagram and (7) Collaboration diagram, and Implementation diagrams, which include (8) Component diagram and (9) Deployment diagram.

Class diagrams are a melding of OMT (Rumbaugh et al., 1991), Booch (1991), and class diagrams of most other OO methods. Use-case diagrams are similar in appearance to those in OOSE (Jacobson et al., 1992). Statechart diagrams are substantially based on Statecharts of Harel (1987; 1988; 1992). Activity diagrams are similar to the workflow diagrams developed by pre-OO sources. Sequence diagrams were found in a variety of OO methods under a variety of names (interaction, message trace, and event trace) and date to pre-OO days. Collaboration diagrams, which form the basis of patterns, were adapted from object diagrams of Booch (1991), Fusion's object interaction graphs (Coleman et al., 1994) and other sources. The implementation (component and deployment) diagrams are derived from Booch's module and process diagrams. Stereotypes are one of the extension mechanisms and extend the semantics of the metamodel. User-defined icons can be associated with given stereotypes for tailoring the UML to specific processes. UML uses Object Constraint Language, or OCL (Object Management Group, 2000; Warmer and Kleppe, 1999) to specify constraint expressions that cannot be modeled graphically.

One implication of UML being "purely" object-oriented is the exclusion of the DFD model from the UML set of diagrams, which was met with many eye brow raising. Many professionals are dissatisfied with the lack of a "DFD-like" mechanism to explicitly show the dynamic aspect of a system. For example, Maciaszek (2001) has noted that since UML does not provide a good visual model to define the scope of the system, the old-fashioned context diagram of DFDs is frequently used for this task.

Comparing complexity metrical values of UML with other object-oriented techniques, Siau and Qing (2001) found that while each diagram in UML is not distinctly more complex than techniques in other OO methods, as a whole UML is 2 to 11 times more complex than other OO methods. Writing about the lack of precise UML semantics, Cook et al. (1999) noted that since UML embraces a variety of syntactical specializations and extensions through stereotypes and tags, it is difficult to define a precise semantics of UML for all cases. Referring generally to standardization efforts, such as UML, Kobryn (1999) noted that in this sort of environment, sound technical tradeoffs are often overridden by inferior political compromises, resulting specifications that are bloated with patches. D'Souza et al. (1999) indicted that the UML 1.3 has cyclic dependencies between Core and Extension Mechanisms. Melewski (2000) noted that with UML's ambiguities, size and complexity many organizations find that UML is complex and takes long time to work with. Simons (2001) claimed that UML's mix of notations from different design approaches yields a confused picture that is not really productive. Allowing UML to drive the development process results in inadequate object conceptualization, poor control structures and loosely coupled subsystems (Simmons et al., 1999).

15.4 Metamodeling in OO Methods

A metamodel approach was proposed by the Common Object Methodology Metamodel Architecture, or COMMA (Henderson-Sellers, 1994) project, which was elaborated later on (Henderson-Sellers and Bulthuis, 1998). Like OO methods, these works advocate separating the structure from behavior at the meta level as well, noting that "one metamodel will be required for the static structure, one for the dynamic behavior, and one for the development process."

In a more recent effort, OMG (2001) has standardized Meta-Object Facility (2000) as a 4-layer architecture. MOF's meta-metamodel level is the foundation on which several metamodels can be defined and standardized. The first and most notable example is the UML metamodel. The second one is Common Warehouse Metadata (CWM). Model-Driven Architecture is an OMG initiative for an overarching, technology-independent standard for lifecycle support of software systems.

15.5 OO Methods – A Summary

Table 19 summarizes the OO methods surveyed. It lists the models that the methodology includes within its analysis phase and how transition from analysis to design proceeds.

Table 19. Analysis models and transition to design of various object-oriented methods

Method Name (Abbr.)	First Author(s)	Structure model			Transition to Design
		Structure model	Behavior model	Additional model(s)	
Object Modeling Technique (OMT)	Rumbaugh et al. (1991)	Object Model	Dynamic Model	Functional Model	Expansion of the analysis model
Object-Oriented Software Engineering (OOSE)	Jacobson et al. (1992)	Entity objects	Control objects	Interface objects	Adapting and formalizing the analysis model

Table 19. (continued)

Method Name (Abbr.)	First Author(s)	Structure model			Transition to Design
		Structure model	Behavior model	Additional model(s)	
Object-Oriented Analysis (OOA)	Coad and Yourdon (1991)	Objects with attributes	Services and message passing among objects	-	design of the static aspect only
Object-Oriented Systems Analysis (OSA)	Embley et al. (1992)	Object Relation-ship model (ORM)	Object Behavior model (OBM)	Object Interaction model	not addressed
Object-Oriented Analysis & Design (OOAD)	Martin and Odell (1992)	Object Structure Analysis (OSA)	Object Behavior Analysis (OBA)	Object Flow Diagram (OFD)	Supposed to be a direct mapping of analysis
Object Lifecycles	Shlaer and Mellor (1992)	Informa-tion Models	Process and State Models	-	a direct mapping of analysis using Recursive Design
Object-Oriented Analysis and Design with Applications	Booch (1991)	Class Diagrams + Object Diagrams	State Transition Diagrams, Interaction Diagrams	Module Diagrams, Process Diagrams	The same notations are used for analysis and design
The Fusion Method	Coleman et al. (1994)	Object Model	Interface Model	Data Dictionary	Additional models based on analysis results
MOSES	Henderson-Sellers, Edwards (1994)	O/C (Object/Class) model	Event models, Object-charts	-	-
OO Process, Environment and Notation (OPEN)	Firesmith, Henderson-Sellers, Graham (1996)	Semantic Nets	Interaction Diagrams, State Tran-sition Dia-grams	Use Case Diagrams	-
Unified Modeling Language (UML)	Rumbaugh et al. (1998)	Class Diagrams (D.)	Statechart, Activity D., Sequence D., Colla-boration D.	Use-case D., Component D., Deployment D.	-

As Table 19 shows, all the methods surveyed above are multi-model methods: they employ between three to nine models to show different views of the systems being developed. Each method contains at least one model for structure specification and another for behavior specification, and at least one additional model. OPM is contrasted with the other methods, as it is the only method that integrates the structure and behavior in a single model.

15.6 Software Development Approaches and Trends

A number of software development approaches and trends have been suggested in recent years to address the problem of software development, which are related to OO in one way or another, and hence also to OPM. These include Aspect-Oriented Programming, the Rational Unified Process, Extreme Programming, and Agile Modeling. RUP has been classified as heavyweight, while XP and AM, as lightweight, or agile.

15.6.1 Aspect-Oriented Programming

Aspect-Oriented Programming (AOP) focuses on the idea of aspects that stem from separation of concerns. Separation of concerns is regarded as a key goal of software design, as it enables the design and implementation of clean, reusable, and understandable modules. A clean separation of concerns reduces the coupling between modules and inherently increases their cohesion. While objects have been a great success at facilitating the separation of concerns, they are considered to be limited in their ability to modularize systemic concerns that are not localized to a single module's boundaries and tend to crosscut the system's class and module structure. Much of the complexity and brittleness in existing systems is attributed to the way in which the implementation of these kinds of concerns comes to be intertwined throughout the code. AOP proposes a new unit of software modularity, an aspect, to provide a better handle on managing crosscutting concerns. Like objects, aspects are intended to be used in both design and implementation. During design, aspects facilitate thinking about crosscutting concerns. During implementation, AOP languages make it possible to program directly in terms of design aspects, just as object-oriented languages have made it possible to program directly in terms of design.

15.6.2 The Rational Unified Process

The Rational Unified Process (RUP) recognizes a "software lifecycle" as a cycle that covers four phases: inception, elaboration, construction, and transition. Each phase spans the time between two major milestones, which are synchronization points. During inception, stakeholders should come to agreement on the objectives, rough architecture, and rough schedule of the project. During elaboration, architecture is tightened, use cases are discovered and a minimal set is implemented. During construction the product is being developed, and use-cases may change as the customer provides feedback to the developer. During transition, the customer takes ownership of the system, and the functionality of the system is applied, incorporating customer feedback. A project plan is a malleable, flexible, and responsive plan of contingencies. As the project proceeds, developers learn how quickly iterations are developed. The customer sees the developing project and changes requirements.

15.6.3 Extreme Programming

Extreme Programming (XP) is a deliberate and disciplined approach to software development. XP emphasizes customer satisfaction and promotes teamwork. Risky projects with dynamic requirements are considered good candidates for XP, as they are expected to experience greater success and increased developer productivity. XP has a simple set of rules and practices, which make the customers partners in the software process and developers actively contribute regardless of experience level.

15.6.4 Agile Modeling

While RUP has been classified as a heavyweight method, XP and Agile Modeling (AM) are considered lightweight, or agile. AM is a collection of values, principles, and practices for modeling software that can be applied on a software development project in an effective and lightweight manner. Agile models do not require perfection and may be applied to requirements, analysis, architecture, and design. The Agile Alliance promotes individuals and interactions over processes and tools, working software over comprehensive documentation, customer collaboration over contract negotiation, and responding to change over following a plan.

15.7 Challenges for OO Methods

The shift of emphasis in system analysis paradigms from processes to objects has been accompanied by degradation in the power of object-oriented methodologies to express systems' behavior without auxiliary "non-object-oriented" tools. The basic OO model in virtually any OO method or language is the static class/object model. All the other models are add-ons that are needed to account for system dynamics and function. As Abadi and Cardelli (1996) noted, while procedural language constructs are generally well understood, an analogous understanding has not yet emerged for OO languages, and there seems to be a mismatch when one tries to model objects with functions.

The restrictions that the Object paradigm imposes on modeling system dynamics account for its unsatisfactory behavior modeling capability. This drawback is a consequence of suppressing the role of processes by forcing them to be attached to, or "encapsulated" within particular classes of objects. This is not natural, because the occurrence of processes in real-life, non-software systems requires the coexistence of a number of objects from different classes. The result of trying to apply a programming paradigm such as OO, as good as it may be, to real-world problem domains often results in awkward problem domain models. Kovitz (1998) clearly reflected the problem of capturing software requirements using the object-oriented approach:

> In my experience, object orientation and requirements often make a bad marriage. ... For example, the idea of a class and a set of methods that operate on it is a very useful discipline for keeping programs simple and maintainable. But ... a very common and important design decision is: which class gets which methods? Do you give the "bake" method to the "cookie" class or to the "oven" class? In reality, cookies bake when you put them in a hot oven. Neither one of them has the "bake" operation all to itself.

15.7.1 A Historic Perspective

In response to the question "was object-oriented programming invented so that it would be possible to write programs that model the real world?" Logan (2000) wrote that this is a widely held misconception, since Alan Kay had coined the term "object-oriented programming" (OOP) by applying it to his language, Smalltalk, in the early 1970s, based on the metaphor of a biological cell. This metaphor was based on the idea that information can pass through the cell wall, but what is on the inside is hidden from what is on the outside. In the 1960s, Kay observed several software systems that exhibited "cell-like" behavior in that they encapsulated data: Sketchpad graphics system, the Simula simulation system and a Burroughs operating system. Alan Kay did not propose OO as a means to model the real world. Historically, that aspect of OO programming came later. Simula was originally intended to be a simulation programming language (Holmevik, 1995). The encapsulation/message like mechanism, which helped write complex simulations, made Simula a good general purpose programming language, because all programs benefit from managing complexity.

Three major inter-related issues, encapsulation, model multiplicity and complexity management, prevent OO methods from fully realizing the potential they are expected to deliver. Overcoming them poses real challenges, as they question the very foundations of OO. In the following subsections we discuss each challenge separately and how the three are related.

15.7.2 The Encapsulation Challenge

The encapsulation principle, a major OO cornerstone, requires that any process be affiliated with (or "owned" by) some object, within which it is defined. This restriction implies that it is not possible to define a "stand alone" process class in a manner analogous to the way an object class is defined. Thus, in OO, a process can be defined only within a particular object class. Therefore, every process is forced to be an operation. While this encapsulation constraint may be a suitable and effective programming paradigm, it has unnecessarily been a source of endless confusion and counter-intuitive modeling of real life situations. Mandating the association of each process with some object class regardless of the fact that the process may involve more than one object class, an apparently useful programming practice, inadvertently found its way into the analysis and design domains.

When we, humans, try to understand a complex system, we naturally think of the objects that comprise the system and how they relate to each other in the long run. This is the structure of the system. To understand what these objects are for, as soon as we identify an object, we examine how it interacts with other objects through processes and how it may change over time. This is the behavior of the system. We naturally tend to model processes that groups of objects from various classes take part in, without being forced to assign an "owner" for the process.[1] But since OO's encapsulation forbids stand-alone processes, one is bound to model **Baking** as a method of **Cookie, Cashing** as a method of **Check, Manufacturing** as a method of **Raw Material**, and so forth. The designer is preoccupied with associating processes with their owner objects, forcing him or her to think in terms of the solution space rather than the problem space. Overcoming the fact that some object is the owner of each process requires the development of an elaborate messaging mechanism that has little to do with reality.

Some researchers have identified this problem and suggested to break away from the encapsulation restriction when necessary. For example, Kappel (1995) noted that "complex operations involving several classes should be modeled independently of a particular object class." The Datasim Development Process (2000) also pointed out the inadequacy of encapsulation to analysis and design, as it forces developers to think about data in the early stages of software development. However, the OO encapsulation principle has been so dominant and influential that little attention has been paid to those views.

15.7.3 The Model Multiplicity Challenge

As we have seen, OO methods usually employ at least three distinct models for specifying various system aspects – mainly structure, function, and behavior. In real life, however, structure and behavior are intertwined rather than compartmen-

---

[1]   OO veterans, who have been trained to think in terms of encapsulation, tend to define processes as objects or as methods, but not as stand-alone things.

talized. The straightforward, commonsense intuition of thinking concurrently about structure and behavior is severely impeded by the separation into different structure and behavior models, or "views," which object-oriented methods advocate. The *model multiplicity problem* arises from the difficult task of mental integration among the various models within a typical OO method, which the analyst is expected to carry out.

Advocates of the popular multiple-view approach argue that "separation of concerns" mandates the use of different views to examine different concerns, since showing everything on one diagram is too hard to digest. This is true, but the solution lies not in separating the model into distinct views by aspects, but into views at different levels of granularity, using refining and abstracting mechanisms such as those discussed in Chapter 9. Rather than dictating separation into various aspects, each OPD ("OPM view") allows (but does not enforce) mixing and matching of structure and behavior as the architect sees fit.

Responding to the question "How do you find the objects for an object-oriented design?" Martin wrote:

Neither class diagrams nor collaboration diagrams take precedence. It does not matter which ones you start with. The important thing is that you should avoid doing too much of one kind, without balancing them with the other kind. Class diagrams cannot stand alone. Neither can collaboration diagrams. They reinforce each other, and need to be developed concurrently with each other. Failure to develop these diagrams concurrently will result in dynamic models that cannot be supported statically, or static models that cannot be implemented dynamically.

This response points to the need to work concurrently on at least two models and "balance" the load. How much is "too much of one kind" is hard to prescribe. Indeed, in a multi-model or multi-view environment, understanding a system and the way it operates and changes over time requires concurrent reference to the various system views and the creation of abstract associations that link them. Rather than being built into the method, the mental burden of integrating the collection of models or views is put on the shoulders of the developers who are required to grapple with a system that is complex in itself, and they are mentally overloaded for no good reason.

Mentally integrating the structure and behavior models alone is difficult enough in any method that advocates the separation of these two major system's aspects into two different models. Concurrent development of two models is practically difficult, as one needs to constantly switch from one model to the other, which implies switching between sets of graphic notations and conventions, and making sure that each construct in one view is correctly represented in the other. Under these circumstances, keeping all the views synchronized and avoiding contradictions and mismatches between the different views becomes very difficult. The larger the number of models, the more difficult it becomes to carry out the integration and avoid conflicts. The difficulty of maintaining integrity increases exponentially with the number of views. As we add the n-th model to the $n - 1$ models already in existence, we have to check consistency between this additional model and the rest. The number of interactions among the n models, if $n > 1$, is $n!/2$. For UML, where $n=9$, this number is already in the hundreds. Technical solutions that involve computer-aided software engineering tools to impose consistency alleviate manual checking, but they do not address the more serious problem of excessive mental burden. With multi-model methods, the cognitive problem of the human system architect is here stay: a complete mental picture of the system needs to be created from information that is distributed across multiple views with different graphical syntax.

15.7.4 Empirical Evidence of the Model Multiplicity Problem

In a double blind controlled experiment (Peleg and Dori, 2000), which involved 88 senior undergraduate information-engineering students, we compared OMT (Rumbaugh, 1991), extended with Timed Statecharts, with OPM/T, a real-time extension of OPM. The objective was to investigate two major questions related to model multiplicity vs. model singularity: (1) Does a single model enable the synthesis of a better system specification compared with a combination of several models? (2) Which of the two alternative approaches yields a specification that is easier to comprehend?

Table 20. Overall number of errors in the specification quality part of the experiment

OMT w/Timed Statcharts		OPM/T		T-value (two-tailed)	P-value
Average	S.D.	Average	S.D.		
5.81	2.95	2.80	2.11	−5.36	0.01

Our conjecture was that since fusion of the various models within one's mind is difficult, using multiple views to model a system might adversely affect the two transformations involved in the development and use of information systems (Wand and Wang (1996). The first transformation is modeling, i.e., specifying the real-world system using a particular single-model or multi-model methodology. The second transformation is comprehension (interpreting) of the system specified in a particular methodology. The experiment has established that OPM was more effective than OMT in terms of synthesis (see Table 20). The specification comprehension resulted significant differences between the two methods in specific issues.

15.7.5 The Complexity Management Challenge

In Section 9.1 we discussed in some detail the fact that OO system development methods have adopted the decomposition principle by breaking the system into a number of models, each dealing with a different aspect of the system, such as structure, behavior, and function. This aspect decomposition, which links back to the model multiplicity problem, is at the heart of UML. Whether by design or by default, OO methods achieve the objective of simplifying complexity or managing it by breaking the system into different models (or the model into different views, the difference is merely semantic). However, breaking systems that are really complex into three, six or nine views is not guaranteed to be an adequate solution to the complexity management problem. Neither is the introduction of aggregating concepts like packages, which are used in OMG's Meta-Object Facility (2000), abbreviated MOF:

> *The Package is the MOF Model construct for grouping elements ... Packages may be generalized by (inherit from) one or more other Packages in a way that is analogous to Class generalization. When one Package inherits from another, the inheriting (sub-) Package acquires all of the meta-model elements belonging to the (super-) Package it inherits from.*

Based on this definition, the advantage of using a specially designated construct such as Package rather than using an object class, which is at a more abstract level, is not clear. Rather than decomposing a system according to its various aspects, OPM's decomposition is based in an orthogonal fashion on the system's levels of abstraction and is achieved through the various refinement and abstraction mechanisms described in Chapter 9.

15.8 OPM and OO

OPM has borrowed many good ideas from the realm of object-orientation (OO).[2] An intriguing question concerns the relationships between OPM and OO. The question can be addressed from different perspectives. Perhaps the most important one is the scope perspective: what objective do OO and OPM aim at? From this perspective, OPM is geared towards modeling systems in general, while OO is focused on software systems. From the evolution perspective, OO started bottom-up as a programming paradigm that made its way to the worlds of analysis and design. OPM was conceived top-down without being influenced by capabilities or limitations of programming languages. From an ontological perspective, OO puts objects at the center stage, while OPM's primary building blocks are objects and processes. From complexity management perspective, OO uses several views to separate concerns, while OPM handles complexity by refining and abstracting. From a number of models or views perspective, OPM uses a single model, while OO methods use several. From a modality perspective, OO uses graphical models at the analysis and design level and programming languages at the execution level. OPM uses two equivalent modalities, graphics (OPDs) and text (OPL), to express the single OPM model. From a metamodeling perspective, using reflective meta-modeling, OPM can model itself using its own ontology. OO methods (such as UML) require higher-level, more abstract facilities (such as MOF) to be meta-

---

[2] Depending on the context, OO stands for object-oriented or object-orientation.

modeled. From constraint specification perspective, OO require add-ons such as OCL (Object Management Group, 2000; Warmer and Kleppe, 1999), while OPM is self-contained.

15.8.1 The UML 2.0 Initiative

UML 2.0 (2000) Request for Proposals is a major OMG revision in UML 1.X. OPM has been proposed as a possible UML 2.0 infrastructure (Sight Code, 2001).[3] The proposed enhancement to UML, if accepted, would make UML suitable as a universal modeling tool. Figure 15.1 is an OPD that describes the current state of affairs. A main problem here is circularity: MOF is described using UML, while UML is metamodeled through MOF. As the MOF document states, UML and MOF are very similar. In the spirit of reusability and simplification, it makes sense to unify the two, as specified in the OPD in Figure 15.1.

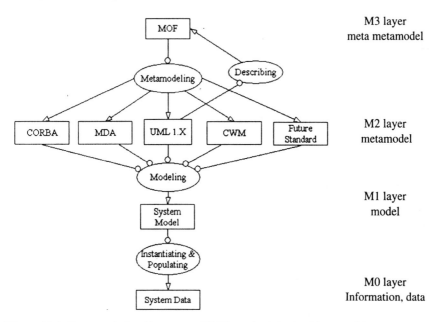

Figure 15.1. Current relationships among OMG standards in the 4-layer architecture

Figure 15.2 shows UML 2.0 as the only instrument for **Modeling**, which specializes into **Reflective Metamodeling**, **Standard Modeling**, and **System Modeling**. **Reflective Metamodeling** is the only real metamodeling process. It defines UML 2.0 in terms of UML 2.0. **Standard Modeling** is what has been called "metamodeling," but since a standard can be considered a system, there is no difference in

---

3   The proposal was supposed to be presented at the OMG Technical Meeting in Toronto on September 13, 2001, but the closing of air traffic across America following the tragic terrorist attack on WTC prevented the author from getting there.

principle between modeling any system and modeling a standard. Finally, **System Modeling** is the "regular" modeling activity that system architects do when developing systems. All modeling process specializations use UML 2.0 as their modeling instrument, making MOF redundant.

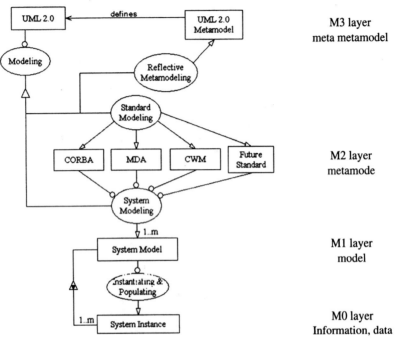

Figure 15.2. Proposed relationships among OMG standards in the 4-layer architecture, where UML 2.0 and MOF are unified

15.8.2 Systemantica: an OPM Supporting Tool

Systemantica® Version 1.0 (Sight Code, 2001), which is supplied with this book in a limited version, is a software tool that supports developing systems in OPM. Developed by Saggi Nevo, it implements the majority of OPM features specified in this book, and is capable of generating OPL sentences from OPD sentences and vice versa. The complete ATM system, specified in chapters 2, 3 and 9, is provided along with the software as a worked-out case study. Systemantica® is capable of automatically converting any OPD to two types of UML diagrams: Class diagrams and use case diagrams. Figure 15.3 shows a screenshot of Systemantica® 1.0 with SD of the ATM case study.

Figure 15.4 shows the automatically derived UML class diagram of the ATM case study, while Figure 15.5 shows the corresponding use-case diagram, where the stick figures were added manually.

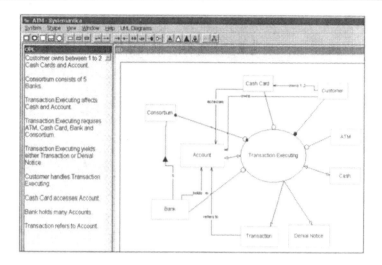

Figure 15.3. A screenshot of Systemantica® Version 1.0 showing SD of the ATM case study

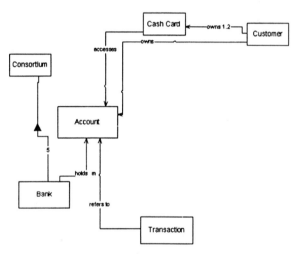

Figure 15.4. A screenshot of Systemantica® Version 1.0 showing the derived UML class diagram of the ATM case study

15.8.3 OPM Applications and Research: Present and Future

The generality and width of scope of OPM, along with its combination of intuitive graphics and natural language, makes it a candidate for infrastructure of applications in various disciplines and development directions. OPM has already been applied and is being applied in a number of large manufacturing corporations,

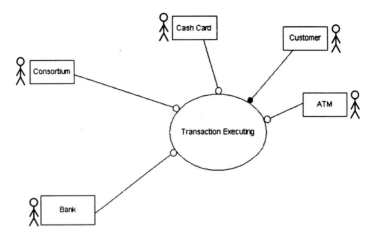

Figure 15.5. A screenshot of Systemantica® Version 1.0 showing the derived UML use-case diagram of the ATM case study. Stick figures were added manually.

whose businesses are manufacturing of jet engines, metal cutting tools, semi-conductors, and manufacturing of robots for aircraft part manufacturing.

Academically, a number of master's theses and Ph.D. dissertations have been completed while other are under way. Goodman (1996) showed how C++ code can be automatically generated from OPM analysis and design results. His work provided a basis for the Ph.D. dissertation of Peleg (1999), who modeled system dynamics through OPM and experimentally showed the benefit of OPM as a single-model methodology. Her work, in turn, provided a basis for using OPM for real-time systems (Peleg and Dori, 2000). Sturm (1999) applied relational and object-relational database models to OPM analysis and design results, providing a basis for automated database schema generation. Reinhartz-Berger (1999) showed how to generate Java code from OPL Script. Christine Miyachi (2001) has modeled the Capability Maturity Model with OPM. Benjamin Koo (2001) has shown how to improve product development capability maturity through OPM.

As of writing of this book, Dagan Gilat is completing his Ph.D. dissertation, which extends OPM for simulation of discrete events. Soffer is completing her Ph.D. dissertation on applying OPM for adapting an off-the-shelf Enterprise Resource Planning software package for an organization (Soffer et al., 2001). Reinhartz-Berger is conducting her Ph.D. research on Web application development through OPM, and Soderborg is integrating OPM with Axiomatic Design.

OPM offers a wealth of research issues in various directions. A partial list includes automated processing of natural language script, automated software application development, system lifecycle and evolution support, automated translations among natural languages, automated reasoning and fact discovery from given OPM models, knowledge extraction and management, applications as a system modeling language, OPM as an evolving and learning system.

Summary

- Systems analysis methods evolved from being oriented towards process into object-oriented.
- Data Flow Diagrams (DFDs) are process-oriented and comprise processes, data stores, data flows, and external entities.
- Entity-Relationship Diagrams (ERDs) represent data elements and static associations among them.
- The DFD-ERD combination was a prevalent way of designing information systems in the 1970's and 1980's, prior to the popularization of Object-Orientation.
- The shift of OO from procedural to structural modeling was accompanied by a degradation of OO methods to adequately model the dynamic aspect of systems.
- Many OO methods proliferated in the early and mid-1990's. By and large they comprise a static object-and-class model, which was accompanied by one or more procedural, dynamic model and other complementary models.
- Object Management Group's adoption in 1997 of Unified Modeling Language (UML) as the software industry standard has been an important milestone, as the software developer community can now share a common language.
- UML has a number of known problems that are being addressed in UML 2.0.
- Encapsulation, model multiplicity and complexity management are challenges that question foundations of OO.
- OPM has adopted and extended many OO concepts and ideas. It has also incorporated a number of fundamental ideas that go beyond and extend OO principles, notable the possible definition of processes independently of objects and the way objects interact with each other via processes.
- OO and OPM are similar in many respects and differ in many others. Perhaps the most fundamental one is that while OO aims at developing software systems, OPM has been conceived as an approach to modeling systems in general. This difference is a source of many others.

Problems

1. Metamodel DFDs in OPM.
2. Metamodel ERDs in OPM.
3. Using the metamodels from problems 1 and 2, show how ERD entities and DFD elements relate to each other in the model of an information system.
4. Use OPM to model the top-level OPD of the way system developers used to develop systems with a combination of DFDs and ERDs.
5. Based on the OPM model of **SD** built in Problem 4, drill down to the level of **SD1**, where you can use the model of DFD-ERD from Problem 3.
6. Metamodel the top level of UML using OPM.
7. Construct a top-level metamodel of UML in OPM.
8. Create an OPM model that is based on the specification of the Class model in Section 3.19 of the formal document of UML 1.3 (Object Management Group, 2000).
9. Create an OPM model that is based on the specification of the Use Case model in Section 3.52 of the formal document of UML 1.3 (Object Management Group, 2000).
10. Discuss the challenges of OO. Explain whether OPM meets any of them, and if so, how?
11. Summarize the main differences between OO and OPM.

Appendix A: The ATM System

The ATM system was partly specified in Chapters 2 and 3. The entire system, to an appropriate level of detail, is shown below.

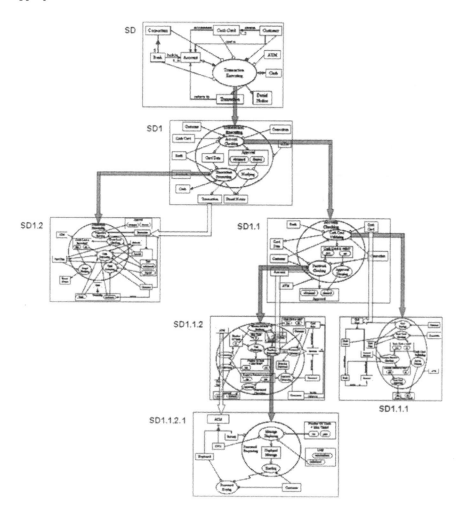

Figure A.1. The system map of the ATM system

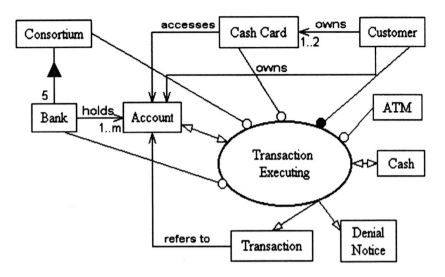

Figure A.2. SD, the system diagram of the ATM system

SD Paragraph:
Consortium consists of **5 Banks**.
Bank holds at least one **Account**.
Customer owns Account.
Customer owns 1 to **2 Cash Cards**.
Cash Card accesses Account.
Transaction refers to Account.
Customer handles **Transaction Executing**.
Transaction Executing requires **ATM, Cash Card, Consortium,** and **Bank**.
Transaction Executing affects **Account** and **Cash**.
Transaction Executing yields either **Transaction** or **Denial Notice**.

Transaction Executing from **SD** zooms in **SD1** into **Account Checking,
Transaction Processing,** and **Notifying,** as well as **Approval** and
Card Data.

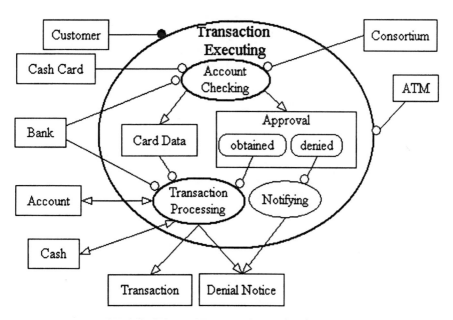

Figure A.3. SD1 of the ATM system

SD1 Paragraph:
Customer handles **Transaction Executing.**
Transaction Executing requires **ATM.**
Account Checking yields **Approval** and **Card Data.**
Account Checking requires **Bank, Cash Card** and **Consortium.**
Approval can be **obtained** or **denied.**
Transaction Processing occurs if **Approval** is obtained.
Transaction Processing requires **Bank** and **Card Data.**
Transaction Processing affects **Account** and **Cash.**
Transaction Processing yields either **Transaction** or **Denial Notice.**
Notifying occurs if **Approval** is denied.
Either **Notifying** or **Transaction Processing** yields **Denial Notice.**

Account Checking from SD1 zooms in SD1.1 into Cash Card Validating, Password Checking, and Approval Denying, as well as "Cash Card is Valid?"

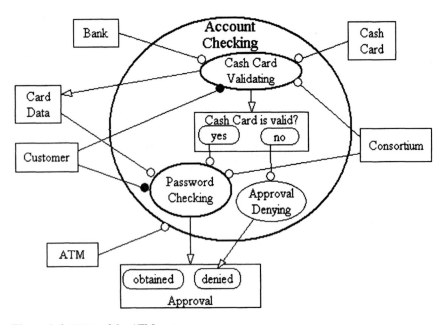

Figure A.4. SD1.1 of the ATM system

SD1.1 Paragraph:

Account Checking requires ATM.

Customer handles Cash Card Validating and Password Checking.

Cash Card Validating requires Consortium, Bank, and Cash Card.

Cash Card Validating yields Card Data.

Cash Card Validating determines whether Cash Card is valid.

Cash Card Validating and Password Checking require Consortium.

Password Checking occurs if Cash Card is valid, otherwise Approval Denying occurs.

Password Checking requires Consortium, and Card Data.

Password Checking yields Approval.

Approval can be obtained or denied.

Approval Denying yields denied Approval.

Cash Card from **SD1.1** unfolds in **SD1.1.1** to exhibit **Bank Code** and **Account Number**.

Cash Card Validating from **SD1.1** zooms in **SD1.1.1** into **Card Reading, Bank Code Checking, Account Number Checking, Access Denying,** and **Cash Card Approving,** as well as **"Bank Code is Valid?"** and **"Account Number is Valid?".**

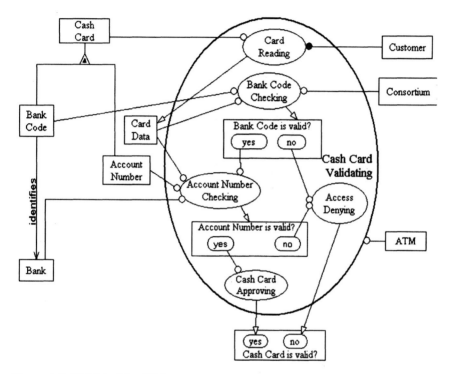

Figure A.5. SD1.1.1 of the ATM system

SD1.1.1 Paragraph:

Bank Code identifies Bank.

Cash Card Validating requires **ATM**.

Customer handles **Card Reading**.

Card Reading requires **Cash Card**.

Card Reading yields **Card Data**.

Bank Code Checking requires **Consortium, Bank Code,** and **Card Data**.

Bank Code Checking determines whether **Bank Code is valid**.

Bank Code Checking and **Account Number Checking** require **Card Data**.

Account Number Checking occurs if **Bank Code is valid,** otherwise **Access Denying** occurs.

Account Number Checking requires **Bank, Account Number,** and **Card Data**.

Account Number Checking determines whether **Account Number is valid**.

Cash Card Approving occurs if **Account Number is valid,** otherwise **Access Denying** occurs.

Cash Card Approving determines that **Cash Card is valid.**

Access Denying occurs either if **Bank Code** is not **valid** or if **Account Number is not valid.**

Access Denying determines that **Cash Card** is not **valid.**

Account from **SD1.1** unfolds in **SD1.1.2** to relate to **Password.**

Password Checking from **SD1.1** zooms in **SD1.1.2** into **Number Of Trials Initializing, Password Requesting, Password Keying, Password Comparing, Trial Comparing, Approving, Confiscating,** and **Incrementing,** as well as **Number Of Trials, Max Trials, Loop, Keyed in Password, "Number Of Trails is greater than Max Trials?",** and **"Keyed in Password is correct?".**

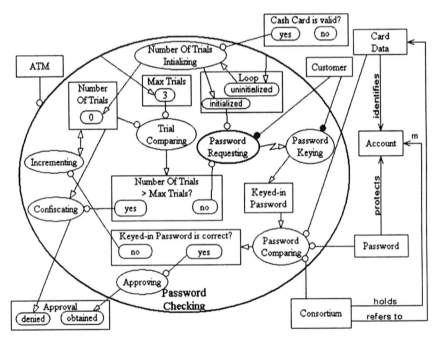

Figure A.6. SD1.1.2 of the ATM system

SD1.1.2 Paragraph:

Consortium holds many **Accounts.**

Consortium refers to Card Data.

Card Data identifies Account.

Password protects Account.

Password Checking requires **ATM.**

Customer handles **Password Requesting** and **Password Keying.**

Password Checking yields **Max Trials** with value **3,** and **uninitialized Loop.**

Number Of Trials Initializing occurs if **Cash Card is valid** and if **Loop** is **uninitialized.**

Number Of Trials Initializing yields **Number Of Trials** with value **0.**

Number Of Trials Initializing changes **Loop** from **uninitialized** to **initialized.**

Password Requesting occurs if **Loop** is **initialized** and if **Number Of Trials** is not greater than **Max Trials.**

Password Requesting invokes **Password Keying.**

Password Keying yields **Keyed-in Password.**

Password Comparing requires **Consortium, Card Data, and Password.**

Password Comparing consumes **Keyed-in Password.**

Password Comparing determines whether **Keyed-in Password is correct.**

Approving occurs if **Keyed-in Password is correct,** otherwise **Incrementing** occurs.

Approving yields **obtained Approval.**

Incrementing affects **Number Of Trials.**

Trial Comparing requires **Max Trials** with value **3** and **Number Of Trials.**

Trial Comparing determines whether **Number Of Trials** is greater than **Max Trials.**

Confiscating occurs if **Number Of Trials** is greater than **Max Trials,** otherwise **Password Requesting** occurs.

Confiscating consumes **Number Of Trials** and **Cash Card.**

Confiscating yields **denied Approval.**

Password Requesting from **SD1.1.2** zooms in **SD1.1.2.1** into **Message Displaying** and **Reading,** as well as **Displayed Message.**

ATM from **SD1.1.2** unfolds in **SD1.1.2.1** to consist of **Keyboard, CPU, Screen,** and additional parts.

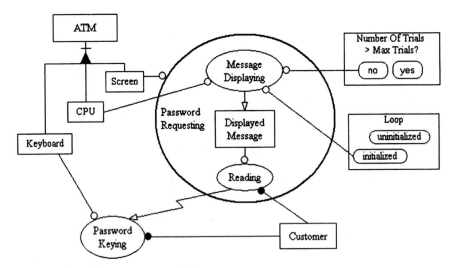

Figure A.7. SD1.1.2.1 of the ATM system

Customer handles **Password Keying** and **Reading.**

Password Requesting requires **Screen.**

Loop can be **uninitialized** or **initialized.**

Message Displaying occurs if **Loop** is **initialized,** and if **Number Of Trials** is not greater than **Max Trials.**

Message Displaying requires **CPU.**

Message Displaying yields **Displayed Message.**

Reading requires **Screen.**

Reading invokes **Password Keying.**

Transaction from **SD1** unfolds in **SD1.2** to exhibit **Amount** and **Type**.
Cash from **SD1** unfolds in **SD1.2** to exhibit **Ownership**.
Transaction Processing from **SD1** zooms in **SD1.2** into **Transaction Querying, Credit Limit Checking, Cash Dispensing, Cash Accepting,** and **Denial Notifying,** as well as **"Credit Limit is Exceeded?"**

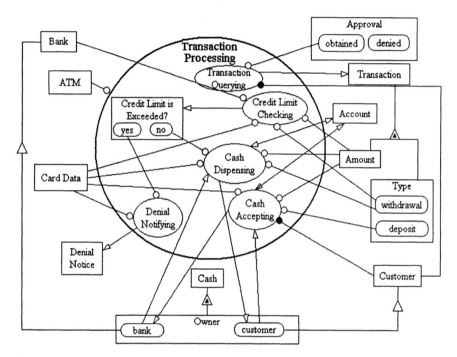

Figure A.8. SD1.2 of the ATM system

SD1.2 Paragraph:
Type of **Transaction** can be **withdrawal** or **deposit**.
Cash exhibits **Owner**.
Customer and **Bank** play the role of **Owners** for **Cash**.
Transaction exhibits **Type,** which can be **withdrawal** or **deposit,** and **Amount**.
Customer handles **Transaction Querying** and **Cash Accepting**.
Transaction Querying occurs if **Approval** is **obtained**.
Transaction Querying yields **Transaction**.
Credit Limit Checking occurs if **Type** of **Transaction** is **withdrawal**.
Credit Limit Checking requires **Card Data, Bank,** and **Amount**.
Credit Limit Checking determines whether **Credit Limit is exceeded**.
Denial Notifying occurs if **Credit Limit is exceeded,** otherwise **Cash Dispensing** occurs.
Denial Notifying requires **Card Data**.
Denial Notifying yields **Denial Notice**.

Cash Dispensing occurs if **Type** of **Transaction** is **withdrawal** and if **Credit Limit is not exceeded.**

Credit Limit Checking and **Cash Dispensing** occur if **Type** is **withdrawal.**

Cash Dispensing requires **Card Data** and **Amount.**

Cash Accepting and **Cash Dispensing** affect **Account.**

Cash Dispensing changes **Ownership** of **Cash** from **bank** to **customer.**

Cash Accepting occurs if **Type** of **Transaction** is **deposit.**

Cash Accepting requires **Card Data** and **Amount.**

Cash Accepting changes **Owner** of **Cash** from **customer** to **bank.**

References

Abadi, M. and Cardelli, L. A Theory of Objects. Monographs in Computer Science, Springer-Verlag, New York, 1996. http://www.luca.demon.co.uk/TheoryOfObjects/Prologue.html

Archaeology World. School of Archaeology and Anthropology at the Australian National University, 2001. http://artalpha.anu.edu.au/web/arc/aboutus/studs/roddom/research2.htm

Ashby, W.R. Design for a Brain. Chapman and Hall, London, 1956.

Ashby, W.R. Concepts of Operand, Operator, Transform. Washington University, St. Louis, MO, 2001. http://www.gwu.edu/~asc/biographies/ashby/MATRIX/SG/sg_1.html

Bar-Yam, Y. Dynamics of Complex Systems. Perseus Books, Reading, MA, 1997.

Bauer, F.L. and Wössner, H. Algorithmic Language and Program Development. Texts and Monographs in Computer Science, Springer-Verlag, Berlin, 1981.

Beer, S. On the Nature of Models: Let Us Now Praise Famous Men and Women, Too. Informing Science 2(3), pp. 69–82, 1999.

Ben-Dov, Y. Quantum Theory: Reality and Mystery, In Ma?Da! M. Dascal (Ed.), Dvir Publishing House, Tel Aviv, Israel, 1997 (in Hebrew).

Bertalanffy, L.V. General Systems Theory: Foundations, Development, Applications. George Braziller, New York, 1968.

Bertalanffy, L.V. Perspectives on General Systems Theory. George Braziller, New York, 1975.

Booch, G. Object-Oriented Design with Applications. Benjamin Cummings, Redwood City, CA, 1991.

Boulding, K.E. General Systems Theory: The Skeleton of Science. Management Science 2(3), pp. 197–208, 1956.

Bouvier, E., Cohen, E. and Najman, L. From Crowd Simulation to Airbag Deployment: Particle Systems, a New Paradigm for Simulation. Journal of Electronic Imaging 6(1), pp. 94–107, 1997.

Bubenko, J.A. Jr. Information System Methodologies – A Research Review. In Olle et al. 1986.

Bunge, M. Treatise on Basic Philosophy, Vol. 3, Ontology I, The Furniture of the World. Reidel, Boston, MA, 1977.

Bunge, M. Treatise on Basic Philosophy, Vol. 4, Ontology II, A World of Systems. Reidel, Boston, MA, 1979.

Carruthers, P. Language, Thought and Consciousness: An Essay in Philosophical Psychology. Cambridge University Press, Cambridge, MA, 1996.

Central Artery/Tunnel Project, Boston, MA, 2001. http://www.bigdig.com/

Chen, D. and Stroup, W. General System Theory: Toward a Conceptual Framework for Science and Technology Education for All. Journal of Science Education and Technology 2(3), pp. 447–459, 1993.

Chen, P.P. The Entity Relationship Model – Toward a Unifying View of Data. ACM Trans. on Data Base Systems 1(1), pp. 9–36, 1976.

Coad, R. and Yourdon, E. Object-Oriented Analysis. Prentice-Hall, Englewood Cliffs, NJ, 1991.

Coad, R. and Yourdon, E. Object-Oriented Design. Prentice-Hall, Englewood Cliffs, NJ, 1991A.

Coleman, D., Hayes, F. and Bear, S. Introducing Objectcharts and How to Use Statecharts in Object-Oriented Design. IEEE Transactions on Software Engineering 18(1), pp. 9–18, 1992.

Coleman, D., Arnold, A., Bodoff, S., Dollin, C., Gilchrist, H., Hayes, F. and Jeremaes, P. Object-Oriented Development: The Fusion Method. Prentice-Hall, Englewood Cliffs, NJ, 1994.

Computer Desktop Encyclopedia. Computer Language Company, Point Pleasant, PA, 2001. http://www.computerlanguage.com/at.html

Cook, S. Foreword to Warmer and Kleppe 1999.

Cook, S., Kleppe, A., Mitchell, R., Rumpe, B., Warmer, J., and Wills, A.C. Defining UML Family Members Using Prefaces. In Mingins, C. and Meyer, B. (Eds.) Proc. Technology of Object-Oriented Languages and Systems, TOOLS-Pacific. IEEE Computer Society, 1999.

Crawley, E. Lecture Notes in System Architecture, Systems Design and Management Course. MIT, Cambridge, MA, January 2000.

D'Souza, D., Sane, A. and Birchenough, A. UML Profiles Considered Redundant. 1999. home.earthlink.net/~salhir/TheUML-TwoYearsAfterAdoptionOfTheStandard.PDF

Datasim Development Process. White Paper, 2000. www.datasim-education.com

Davidson, E.J. Jerry's Discussion Icebreakers. iscd@uindy.edu mailing list, 1998.

Dawkins, R. The Selfish Gene. Oxford University Press, London, 1989.

De Marco, T. Structured Analysis and System Specification. Yourdon Press, New York, 1978.

Defay, R. Introduction à la Thermodynamique des Systèmes Ouvertes. Académie Royale de Belgique, Bulletin de la Classe des Sciences, Vol. 53, 1929.

Defense Systems Management College. Systems Engineering Fundamentals. Defense Systems Management College Press, Fort Belvoir, VA, 1999.

Dirks, T. The Greatest Films, 2001. http://www.filmsite.org/gone.html

Dori, D. Object-Process Analysis: Maintaining the Balance Between System Structure and Behavior. Journal of Logic and Computation 5(2), pp. 227–249, 1995.

Dori, D. Arc Segmentation in the Machine Drawing Understanding Environment. IEEE Transactions on Pattern Analysis and Machine Intelligence, T-PAMI 17 (11), pp. 1057–1068, 1995A.

Dori, D. Unifying System Structure and Behavior Through Object-Process Analysis. Journal of Object-Oriented Programming, July-August 1996, pp. 66–73.

Dori, D. Object-Process Analysis of Computer Integrated Manufacturing Documentation and Inspection Functions. International Journal of Computer Integrated Manufacturing 9(5), pp 339–353, 1996A.

Dori, D. Analysis and Representation of the Image Understanding Environment Using the Object-Process Methodology. Journal of Object-Oriented Programming 9(4), pp. 30–38, 1996B.

Dori, D. and Goodman, M. On Bridging the Analysis-Design and Structure-Behavior Grand Canyons with Object Paradigms. Report on Object Analysis and Design 2(5), pp. 25–35, 1996.

Dori, D. Object-Process Methodology Applied to Modeling Credit Card Transactions. Journal of Database Management 12(1), pp. 2–12, 2001.

Dori, Y.J. Cooperative Development of Organic Chemistry Computer Assisted Instruction by Experts, Teachers and Students. Journal of Science Education and Technology 4(2), pp. 163–170, 1995.

Downton, C. In Smolan, R. and Erwitt, J. One Digital Day. Time Book/Random House in association with Against All Odds Production, 1998.

Embley, D., Kurtz, B. and Woodfield, S. Object-Oriented Systems Analysis. Prentice-Hall, Englewood Cliffs, NJ, 1992.

Encyclopedia of World Problems and Human Potential. 4th Edition, Union of International Associations, Brussels, 1994-95. http://www.uia.org/uiapubs/pubency.htm

Encyclopedia.com, Infonautics Corporation, 2001. http://encyclopedia.com/printable/04040.html

Essink, L.J.B. A Modeling Approach to Information Systems Development. In Olle et al. 1986, pp. 55–86.

Evans, A. Dependencies and Associations. Precise UML Mailing List, June 8, 2001, puml-list@cs.york.ac.uk

Farias, R. and Recami E. Introduction of a Quantum of Time (Chronon) and Its Consequences for Quantum Mechanics, Quantum Physics, abstract quant-ph/9706059, 2001. http://xxx.lanl.gov/abs/quant-ph/9706059

Firesmith, D., Henderson-Sellers, B. and Graham, I. The OML Reference Manual. SIGS Books, New York, NY, 1997.

Firlej, M. and Hellens, D. Knowledge Elicitation: A Practical Handbook. Prentice-Hall, New York, 1991.

Forrester, J.W. Industrial Dynamics. Productivity Press, Cambridge, MA, 1961.

Forrester, J.W. World Dynamics. Productivity Press, Cambridge, MA, 1973.

Forrester, J.W. Collected Papers of Jay W. Forrester. Wright Allen Press, Cambridge, MA, 1975.

Fowler, M. UML Distilled. 2nd Edition, Addison-Wesley, Reading, MA, 1999.

Fox, M. and Gruninger, M. Enterprise Modeling. AI Magazine, Fall 1998, pp. 109–121.

Gane, C. and Sarson, T. Structured Systems Analysis, Tools and Techniques. Prentice-Hall, Englewood Cliffs, NJ, 1979.

Goodman, M. The Transition from Analysis to Design in the Object-Process Methodology. M.Sc. Thesis, Faculty of Industrial Engineering and Management, Technion, Israel Institute of Technology, Haifa, Israel, 1996.

Harel, D. Statecharts: A Visual Formalism for Complex Systems. Science of Computer Programming 8, pp. 231–274, 1987.

Harel, D. On Visual Formalisms. Communications of the ACM 31(5), pp. 514–530, 1988.

Harel, D. Biting the Silver Bullet: Toward a Brighter Future for System Development. Computer, Jan. 1992, pp. 8–20.

Hatley, D.J. and Pirbhai, I.A. Strategies for Real-Time System Specification. Dorset House, New York, 1988.

Heidegger, M. Being and Time. Harper & Row, New York, 1962.

Henderson-Sellers, B. COMMA: An Architecture for Method Interoperability. Report on Analysis and Design 1(3), pp. 25–28, 1994.

Henderson-Sellers, B. and Bulthuis, A. Object-Oriented Metamethods. Springer-Verlag, New York, 1998.

Henderson-Sellers, B. and Graham, I.M. OPEN: Toward Method Convergence? IEEE Computer 29(4), pp. 86–89, 1996.

Henderson-Sellers, B., Simons, T. and Younessi, H. The OPEN Toolbox of Techniques. Addison-Wesley, Reading, MA, 1998.

Henry Ford Museum & Greenfield Village, Dearborn, MI, 2001. http://www.hfmgv.org/histories/wright/wrights.html#airplane

Heylighen, F. Principia Cybernetica Web, 2001. http://pespmc1.vub.ac.be/HEYL.html

Holmevik, J.R. The History of Simula, Center for Technology and Society, University of Trondheim, N-7055 Dragvoll, Norway, 1995.

IDEF Family of Methods. A Structured Approach to Enterprise Modeling and Analysis, 2001. www.idef.com

Johnson, R.A., Kast, F.E. and Rosenzweig, J.A. The Theory and Management of Business. McGraw-Hill, New York, 1967.

Kant, I. Critique of Pure Reason. Transl. Werner Pluhar, 1787. In McCormick, M. The Internet Encyclopedia of Philosophy. Hackett, Indianapolis, 1996. http://www.utm.edu/research/iep/k/kantmeta.htm

Kappel, G. The Advocatus Diaboli of Object-Oriented Development. Dagstuhl Seminar Report 9434, p.19, 1995.

Kerzner, H. Project Management: A Systems Approach to Planning, Scheduling and Controlling. 5th Edition, Van Nostrand Reinhold, New York, 1995.

Kilov, H. and Simmonds, I. D. Business Patterns: Reusable Abstract Constructs for Business Specification. In Humphreys, P., Bannon, K., McCosh, A., Migliarese, P. and Pomerol, J.S. (Eds.), Implementing Systems for Supporting Management Decisions. Chapman and Hall, London, 1996.

Knowledge Interchange Format, 2001. http://logic.stanford.edu/kif/

Kopczak, L. and Lee, H. Hewlett Packard: Deskjet Printer Supply Chain (A). Board of Trustees of the Leland Stanford Junior University, Palo Alto, CA, 1994.

Kobryn, C. UML 2001: A Standardization Odyssey. Communications of the ACM 42(10), pp. 29–37, 1999.

Kovitz, B.L. Using OO Modeling for Requirements Analysis. Object-Orientation Tips, 1998. http://ootips.org/oo-for-analysis.html

Kung, C.H. and Sølvberg, A. Activity and Behavior Modeling. In Olle et al. 1986.

Laszlo, E. Introduction to Systems Philosophy: Toward a New Paradigm of Contemporary Thought. Gordon and Breach, New York, 1986.

Latimer, C. and Stevens, C. Some Remarks on Wholes, Parts and Their Perception. Psycoloquy 8(13), Part Whole Perception (1), 1997. http://www.cogsci.soton.ac.uk/psyc-bin/newpsy?8.13

Lillesand, T.M. and Kiefer, R.W. Digital Image Processing, Remote Sensing and Image Interpretation. John Wiley & Sons, New York, 1994.

Lloyd, S. Physical Measures of Complexity. In Jen, E. (Ed.), Lectures in Complex Systems, pp. 67–73, Addison-Wesley, Redwood City, CA, 1989.

Logan, P. The Origins of Object Orientation. Object-Orientation Tips, 2000. http://ootips.org/history.html

Lohr, S. Pioneers of the Fortran Programming Language. The New York Times, June 13, 2001. http://www.nytimes.com/2001/06/13/technology/13LOHR.html

Long, J.G., George Washington University, Washington, DC, 2001. http://www.seas.gwu.edu/~nelwww/ason.html

Lotka, A.J. Elements of Mathematical Biology. Dover, New York, 1920, 1956.

Lowe, E.J. In Goodman, N. (Ed.), Fact, Fiction, and Forecast. 4th Edition, Cambridge, MA, 1983. http://www.xrefer.com/entry/552150

Maciaszek, L.A. Requirements Analysis and System Design, Developing Information Systems with UML. Addison-Wesley, Harlow, England, 2001.

Martin, J. and Odell, O. Object-Oriented Analysis & Design. Prentice-Hall, Englewood Cliffs, NJ, 1992.

Maturana, H.R. and Varela, F. Autopoiesis and Cognition: The Realization of the Living. Reidel, Dordrecht, 1980.

Meinhardt, H. The Algorithmic Beauty of Sea Shells. Springer-Verlag, Berlin, 1995.

Melewski, D. UML Gains Ground. Computer Associates, 2000. http://www.platinum.com/products/reprint/uml_adt.htm

Meta-Object Facility, Version 1.3. Object Management Group, 2000. http://www.omg.org/technology/documents/formal/mof.htm

Meyer, B. Eiffel: The Language. Prentice-Hall, New York, 1990.

Meyer, B. Object-Oriented Software Construction. Prentice-Hall, New York, 1997.

Meyersdorf, D. and Dori, D. The R&D Universe and Its Feedback Cycles: An Object-Process Analysis. R&D Management 27(4), pp. 333–344, 1997.

Miyachi, C. Modeling the Capability Maturity Model with Object-Process Methodology. M.S. Thesis, Engineering Systems Division, MIT, Cambridge, MA, 2001.

Moore, F.G. (Ed.) A Management Sourcebook. Harper and Row, New York, 1964.

Object Management Group. Unified Modeling Language (UML) 1.3 Documentation, 2000. http://www.omg.org/cgi-bin/doc?formal/2000-03-01

Olle, T.W., Sol, H.G. and Verrijn-Stuart, A.A. (Eds.) Information Systems Design Methodologies – Improving the Practice. Elsevier Science Publishers (North Holland), IFIP, 1986.

OMG, Object Management Group website, 2001. www.omg.org

Ontology Markup Language, 2001. http://wave.eecs.wsu.edu/CKRMI/OML.html

OPEN – Object-Oriented Process, Environment and Notation, 2001. http://www.open.org.au/

Open Group Architectural Framework, 2001. http://www.opengroup.org/public/arch/

Oregon State University Archives, 2001. http://www.orst.edu/Dept/archives/definitions/silver.halide.film.html

Osburn, T. Three Gorge Dam, 2001. http://www.wsu.edu/~hallagan/THREEG.HTML

Pahl, G. and Beitz, W. Engineering Design – A Systematic Approach. 2nd Edition, Springer-Verlag, Berlin, 1996.

Pedrycz, W. and Zadeh, L. Fuzzy Sets Engineering, CRC Press. 1995.

Peleg, M. Modeling System Dynamics Through the Object-Process Methodology. Ph.D. Dissertation, Faculty of Industrial Engineering and Management, Technion, Israel Institute of Technology, Haifa, Israel, 1999.

Peleg, M. and Dori, D. Representing Control Flow Constructs in Object-Process Diagrams. Journal of Object-Oriented Programming 11(3), pp. 58–71, 1998.

Peleg, M. and Dori, D. Extending the Object-Process Methodology to Handle Real-Time Systems. Journal of Object-Oriented Programming 11(8), pp. 53–58, 1999.

Peleg, M. and Dori, D. The Model Multiplicity Problem: Experimenting with Real-Time Specification Methods. IEEE Transactions on Software Engineering 26(8), pp. 742–759, 2000. http://iew3.technion.ac.il:8080/Home/Users/dori/Model_Multiplicity_Paper.pdf

Reinhartz-Berger, I. Generating Java Code from Object-Process Language Script. M.Sc. Thesis, Faculty of Industrial Engineering and Management, Technion, Israel Institute of Technology, Haifa, Israel, 1999.

Reinhartz-Berger, I., Dori, D. and Katz, S. OPM/Web – Object-Process Methodology for Developing Web Applications Annals of Software Engineering, 2002 (to appear).

Rescher, N. and Oppenheim, P. Logical Analysis of Gestalt Concepts. British Journal for the Philosophy of Science 6, pp. 89–106, 1955.

RTCA Select Committee for Free Flight Implementation, National Airspace System – Concept of Operations. RTCA, Washington, DC, 2000.

Ruckelshaus, W.D. Risk, Science, and Democracy. In Glickman, T.S. and Gough, M. (Eds.), Readings in Risk. Resources for the Future, Washington, DC, 1990.

Rumbaugh, J., Blaha, M., Premerlani, W., Eddy, F. and Lorenson, W. Object-Oriented Modeling and Design. Prentice-Hall, Englewood Cliffs, NJ, 1991.

Rumbaugh, J., Jacobson, I. and Booch, G. The Unified Modeling Language Reference Manual. Object Technology Series, Addison-Wesley, Reading, MA, 1998.

Schrödinger, E. What Is Life? Cambridge University Press, London, 1944, 1967.

Senge, P.M. The Fifth Discipline: The Art and Practice of the Learning Organization. Doubleday Currency, New York, 1990.

Shannon, C.E. and Weaver, J. The Mathematical Theory of Communication. University of Illinois Press, 1949.

Shoval, P. ADISSA: Architectural Design of Information Systems Based on Structured Analysis. Information Systems 13, pp. 193–210, 1988.

Shlaer, S. and Mellor, S.J. Object Lifecycles – Modeling the World in States. Yourdon Press, PTR Prentice-Hall, Englewood Cliffs, NJ, 1992.

Siau, K. and Qing, C. Unified Modeling Language (UML) – A Complexity Analysis. Journal of Database Management 12(1), pp. 26–34, 2001.

Sight Code, 2001. www.sightcode.com

Simons, T. Dependencies and Associations. puml-list@cs.york.ac.uk email forum, June 21, 2001

Simons, A.J.H. and Graham, I. 30 Things that Go Wrong in Object Modeling with UML 1.3. In Kilov, H., Rumpe, B., and Simmonds, I. (Eds.), Behavioral Specifications of Businesses and Systems, pp. 237–257. Kluwer Academic Publishers, 1999.

Sowa, J.F. Principles of Ontology, 2001. http://www-ksl.stanford.edu/onto-std/mail-archive/0136.html

Soffer, P., Golany, B., Dori, D. and Wand, Y. Modeling Off-the-Shelf Information Systems Requirements: An Ontological Approach. Requirements Engineering 6, pp. 183-199, 2001.

Sturm, A. Applying an Object-Relational Database Model to OPM Analysis and Design Results. M.Sc. Thesis, Faculty of Industrial Engineering and Management, Technion, Israel Institute of Technology, Haifa, Israel, 1999.

Suh, N.P. Axiomatic Design of Mechanical Systems. Journal of Mechanical Design 117, pp. 2–10, 1995.

The American Heritage Dictionary of the English Language. 3rd Edition, Houghton Mifflin Company, 1996.

The State Hermitage Museum, St. Petersburg, Russia. Pazyryk Burial, Mound 5, 5th–4th centuries BC, 2001. http://www.hermitagemuseum.org/html_En/03/hm3_2_7e.html

Ulrich, K.T. and Eppinger, S.D. Product Design and Development. 2nd Edition, McGraw-Hill, Boston, MA, 2000.

UML 2.0 RFP, 2000. http://cgi.omg.org/cgi-bin/doc?ad/00-09-02

University of Arizona. NATS-nline, 2001. http://www.ic.arizona.edu/~nats101/n1.html

University of Maryland. Environmental Safety, 2001. http://www.inform.umd.edu/CampusInfo/Departments/EnvirSafety/rs/material/tmsg/rs3.html

Verheijen, G.M.A. and Van Bekkum, J. NIAM: An Information Analysis Method. In Olle et al. 1986, pp. 289–318.

Wand, Y. and Wang, R.Y. Anchoring Data Quality Dimensions in Ontological Foundations. Communications of the ACM 39(11), pp. 86–95, 1996.

Wand, Y. and Weber, R. An Ontological Evaluation of Systems Analysis and Design Methods. In Falkenberg, E.D. and Lindgreen, P. (Eds.), Information System Concepts: An In-Depth Analysis, pp. 145–172. Elsevier Science Publishers (North Holland), IFIP, 1989.

Wand, Y. and Weber, R. On the Ontological Expressiveness of Information Systems Analysis and Design Grammars. Journal of Information Systems 3, pp. 217–237, 1993.

Warmer, J. and Kleppe, A. The Object-Constraint Language: Modeling with UML. Addison-Wesley, Reading, MA, 1999.

Weber, R. Towards a Theory of Artifacts: A Paradigmatic Base for Information Systems Research. Journal of Information Systems, Spring 1987, pp. 3–19.

Webster's Encyclopedic Unabridged Dictionary of the English Language. Portland House, New York, 1984.

Webster's New Dictionary. Promotional Sales Books, 1997.

Wenyin, L. and Dori, D. A Generic Integrated Line Detection Algorithm and Its Object-Process Specification. Computer Vision – Image Understanding (CVIU) 70(3), pp. 420–437, 1998.

Wenyin, L. and Dori, D. Object-Process Diagrams as an Explicit Algorithm Specification Tool. Journal of Object-Oriented Programming 12(2), pp. 52–59, 1999.

Wiener, N. Cybernetics, or Control and Communication in the Animal and the Machine. MIT Press, Cambridge, MA, 1948, 1961.

Winograd, T. and Flores, F. Understanding Computers and Cognition. Addison-Wesley, Reading, MA, 1987.

Witcombe, C. Archaeoastronomy at Stonehenge. Sweet Briar College, Sweet Briar, VA, USA, 2001. http://witcombe.sbc.edu/earthmysteries/EMStonehengeD.html

Index

Bold page numbers refer to definitions. Bold italics refer to OPL sentence types.

Links: The Mortar (continued)

Procedural Links

These links are generally used between an object and a process. They cannot be used to link objects together.

Link Name	OPD Symbol	OPL Sentence	Description
Consumption	Processing ◁—▪ Object	**Processing** consumes **Object.**	Process uses object up entirely during its occurrence.
Result	Processing —▷ Object	**Processing** yields **Object.**	Process creates an entirely new object during its occurrence.
Effect	Processing ◁—▷ Object	**Processing** affects **Object.**	Process changes the state of the object in an unspecified manner.
Input and Output	Object (input state) (output state) — Processing	**Processing** changes **Object** from **input state** to **output state.**	The object is at input state prior to the process occurrence, and at output state as a result of its occurrence.
Agent	Object —●(Processing)	**Object** handles **Processing.**	Object is a human that is not changed by the process; process needs the agent object in order to occur.
Instrument	Object —○(Processing)	**Processing** requires **Object.**	Object is a non-human that is not changed by the process; process needs the instrument object in order to occur.
Invocation	X Processing ▷△▷ Y Processing	**X Processing** invokes **Y Processing.**	First process directly starts up a second process, without an inter-mediate object.

States

State sentences and images

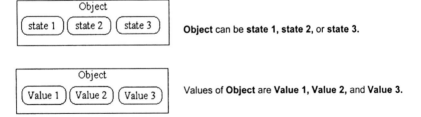

Object can be **state 1**, **state 2**, or **state 3**.

Values of **Object** are **Value 1**, **Value 2**, and **Value 3**.

Printing (Computer to Film): Saladruck, Berlin
Binding: Stürtz AG, Würzburg

States (continued)

State-related Links

Link Name	OPD Symbol	OPL Sentence	Description
Condition	Processing — Object [state 1] [state 2]	**Processing** occurs if **Object** is **state 1**.	Object is an instrument. It must be at a specific state in order for the process to occur.
Agent Condition	Object [state 1] [state 2] — Processing	**Object** must be at **state 2** for **Processing** to occur.	Object is an agent. It must be at a specific state in order for the process to occur.
Qualification	Object — [state 1] / Attribute — Qualified Object	**Qualified Object** is an **Object**, the **Attribute** of which is **state 1**.	Qualified Object is a type of Object. It must be at a particular state of Object's Attribute.
Instance Qualification	Object — [state 1] / Attribute — Qualified Object	**Qualified Object** is an instance of an **Object**, the **Attribute** of which is **state 1**.	Qualified Object is an instance of class Object. It must be at a particular state of Object's Attribute.
State Specificied Consumption	Object [state] — Processing	**Processing** consumes **state Object**.	Process consumes object only if it is at a certain state.
State Specificied Result	Object [state] — Processing	**Processing** yields **state Object**.	Process creates object at a certain state.

Boolean Objects

Specialized informatical objects. Boolean objects are questions, and they always have two states (the answers): yes and no.

Link type	OPD Symbols	OPL Sentence	Description
Determination (a Result Link)	Determining ↓	**Determining** determines whether **Object** is **proper**.	Process yields a Boolean object that poses a "yes or no" question. The process then determines the answer.
Condition link	Object is proper? [yes] — A Processing	**A Processing** occurs if **Object** is **proper**.	If the answer is "yes," a certain process occurs. If the answer is "no", a different process occurs.
Negative condition link	[no] — B Processing	**B Processing** occurs if **Object** is not **proper**.	
Both condition links	Object is proper? [yes] — A Processing [no] — B Processing	**A Processing** occurs if **Object** is **proper**, otherwise **B Processing** occurs.	Compound sentence: if the answer is "yes," a certain process occurs, otherwise a different process occurs.

CPSIA information can be obtained at www.ICGtesting.com
Printed in the USA
LVOW010422120911

245854LV00006B/1/P